SERICULTURAL ENTOMOLOGY

SERICULTURAL ENTOMOLOGY

Dr. Ravindra Nath Singh
Dr. Beera Saratchandra

A.P.H. PUBLISHING CORPORATION
4435-36/7, ANSARI ROAD, DARYA GANJ
NEW DELHI-110 002

Published by
S.B. Nangia
A P H Publishing Corporation
4435-36/7, Ansari Road, Daryaganj
New Delhi 110002
Ph.: 23274050
E-mail : aphbooks@gmail.com

2026

Printed at
Balaji Offset
Navin Shahdara, Delhi 110032

FOREWORD

Plant protection and silkworm care are the most important components to maintain the momentum of sericultural progress. Efficient management of technological advancements, systematic adoption and meticulous practices can increase sericulture productivity. In view of the increasing demands for silk and considering the global concern for environmental protection, sericulture must not only be productive but highly sustainable to the existing natural resource base.

Therefore, effective strategies are required to reduce present use of toxic chemicals. Chemical control forms the prime and foremost method for the management of insect pests. Prolific use of chemical insecticides significantly curtailed the insect pests in the past but in due course it resulted in the development of resistance to insecticides in insects, environmental degradation and increase in the cost of cultivation.

To overcome these unfavourable situations, Integrated Pest Management (IPM) strategies were advocated. Integrated Pest Management (IPM) approach has helped in achieving desired successes in reducing pesticide consumption. For its success, it is necessary that detail information on insect pests is available for the students, researchers and extension workers. The present book entitled "Sericultural Entomology" is a good effort in this direction and provides detail information on pests and their control, measures pertaining to mulberry and Vanya silkworms and its host plants. I congratulate the authors for their untiring efforts in bringing out this useful publication. I am sure this publication will be useful to all concerned with pest management in sericulture.

Sathyavathy 14/10/2010.

Bangalore
October 2010

(M. Sathiyavathy, IAS)
Member Secretary
Central Silk Board
Ministry of Textiles, Govt. of India

Dedicated

To

My Mother

Late Gulabo Devi

Dr. R. N. Singh

PREFACE

Intensive sericulture by improving systematic plantation, water management, crop varieties, fertilization and other agronomic practices has greatly enhanced pest incidence coupled with more and more deliberate use of pesticides. The complexity in the behaviour and life cycle of insects warrant a special attention for their effective management especially in the changing scenario of our modern sericulture. For a sustained development in all spheres of our life, sustainable sericulture has become a significant topic of discussion. We depend on sericulture for silk, employment, and foreign exchange, making it an essential part of our economy. Many people take sericulture for granted while voicing concern over adverse effects of sericultural production practices on the environment. It is because insect pests have mainly been controlled with synthetic insecticides in the last fifty years. Most insecticidal compounds fall within four main classes, the organochlorine insecticides, organophosphorus insecticides, the carbamate and pyrethroid insecticides. Out of these the major classes in the uses today are organophosphates and carbamates. There are problems of pesticides resistance and negative effects on non-target organisms including man and environment. Environmental contamination of air, water, soil and food with pesticides may alter the structure (species richness, biological density and diversity) and functional activities of eco system and may ultimately threaten the very survival of the human race. The greater awareness of the dangers of environmental pollution arising from the wide spread use of chemical pesticides is reflected in the increasing emphasis placed on safer and most selective chemicals. In many countries candidate chemicals have to satisfy stringent legislative criteria before they can be marketed as pesticides; the effect of pesticides on the environment are discussed together with future development.

Scientists for the control of pests have developed many devices, principles and practices. The physical devices meant for destroying pests have the scientific basis that the greater the number killed, the lesser the damage done. Taking advantage of natural barriers around islands and continents and by creating artificial restriction, the entry of pests from other lands can be stopped physically. As sericulture progressed in the past, there was accumulation of knowledge based on common observations and scientific discoveries. In this way many sericultural practices were developed which made living conditions rather harsh for pests and thus damage to crop was reduced. At the same time certain inorganic and organic compounds used were found to be effective in killing the pests. Many of these organic compounds were synthetic, belonging to the group's chlorinated hydrocarbons, carbamates, organophosphates and pyrethroids, etc. It was noticed later on that, by the application of these pesticides, there was a regular resurgence of the pests, which was due to the death of natural predators and parasites. By the suppression of these natural agents of control, the population balance was upset and the pest would appear in severe form almost every year. The importance of ecology was then realized and the integrated approaches to pest control were thought of as more sound strategies to avoid violent fluctuations in pest populations and thus to achieve a more lasting and more economical control of pests.

Insect pest control continues to be a challenge for silk producers and researchers. Insect resistance to commonly used pesticides and the removal of toxic pesticides from the market have taken their toll on the ability of silk producers to produce high quality, pest-free crops within economical means. In addition to this, they must not endanger their workers or the environment. Areas currently of special interest include selective herbicides against the host plants pest, chemical controlling the growth of insects and

Integrated Pest Management techniques are of special importance.

During the last two decades sericulture scientists have reported extensively about the pest control strategy in seri ecosystem but much of the information often focuses on silkworm diseases and parasites. Moreover many of the publications envisage either on chemical control or mechanical control of pests and lack comprehensive approach of plant protection as a component of silk production technology. To find out suitable control measures against the pests of mulberry and non mulberry silk host plants, it is essential to know the behaviour of silkworms and existing pest populations in seri ecosystem. Therefore, brief life cycle of all the commercially exploited silkworms in India has been described. The succession of various pest population in mulberry, tasar, eri and muga plantation and its suitable control measures have been described. Though use of synthetic insecticides has provided effective control of all major pests, yet their undesirable side effects (pests out breaks, resurgence, resistance development, harmful residues, etc.) limit their continued use. To minimize such problems, farmers must diversify their pest control practices – a strategy that scientists now term Integrated Pest Management.

Insect Pest Management is an ecosystem-based pest management strategy that focuses on long-term prevention of pests and their damage through a combination of techniques such as biological control, habitat manipulation, modification of cultural practices, and use of resistant cultivars. Pesticides are used only when needed as determined by established guidelines. It is an economically justified and sustainable system of crop protection that leads to maximum productivity with the least possible adverse impact on the total environment. In silk production technology IPM is a schedule of practices, which starts from field selection till harvest of cocoons. The major components in this approach are cultural, mechanical, biological and chemical methods of insect pests, diseases and rodent control in a compatible manner.

IPM presents a balanced overview of environmentally safe and ecologically sound practices for managing insects. The book approaches its subject in a systematic and comprehensive manner making it a useful resource for professionals in the fields of entomology, sericulture, horticulture, ecology, and environmental sciences, as well as to silk producers, industrial chemists, and people concerned with regulatory and legislative issues. This book has many unique features. It comprises brief description of tasar and Oak tasar culture, muga and eri culture and mulberry culture. Separate chapters have been dealt on pests of mulberry and non-mulberry silk host plants. This book covers specific ecological measures, environmentally acceptable physical, mechanical, cultural and bioecological control measures, use of chemical pesticides, and a detailed account of legal and legislative issues. It also includes a chapter on sampling and forecasting of pest populations and its biological control. Recent and relevant information on prospects of Integrated Pest Management in sericulture have been discussed in detail.

During writing this book, we felt, that to produce a fully up-to-date book is well-nigh impossible, as new informations have been accumulating almost everyday. It was, therefore, decided to restrict details to a level which would be most useful to the Indian student doing post graduate study in sericulture, zoology and entomology. Scientists and professors from various institutions have contributed ideas and furnished suggestions towards the preparation of this book, and we express our sincere gratitude to them. Incorporation of a long list of names may not appear quite feasible, but due acknowledgements have been made at appropriate place.

However, we are extremely thankful to Dr. B. K. Singh and Dr. S. K. Singh, Professor in Zoology, Post Graduate Department of Zoology, Magadh University Bodh Gaya for many useful suggestions and encouragement. Dr. Md. Raziuddin, Professor in Zoology, Vinoba Bhave University, Hazaribagh, Jharkhand for constant help and providing ideas. We express our gratitude to Late Dr. S. Jayaraj, Chairman S. Jayaraj Research foundation, Chennai for his help and constant encouragement.

We feel highly privileged and honoured by Ms. M. Sathiyavathy, Member Secretary, Central Silk Board, Bangalore for kindly agreeing to write the foreword of this book.

We deeply appreciate the help received from Dr. M. Maheshwari, Dr. S. T. Chistiana, Dr. Tribhuwan Singh, Sri Jayarama Raju, Sri. J.C. Mahanta scientist, Central Silk Board, Bangalore during the preparation of the manuscript. We express our gratitude to Sri Rajesh Kumar Sinha, Central Silk Board, Bangalore for helping in critically reading the manuscript and also in preparing the bibliography; Sri. K.K. Shetty, Deputy Secretary CSB, for vetting the manuscript from the language point of view; Mr. R. Punniyamoorthy for typing part of the manuscript.

We are grateful to Dr. K. B. Nangia, APH Publishing Corporation, New Delhi whose interest and enthusiasm in putting the work into print has been admirable. Finally, we must record our thanks to our all colleagues who have encouraged or assisted us in this great task. Suggestions for improvement will be most welcome.

2010 Dr. R. N. Singh
Bangalore Dr. B. Saratchandra

CONTENTS

List of the Figures

About the Authors

DR. R.N. SINGH

Dr. R.N. Singh is scientist with Central Silk Board, Ministry of Textiles, Govt. of India, Bangalore. He obtained his M. Sc degree in Zoology in 1975 and Ph.D degree in 1982 from Magadh University, Bodh Gaya. He was Post Doctorate Research Fellow of the Indian Council of Medical Research from 1982 to 1984. He joined Central Silk Board in 1984 and worked in various capacities at different places. He has contributed mostly in integrated pest management, biological control, insect behaviour, particularly the pests and predators of silkworm and its host plants. Dr. Singh is a member of various professional societies in India and abroad, and is a Fellow of Zoological Society of India. He has received Dr. B.S. Chauhan Gold medal award by Zoological Society of India for contribution in pest management in sericulture during 2001 and senior scientist award during 2003. At present he is associated with co-ordination and monitoring of Research projects being undertaken by various sericultural research and training institutes of Central Silk Board. He has published more than 150 Research papers including popular articles in various National and International Journals and is author/ editor of three books. He is an expert in the field of pest management. He has delivered several lectures in the National and International conferences and symposia.

DR. B. SARATCHANDRA

Dr. Beera Saratchandra is the Director (Technical), Central Silk Board, Ministry of Textiles, Govt. of India, Bangalore. He obtained his M. Sc degree in Zoology in 1975 and Ph. D degree in 1980 from Andhra University, Visakhapatnam, where he continued as Post Doctorate Research Fellow. He joined Central Silk Board during 1983, worked as Joint Director and Director for over 27 years in various research establishments from the North-east to the south of India gaining rich knowledge and experience in the field of vanya and mulberry silk production, and made good contribution to the subject. He is a member of various academic and scientific bodies and Fellow of the Zoology Society of India. He has visited several countries for various assignments. He is a founder Director of Central Sericultural Germplasm Resource Centre, Hosur, Tamil Nadu and later served as Director, Central Sericultural Research and Training Institute, Berhampore, West Bengal. He is an expert in the field of silkworm pathology, silkworm rearing technology, entomology and seed technology. He has published one book and more than 125 research papers in reputed journals and participated in number of National and International conferences, seminars and workshops.

About the Book

This book covers the problems of pest management in sericulture through research and development of different integrated pest management tactics. The book has been written primarily for students of sericulture and applied biology, but it can also be used with equal benefit by researchers, extension workers and educated farmers. The distribution of silkworm host plants and brief life cycle of all the commercially exploited silkworms in India have been described. The biology, life cycle, nature of damage and recommendations for the control of major pests of mulberry, tasar, oak tasar, muga and eri silkworm host plants have been extensively discussed. Life history and control of major parasites and predators of silkworm have been discussed in detail. The principles and methods of pest control have been described with ecological bias. Efforts have made to include cultural, mechanical, legislative, biological and chemical methods of pest control so that student can draw help from them while devising integrated control practices. This book also discusses the general principles underlying chemical pesticides along with a brief history of their development. It lists the main types of insecticides, and herbicides with discussion of their chemical preparation, uses and biochemical mode of action. Some novel methods of pest control viz. use of attractants, repellants chemo-sterilants and herbal methods have been discussed in detail. The adverse impact of the agrochemicals on the ecosystem has been described along with mitigating measures such as biological control. Conservation and augmentation of predators and parasitoids for control of parasites have also been included. The principles, concepts, mechanisms and current practices of biological control of insect pests have been described in detail. Classical biological control includes augmentation of natural enemies and restoration of ecology. All these aspects have been discussed in detail. The book ends with a chapter on integrated pest management which carries the latest knowledge in terms of practical hints for effective and economical pest control.

CHAPTER I

INTRODUCTION

The struggle between the insect pests and the host plants has been going on since time immemorial. The emergence of chemotypes among plants and races or biotypes among insects is the result of the struggle for existence and the nature's selection pressure exerted on insect pests and host plants. Plants develop newer toxic chemicals to combat their insect pests while the insects develop detoxification mechanisms in their defence. Unlike the other host, the pest depends on its host for survival and hence prefers to co-exist thus entering into an adaptive mechanism maintaining its ecological balance in nature. When an insect population increases to a level causing economic injury to the desired plant, directly inflicting human welfare and his profits, the insect is called as pest. Almost all the harmful insects are known as pests. There are a number of processes by which insects have been causing harm over the past several years. Likewise man's effort in containing the damage from the harmful insects too has a long and varied history. A knowledge of this history adds important dimensions to the study of pest management which provide insights into the driving forces (technical, economic, social) that have forged current pest management practices which in turn provides some idea of the forces likely to be acting in the future (Norton and Mumford, 1993; Dent, 1995).

Since time immemorial man has been in constant strife with insect pests competing for food, fibre and even shelter. Recent publications (McEwen, 1978; Pimentel, 1978) estimate that despite our best efforts world crop losses to pests approximate 35% of production. Insect's account for a significant portion of this staggering figure. These reduced yields are further compounded by post-harvest losses to insects and other pests, so that in some tropical countries over 50% of the crop may be lost for this reason, and it is not uncommon for susceptibility to pests to limit the types of sericulture which can be practiced. In addition it must be accounted against the insects that they are the vectors of many of the most debilitating diseases, which may afflict productivity potential. The modern sericultural technology not only paid new dividends by increasing silk production but also drastically disturbed the natural ecosystem. The pest complex and disease spectrum as a consequence has increased and caused manifold problems ultimately resulting in heavy losses in silk industry. Pest causes severe damage to silk host plants and transmitted disease from generation to generation, which ultimately kills the silkworm larvae. Mulberry (*Morus alba*) forms the basic food material for mulberry silkworm (*Bombyx mori*) in tropical and temperate region. Large number of tree species in the dense and humid forests are exclusively utilized for rearing of non mulberry silkworms especially tasar, oak tasar, muga and eri silkworm. Different insect species infest different parts, but all stages of growth are subjected to all types of insect attack. Mulberry like most of the economic plantations and field crops is often attacked by various insect pests. The damage caused by the pests is quick and extensive. Lepidopteran, hemipteran and mites cause considerable damage to the mulberry. The attack by these pests is sporadic and sometimes seasonal. Among the sucking pets, white flies (*Dialeuropora decempuncta* Quaintance & Baker, *Aleurodicus dispersus* Russell), mealybug (*Maconellicoccus hirsutus (Green)*, leaf hoppers, scale insects and thrips are very common. Tasar silkworm (*Antheraea mylitta* Drury) is reared outdoor on *Terminalia arjuna, T. tomentosa* and *S. robusta* whereas oak tasar silkworm (*Antheraea proylei* Jolly) is reared on *Quercus* spp. Muga silkworm (*Antheraea assama* West wood) is reared on som (*Machilus bombycina*) and soalu (*Litsaea polyantha*). Eri silkworm is reared indoor mostly on castor, tapioca and payam. All these silkworm host plants are attacked by various groups of insects. Apart from pest population, the host plants are attacked by large number of diseases especially during rainy season. Monoculture plantation of host plants in large area and suitable climatic condition favour the development of epidemics. The perpetuation of the pests and pathogens is easier due to mono cropping, availability of collateral hosts and land varieties/races. In this context there is, thus urgent need to have detail information on insect pests for developing suitable control measures against various pests of silk host plants.

HISTORY OF PEST MANAGEMENT

The history of pest management dates back to the beginnings of agriculture. It combines important events (discoveries and defining moments), influential people, institutions, organizations and governments in ways that have led us to the current concept of Integrated Pest Management (IPM). IPM has been defined as a 'pest management system that in the context of the associated environment and population dynamics of the pest species, utilizes as suitable techniques and methods in as compatible manner as possible and maintains the pest population levels below those causing economic injury' (Smith and Reynolds, 1966). The concept of utilizing a number of techniques in compatible manner as possible as is, of course, not new and even the different control techniques available to us today were utilized in same shape or form many years BC. The earliest record of pest control seems to be the use of sulfur (an insecticide still in use) by ancient Sumerians about 2500 B.C. In pre-biblical times both the Egyptians and Chinese used insecticides formulated from herbs and oils, particularly for protection of seeds and store grains. First descriptions of cultural controls, especially manipulation of planting dates, were recorded around 1500 B.C., while burning as a cultural control method was first described in 950 B.C. A very important advancement occurred around 300 B.C., was the recognition of 'phenology' as a science.

This led to the principle of timely planting of crops to avoid losses from pests. During this time, the Chinese discovered the use of natural enemies to control insect pests: for example, they placed ants on citrus to reduce pest infestations (Coulsen *et al.*, 1982). These practices date back to over 1,000 years. A few developments in pest control were recorded during the middle ages (end of the Roman Empire until about 1,000 A.D.) However, by 1101 the Chinese had discovered the use of soap to control pests, and by the late 1600s, tobacco infusions and insecticides from other herbs, as well as arsenic, were commonly used. With expansion of agriculture in the 1700s, Reaumur published on the importance of temperature summation in determining insect phenology. This idea was used, almost two centuries later, to predict insect events and determine timing of pest control activities. Also in the 1700s, the concept of plant resistance to insects was advanced for the Hessian fly *Mayetiola destructor*, in the United States, as well as the rediscovery or introduction of botanical insecticides (derived from plants). Genetic basis for resistance is one of the oldest recognized concepts of breeding plants for pest resistance (Panda and Khush, 1995). Theophrastus recognized the differences in disease susceptibility among crop cultivars such as early as 3 B.C. (Allard, 1960) and, of course, farmers would have selected for resistance to local pests by the practice of saving seed from healthy plants for sowing. Over many generations of selection, resistant land races developed, which provided sustainable levels of yield for average local conditions. Thus by 500 A.D. all the general types of control measure available today - insecticides, host resistance, biological and cultural control had already been developed and used by one civilization or another.

References to developments in pest management seem to be very few till 18th century A.D. A number of bizarre events such as the excommunication of cutworms in Berne, Switzerland, in 1476 and the banishing in 1485 of caterpillars by the High Canadian Vicar of Valence perhaps depict the lack of understanding of the causes of pest problems at that time. It was not until 1685 that the first correct interpretation of insect parasitism was published by the British physician Martin Lister who noted that the ichneumon wasps emerging from caterpillars were the result of eggs laid by adult female ichneumonids (van Driesche and Bellows, 1996). This was followed by a similar interpretation of parasitism in *Aphidius species* by the Dutchman Antoni van Leeuvenhock recorded in 1700. The work of Linnaeus burgeoned interest in insect descriptions, which promoted greater observation and study of insect biology. It was Linnaeus who first noted in 1752 that 'Every insect has its predator which follows and destroys it' and then commented on how such 'Predatory insects should be caught and used for disinfecting crops' (Horstadius, 1974). The earliest documentation on host plant resistance occurs in 1762 with the description of hessian fly (*Mayetiola destructor*) resistant wheat cultivars, under hill, in the US by Haven (Panda and Khush, 1995). From these beginnings the scientific discipline of entomology began to take shape at the turn of the 19th century. Pioneers such as John Curtis in the UK, Thomas Say, C. V. Riley in the US created the basis for applied entomology with their publications (e.g., Farm Insects, J. Curtis, printed 1841-57 an illustrated account of pest insects arranged by crops for ease of use by farmers; American Entomology, 1824 by T. Say) and their innovative approaches to pest control which formed the foundation for modern pest control. In the US, the Rocky Mountain locust plagues of the 1870s initiated the formation of the US Entomological Commission (Sheppard and Smith, 1997). In 1881 C. V. Riley was appointed Entomologist of the Department of Agriculture and the Division of Entomology was created within the USDA. W. Saunders and C. Bethuse founded the Entomological Society of Canada in 1863 with the first issue of the journal.

The Canadian Entomologist' published in 1868 (Sheppard and Smith, 1997). C. V. Riley (1873) was the first who was responsible for the introduction of the biological control agent, *Phylloxera,* from Canada into France to control grape pest, which unfortunately failed to give the desired results. The first successful example of classical biological control was cited in 1888 with the introduction and release of the Vedalia beetle *Rodolia* (formerly *Vedalia cardinalis*) for control of the citrus pest of cottony cushion scale (*Icerya purchasi*) in California (van Driesch and Bellows, 1996). Within two years the beetle controlled the scale throughout the state. This project has far-reaching effects because it demonstrated the feasibility of utilizing arthropod predators to control serious pest problems and produce large economic gains. Other areas of biological control were also forging ahead during the 19th century. Insect pathogens were first considered as biological control agents by Bassi in 1836, when he proposed that liquids from cadavers of diseased insects could be mixed with water and sprayed on plants to kill insects. However, it was only when the Russian entomologist Elie Metchnikoff (1884) developed a means to mass produce the fungus, *Metarhizium anisopliae*, to control the curculionid beetle, *Cleanus puntiventris*, the potential of entomopathogens as a practical pest control option has come to lime light. The end of the 19th century saw a fundamental change in agriculture and urbanization, particularly in the United States. Agriculture was the linchpin to urbanization; a rapid growth from a mainly subsistence way of life towards greater mechanization (substituting labour) and commercialisation fuelled urbanization. Since insect pests were a limiting factor in agricultural efficiency, the importance of their control increased. Thus, a demand developed for the investigation, identification and control of insect pests. In this way the subject of entomology became identified as a distinctive scientific discipline with the increase in training and employment of State and Federal entomologists. The numbers increased from less than 20 in the mid-1800s to more than 800 by the 1870s (Sorenson, 1995).

In 1889 the first US national professional group, the American Association for Economic Entomology was founded, followed by the publication of the Journal of Economic Entomology, in 1908 (Howard, 1930). The new era, involving the use of toxic substances for insect control, was led by the use of the dye, 'Paris Green', which was found to be effective against the Colorado beetle in the US and was used with Bordeaux mixture against Grape *Phylloxera* in French vineyards during1870-1890. In the 1890s lead arsenate was also introduced for insect control. By 1910, lead arsenate and Paris Green were the most widely used insecticides sold on a commercial scale with sales reaching £10 million each year. In 1917 calcium arsenate joined lead arsenate as the leading insecticides in addition to the increasing list of available products which included tar plant extracts, derris, nicotine and pyrethrum (Ellis, 1993). Thus, the first 40 years of the 20th century witnessed an increased use and reliance on chemical insecticides, including the introduction of new compounds such as ethylene oxide, thiocyanates and phenothiazine. However, the application of chemical products tended to be haphazard and very imprecise rendering the active ingredients often ineffective or hazardous. Hence, in spite of the growing importance of chemical insecticides in pest control, use of chemical insecticides did not dominate other approaches adopted by farmers. Enormous potential for biological control and breeding for resistant crop plants was noticed and over 30 cases of natural enemy establishment were recorded throughout the world between 1920 and 1930 (Pamell, 1935). Plant breeding for agronomic characters had become well established in the 19th century utilizing quantitative genetics, but it was not really until the rediscovery of Mendel's Law of Heredity in 1900 that geneticists understood qualitative breeding approaches and stumbled upon disease resistance and later insect pest resistance (Robinson, 1996 and 1992). The systematic research of R.H. Painter in the 1920s on the resistance to Hessian fly in wheat cultivars laid the foundations for the development of resistance breeding against insects utilizing qualitative genetics (Panda and Khush, 1995). However, successes were few and far between with breeding for resistance to insects (unlike pathogens); the most notable success at the time being cotton resistant to *Empoasca* bred in South Africa (Parnell, 1935) and India (Husain and Lal, 1940). It was not until the 1960s that the full potential of host plant resistance to insects was fully appreciated but by then the dominance of chemical insecticides for insect control had reached spectacular proportions.

Chemical insecticides came to prominence on the back of the most famous insecticide DDT (dichlorodiphenyl trichloroethane) developed by Paul Muller working for the Geigy Chemical Company in 1939. DDT offered persistence, low cost, virtually no plant damage, broad-spectrum activity and low acute mammalian toxicity. Swiss farmers to control Colorado beetle first used it in 1941 and by 1945 DDT production had reached 1,40,000 tons a year. The success of DDT also stimulated the search for other similar chemicals and the subsequent development of aldrin, HCH, dieldrin, heptachlor and chlordane (Ellis, 1993). Other methods of pest management paled in significance with the success achieved with these impressive chemicals and hence, the use of more biologically oriented approaches declined dramatically in the 1940s and 1950s. This reorientation of effort was mitigated by the overwhelming need to test the efficacy of the ever-expanding arsenal of new chemical compounds (Kogan and Mc Grath, 1993). Pesticides became the only method used by many farmers for the control of agricultural pests. Ironically, at the same time that DDT had obtained notoriety as a panacea for pest control in 1946, the philosophy of IPM came into being in the alfalfa fields of California's San Joaquin Valley (Summers, 1992). In 1946 K.S. Hagen was hired as the first supervised control entomologist in California, where he monitored 10,000 acres of alfalfa for the alfalfa caterpillar, *Colias eulythene*, and its parasite, *Apanteles medicanis* (Hagen *et al.*, 1971). From the experience of combining natural enemies, with host plant resistance and rational use of chemicals in the 1950s against the spotted alfalfa aphid, *Therioaphis maculate*, emerged the integrated pest control philosophy presented in Stern *et al.* (1959).

However, the context in which the IPM framework would achieve significance was happening around the world in situations where pesticides were used in excess. Insect resistance to chemical insecticides was first reported in 1946 in houseflies in Sweden, but in the 1950s it became widespread in many agricultural pests. Resurgence of target pests, upsurges of secondary pests (both caused by the suppression of natural enemies by insecticides), human toxicity and environmental pollution (Metcalf and Metcalf, 1992) caused by pest control programmes relying on the sole use of chemicals reeked havoc in many cropping systems. The ecological and economic impact of chemical pest control came to be known as the 'pesticide treadmill' because once farmers set foot on the treadmill it was virtually impossible to try alternatives and remain in viable business (Clunies-Ross and Hildyard, 1992). In addition, despite problems with the chemicals used, the faith in and 'addiction' to the chemical technology by farmers was assured through the continued development and availability of new active ingredients coming onto the market.

A landmark event in the history of pest management was the publication in 1962 of the book "Silent Spring" by Rachel Carson, which was not really important from the point of view of its technical content but rather from its impact on the general public. Silent Spring brought pest management practices into the public domain for first time and provided the springboard for an increase in the awareness of the general public of the problems associated with chemical insecticide use. This growing environmental concern, combined with the philosophy of integrated control advocated by an increasing vocal band of scientists provided openings for the funding and development of alternative, more environmentally friendly approaches such as insect pheromones, sterile insect techniques, microbial insecticides and host plant resistance. The gypsy moth pheromone was isolated, identified and synthesized in 1962; by 1967 the screw worm, *Cochlomyia hominivorax*, was officially declared eradicated from the US through use of the male sterile technique (Drummond *et al.*, 1988); and in 1972 the first commercial release of a microbial insecticide of *Bacillus thuringiensis* based on the isolate HD-1 for control of lepidopterous pests occurred (Burgess, 1981).

Plant breeding using Mendelian genetics had earlier in the century produced very high yielding dwarf wheat cultivars. The need to improve yield in food crops to alleviate 'third world' malnutrition fostered the establishment of the first International Agricultural Research Centre (IARC) in1960 sponsored by the Consultative Group for International Agricultural Research (CGIAR) and the International Rice (IRRI), Bos Banosin, Philippines. This was followed by seven other centres improving crop reproductivity through development of high yielding varieties (HYV). Part of this development included breeding for insect resistance to brown plant hopper (*Nilaparvata lugens*), the green leaf hopper (*Nephotettix virescens*), the yellow stem borer (*Scirpophaga incertulas*) and the gall midge (*Orseolia oryzae*) (Panda and Khush, 1995). This became popular as the 'Green Revolution' with the 'greening' of developing countries with high yielding cultivars, particularly of rice (Panda and Khush, 1995; Dent, 1992). However, the resistance was based on a form of resistance that could be overcome by the insects and particularly in rice, farmers were advised to spray their high yielding crops with chemical insecticides to protect them. Such measures were still necessary until the 1980s when IPM methods were introduced into rice systems for the first time.

The concept of IPM in the 1960s and 1970s was based on restricting pesticide use through use of economic thresholds and utilization of alternative control options such as biological products or methods, bio pesticides, host plant resistance and cultural methods (Thomas and Waage, 1996; Smith and Reynolds, 1966). IPM was initially launched in the US and ultimately around the world with a large and influential research project known as the Huffaker Project (1972–1979) which was later continued till 1980s (as the Atkinson project) focusing primarily on insect pest management in six crops, cotton, soybean, alfalfa, citrus fruits, pome fruits (apples and pears) and some stone fruits (peaches and plums) (Morse and Buhler, 1997). Involving scientists from universities, USDA and private industry the project sought to organize research to transcend the disciplinary, organizational, political, geographical and crop specialist barriers that usually inhibited collaborative efforts between different scientific disciplines. In this effort, IPM was designed to the specific requirement of capital intensive, technologically sophisticated farmers. Ironically this was the same design for which the high tech chemical pest control tradition had been based, something that IPM endeavoured to replace (Perkins, 1982; Thomas and Waage, 1996). A different model for the development of IPM has been developed in agricultural crops and the similar steps may be started in sericulture. The history of pest management is a subset of the history largely of agriculture. While pests have been a chronic problem in crop plants since the beginning, many of today's serious pest problems are the direct consequence of actions taken to improve crop production (Waage, 1993). The intensification of agriculture has created new and greater pest problems in a number of ways. In the beginning, the concentration of a single plant species or variety in larger and extensive monocultures increases their vulnerability to pests. Secondly high yielding crop cultivars can provide improved conditions facilitating colonization, spread and rapid growth of the pest. Further, the reduction of natural enemies around crops and exclusive dependence on chemicals leading to insect resistance to pesticides leading to increased pest problems are the major factors for outbreaks of pest population (Van Den Bosch, 1978).

Similar situation has arisen in sericulture. From the last three decades, much of the attempts have been made to minimise the pest population of silkworms and its host plants and for quick results indiscriminate use of insecticides were conducted which led to large scale out break of new pests and heavy decrease in the beneficial insects population existing in the rearing filed. Several technologies were developed and released at field level to minimise the pest population but still the situation has not improved. The pests are causing consistent damage to silk host plants and silkworms and ultimately the sericulture is suffering from huge loss. Due to continuous human interference of human beings the ecological condition in seri ecosystem especially non mulberry sericulture areas, the natural habitat of the native natural pests and parasitoids are disturbed and it resulted in excessive increase of pest population. But the greatest challenge is to maintain the ecological habitat of the silk producing insects in nature. Most of the time it is disturbed by increasing or decreasing the insect population of other insects group which directly interferes the population ecology of the sericigenous insects. Silk producing insects live in harmony of the same ecological niches where other lepidopteran insects are breeding and feeding.

Therefore, there is need to search for efficient method for augmentation of the sericigenous insect and reduction of harmful insects. Some of the sericigenous insects are considered as the pest for commercially exploited silkworm host plants. The species producing, tasar, eri, muga and mulberry silk rank in economic value. The economic value of cricula and frithi silk producing insects are also increasing and search for new technology for its commercial exploitation is under progress. Therefore, the detail information on the sericigenous insects will provide an idea for commercial exploitation. It will help to increase the production of quality raw silk, especially non mulberry silk, not only to cater the domestic needs but also to the requirements of international market.

SERICULTURE IN RELATION TO MAN

Wild sericulture is a labour-intensive agro-based cottage industry, which can provide employment to millions of people living in remote rural areas close to the forest. The world trend in silk production and consumption is quite favourable. The demand is increasing and the production in major silk producing countries like

China, Japan, Brazil is gradually diminishing due to several reasons such as diversification, affluence, high cost of the labour, rampant collection of cocoon, deforestation, and unscientific practice. Wild silk moth conservation and utilization enhance the productivity and economic returns from small-scale land users, particularly the target women groups. Silkworm farming provides the economic basis for community development suitable for increasing participation of women Hence, the farming of these wild silk species needs to be adopted as an insect based enterprise by rural communities for the integration of biodiversity conservation with the economic development of the rural people and especially women. This can be a means of diversifying their economic base and, therefore, encourage then participating in the current world efforts to promote conservation-based development.

Silk production can provide a strong incentive for rural communities to adopt sound land management practices as an adjunct to subsistence agriculture. This would promote conservation of natural ecosystem, and wilds silkmoth habitat, in face of growing human population. Wild silk moth farming facilitates the access of women to natural resources, technology and income generation. Development of production system in wild to conserve the genetic diversity and production of wild silk moth species in different ecology is highly desirable. The effect may help alleviate poverty, improve the human condition and preserve the natural grown biological systems on which all life depends. Since India has tremendous scope and opportunities to promote wild sericulture, due attention should be given for its conservation to bring about sustainable development in the new millennium. China's dominant role as the world's largest supplier of both raw material and finished silk products remains unchanged. India's position in International market is temporarily slightly deteriorated, but its domestic market remains buoyant. Activities encouraging or imposing the use of environment friendly textiles and garments in western countries continue to be a matter of concern, especially among supplier in developing countries. silk supplier will have to monitor future developments so that they are ready when the times come. Each countries feels the urgent need for silk promotion, but solving their problem seems to be Herculean. One of the important reasons is the need to burnish the image of the silk, somewhat tarnished after the surge in the trade in cheap sand -washed silk products. It is ironic that producer of various other products such as cigarettes, cosmetics, competing textiles and the airlines service sector, use the world silk to describe the best features, of their products, while the silk producers and processors remain more or less silent. Another concern, namely that of the availability of reliable and accurate statistical information, remains unresolved.

Systematic collection, processing and dissemination of data, both quantitative and qualitative, would greatly enhance the global silk development efforts of everybody concerned – producers, converters and traders alike. Silk is produced by the poor, but is the product for the reach. Silk and silk products have always been associated with luxury and have traditionally been expensive. The rearing of silkworm continues to be extremely labour intensive and time consuming. However, the perception of sericulture as a profitable venture is encouraging in countries with available labour, suitable climate and vast stretch of land to start sericulture activities. Unlike cotton and wool, which are typical, export commodities and often processed in other countries, silk has traditionally been processed and consumed in the processing countries. Silk is mainly used as raw material for luxury products. Today, however, silk goods have attracted a new consumer group, a young generation of consumers with more money to spend, partly because the number of women employed has increased dramatically during the last two decades. Another reason for the interest in the silk, is the general rise in the demand for natural fabrics, brought about by the ongoing environmental movement.Natural silk has value, but the introduction of sand-washed silk in the early 1980s demarcated the trade and brought silk within the reach of most consumers in the west. However, the low cost goods did not live up to the traditional image of silk quality product. In the aftermath of that fashion wave, silk now has to compete with other textiles for a place in the international market. Nevertheless growing demand for ready made article in the silk, continues to encourage manufacturers in the developing countries to try to penetrate markets in the west. Silk is highly preferred for the garments, readymade garments, fabrics for interior decoration and other products such as neckties, scarves, cushion covers and other items for the home, because of this luster, longevity and smoothness. Silk has value an in future will have demands as the "queen of Textiles" or "queen fibre" of nature. It is having unique physical properties having high tenacity and high elongation, natural luster and high drape.

DISCOVERY OF SILK

Although we know that silk originated in China, the exact details of its discovery have been lost the mists of time. A likely tale emerges from the period of 2500 BC. When Empress Tsi Ling-shi was enjoying tea in the palace garden, a cocoon fell out of a mulberry tree into her tea cup. A loose thread floated to the surface and as Ling-shi pulled it, the cocoon unravelled, revealing the mystery of silk thread to the empress. Ling-shi's discovery led to the development of sericulture and the production of silk cloth. Until this period, fibre for cloth came from plants (cotton, hemp and ramie) or animals (sheep, goats, camel and yak). No one knew of such a fine and shiny fibre as silk, and would have believed it came from the secretion of an insect. Hundreds of year later, the discovery of the unusual silk fibre brought about a change in economic status of China. As China and other nations began to travel around the world, they sent ambassadors with gifts to other countries. Silk was the most treasured gift from China, and her intelligent emperors soon realized that they had a very important trade commodity. To meet the great demand for silk, the emperors ordered their citizens to pay a portion of their taxes in silk cloth to ensure that they would remain an important trading power. Later, the Government imposed a law that anyone who revealed the secret of silk production would be put to death. The mystery of how silk was produced was not revealed to the western world until the reign of Emperor Justinian in the 6th century AD. According to a legend, two Christian monks were sent to China to discover the centuries old secret of silk production. They returned to Constantinople two years later with silkworm eggs hidden in hollow canes. From here, silk production spread through out the Europe.

The Silk Road refers to the ancient trade route between Central Asia and China. It comprised of many trade routes across Eurasia in various directions. These Silk Routes were important paths for cultural, commercial and technological exchange between traders, merchants, pilgrims, missionaries, soldiers, nomads and urban dwellers from Ancient China, India, Tibet, Persia and Mediterranean countries for almost 3,000 years. It gets its name from the lucrative Chinese Silk trade, which began during the Han Dynasty (206 BCE-220 CE). Originally, the Chinese traded silk internally within the empire. Caravans from the empire's interior would carry silk to the western edges of the region. They were often attacked by small Central Asian tribes for valuable commodities. When the Silk Road was first established, silk was not the principal commodity of trade. But as the rich and noble families of Rome grew fanatic over silk, it became the hottest luxury good moving along the Silk Road. During the Tang dynasty, almost thirty percent of the trade on the Silk Road was comprised of silk. But the caravans along the Silk Road also carried a vast range of other products such as gold, grapes, pomegranates, rugs, glass, jade, gunpowder, paper, compasses, bamboo, chrysanthemums, etc. Goods were transported from Byzantium (Istanbul), the eastern capital of the Roman Empire, across the Middle East and Central Asia to Xian, the capital of China. Camel caravans from the east carried gems and spices from India, furs from Siberia, silk, porcelain and paper from China. From the West came perfumes and cosmetics from Egypt and Arabia, Ivory from Africa, horses from Central Asia, and gold, silver and glass from Constantinople and Rome. But, the Silk Road was not a mere channel of trade - the trade route was the means of spreading different cultures, beliefs, spiritual wealth and religious doctrines. As the traders on the Silk Road originated from different parts of the world, they carried their own cultural lifestyles around with them. They were missionaries of their own personal cultures, and sometimes they adopted other beliefs as well. Thus, the Silk Road was the crossroad between civilizations, peoples, and cultures. The traders and their great caravans who made the tedious journeys across some of the most inhospitable territory on the earth successful - all combined along the Silk Road to make the first global marketplace in the history of civilization. Trade on the Silk Road revived tremendously during the Yuan Dynasty (1271–1368), when China became largely dependent on its silk trade. Genghis Khan conquered all the small states, unified China and built a large empire under his rule. Trade along the Silk Road reached its zenith during this period. The famous traveller Marco Polo (1254–1324) travelled along the Silk Road visiting the Yuan capital city Dadu (today's Beijing), and wrote his famous book about the Orient. In the late Middle Ages, transcontinental trade over the land routes of the Silk Road became increasingly dangerous and travel by sea became more popular, trade along the Silk Road declined.

PRESENT SILK SCENARIO

Various types of fibres such as cotton, synthetics, cellulose, wool and silk are in use worldwide: silk production is quite significant. The share of silk in world production of all textile fibres in 2007 remained unchanged from its 2000 level of about 0.2%. While mulberry silk production has admittedly been steadily rising, silk will never be available in large quantities and international supplies, will, no, doubt remain limited in future. The world production of raw silk has slightly dropped by 1.5% from 1,28,870 tons in 2005 to 1,26,995 in 2009 (Table 1.1). The China and India as first and second largest producers remain almost unchanged. Silk production Japan, Brazil and Republic of Korea continues to decline, largely because of industrialization. Japan remains the world largest market for silk products and its imports of silk products continues to rise. China has become Japan's largest supplier of manufactured silk goods. The bulk of the silk woven in China comes from power-looms; production in Thailand and India is done in handlooms. All weaving in Brazil and the Republic of Korea is carried out on power-looms. The use of power loom is expanding; this is particularly so in India where the increasing domestic demand for sari is justifying the increased use of this type of loom.

Traditionally, more than 90% of the world market for silk garments was geared to women's wear. Silk product for men was in the past limited to shirts, neckties, handkerchiefs, socks and underwear. The situation has changed significantly in the west; as a direct result of the marketing of sand washed silk. China is today's largest supplier of raw silk to the international market. India by contrast has becomes the world largest importer of raw silk. Per capita consumption of silk in Europe has traditionally been highest in Switzerland, followed by Germany and the United Kingdom. Japan has the highest per capita consumption of silk in the world. Although consumption has followed in recent years, particularly because of the drop in demand for kimonos. Japan position is likely to remain unchanged as demand rises for other silk products. Hong Kong (China) taken together, are today the world largest converters of silk fabrics. Nearly 60% of the silk garments produced in Hong Kong is send to the United States; the balance is exported to various European countries. Hong Kong is the second highest exporter of silk blouses to Germany and India and imports raw silk from China, Brazil and Vietnam, reached between 811 and 550 tons in 2009. China overall domination of International silk trade is expected to continue for years to come despite the growing number of aspiring silk producers, such as Bolvia, Columbia, Cote d'Ivoire, Israel, The Islamic Republic of Iran, Peru, Sri Lanka, Turkey and Uganda.

Apart from that; France, Germany, Italy, Switzerland and the United Kingdom are the Europe leading converters of silk. Germany is by far Europe's leading market for textiles and clothing, including silk products. It has also become one of the world most competitive markets in the sector. Italy has over the years been the world's largest importer of waste silk, which is spun into yarn for the production of knitwear. With per capita consumption about a third of that in Germany, Sweden continues to have the lowest consumption rate in Europe. Brazil is the world's third largest silk producer after China and India. China is the world largest producer of "tussah" a type of wild silk. Total raw silk production is approximately 104000 tons annually, takes place mainly in northern province of Liaoning. The out put of China amounts to 84000 tons of mulberry silk and 20000 tons of 'tussah" in 2009.

SILK PRODUCTION IN INDIA

India is the only country in the world to produces all the commercially known varieties – mulberry, tasar (both tropical and temperate) eri and muga. It ranks second to China as mulberry silk producer and accounts for about 15.5% of world's production of raw silk. It is also the second largest producer of tasar silk, gain after China. It monopolizes production of the golden yellow muga silk. Various categories of people, like the cocoon producer, reeler, twister, weaver and trader are involve in sericulture. India has large domestic market for silk goods and only 15% of the production is exported. About 85% of the silk goods sold on the domestic market consists of traditional items such as sarees, grey fabrics, blouse and

Table 1.1. World Mulberry and Non-Mulberry Raw Silk Prodcution

World Mulberry Rawsilk Production *[Unit : Tonnes]*

Country	2001	2002	2003	2004	2005	2006	2007	2008	2009 (P)	% Share
China	62560	64100	76324	85000	87800	93100	78000	70980	84000	81.06
India	15842	14617	13970	14620	15445	16525	16245	15610	16322	15.75
Japan	431	394	287	263	150	150	105	95	90	0.09
Brazil	1485	1607	1563	1512	1285	1387	1220	1177	811	0.78
Korea Republic	157	154	150	150	150	150	150	135	135	0.13
Uzbekistan	1260	1260	950	950	950	950	950	865	750	0.72
Thailand	1510	1510	1500	1420	1420	1080	760	1100	665	0.64
Vietnam	2035	2200	750	750	750	750	750	680	550	0.53
Others	1692	3814	1500	1500	1500	1000	500	350	304	0.29
Total	**86972**	**86972**	**89656**	**106165**	**109450**	**115092**	**98680**	**90992**	**103627**	**100**

World Rawsilk Production

[Unit : Tonnes]

Country	2001	2002	2003	2004	2005	2006	2007	2008	2009(P)	% Share
China	64567	68600	94600	102560	105360	130000	108420	98620	104000	81.89
India	17351	16319	15742	16500	17305	18475	18320	18370	19690	15.50
Japan	431	394	287	263	150	150	105	95	90	0.07
Brazil	1485	1607	1563	1512	1285	1387	1220	1177	811	0.64
Korea Republic	157	154	150	150	150	150	150	135	135	0.11
Uzbekistan	1260	1260	950	950	950	950	950	865	750	0.59
Thailand	1510	1510	1500	1420	1420	1080	760	1100	665	0.52
Vietnam	2035	2200	750	750	750	750	750	680	550	0.43
Others	1692	3814	1500	1500	1500	1000	500	350	304	0.24
Total	**90488**	**95858**	**117042**	**125605**	**128870**	**153942**	**131175**	**121392**	**126995**	**100**

dhotis. India exports consist exclusively sarees, dress fabrics, ready made garments and made up articles for interior decoration (e.g., bed spreads, cushion covers, curtains). During 1950, the total silk production was about 960 MT. Over the years due to several schemes and programmes implemented by respective state and central government, there has been significant growth in terms of silk production. India has produced a total of 19,690 MT raw silk during 2009-10. Among this, mulberry, tasar, eri and muga contributes about 16,322, 803, 2460 and 105 MT of raw silk respectively. The percentage of mulberry silk to the total raw silk produced in the country is around 82.90 per cent while tasar is only 4.07 per cent. The eri and muga contributes about 12.5 and 0.53 per cent of the total silk (Table 1.2.).

Table 1.2: Silk production in India

Production of Silk			
Year	Total Mulberry Production		
	Dfls	Cocoon	Raw Silk
Unit----->	Lakh. Nos	MT	MT
2002-03	2920.08	128181	14617
2003-04	2598.13	117471	13970
2004-05	2565.28	120027	14620
2005-06	2625.48	126261	15445
2006-07	2838.06	135462	16525
2007-08	2647.45	132038	16245
2008-09	2410.90	124838	15610

MULBERRY PRODUCTION STATISTICS			
Year	Bivoltine Production		
	Dfls	Cocoon	Raw Silk
Unit----->	Lakh. Nos	MT	MT
2002-03	168.76	5438	685
2003-04	165.87	4721	609
2004-05	189.18	6254	893
2005-06	195.24	6696	971
2006-07	229.65	7618	1100
2007-08	226.64	8092	1175
2008-09	218.06	8422	1250
2009-10	211.77	8076	1182

MULBERRY PRODUCTION STATISTICS			
Year	CB Production		
	Dfls	Cocoon	Raw Silk
Unit----->	Lakh. Nos	MT	MT
200-03	2751.32	122743	13932
2003-04	2432.26	112750	13361
2004-05	2376.10	113773	13727
2005-06	2430.24	119565	14474
2006-07	2610.20	127844	15425
2007-08	2647.46	123947	15070
2008-09	2410.90	116416	14360
2009-10	2283.54	123609	15140

Source: as reported by DOSs of States

Vanya Silk (Non-mulberry)			
Dfls: Lakh Nos.		Reeling Cocoons Tasar: Lakh Kahan,	
Rawsilk:MT,			
	Tasar		
Year	**Dfls**	**Reeling Cocoons**	**Raw Silk**
2002-03	90.99	2.77	284
2003-04	84.33	2.84	315
2004-05	84.88	3.04	322
2005-06	54.04	2.71	308
2006-07	95.06	3.30	350
2007-08	100.24	4.04	428
2008-09	130.03	5.33	603
2009-10	142.85	7.45	803
Dfls: Lakh Nos.		Reeling Cocoons Eri: Tonnes,	
Rawsilk:MT,			
	Eri		
Year	**Dfls**	**Reeling* Cocoons**	**Raw Silk**
2002-03	216.02	1767	1316
2003-04	230.2	1810	1352
2004-05	238.8	1932	1448
2005-06	211.1	1749	1442
2006-07	226.5	1858	1485
2007-08	205.00	1983	1530
2008-09	280.00	2590	2038
2009-10	308.91	2801	2460
	Muga		
Dfls: Lakh Nos.		Reeling Cocoons Muga: Lakh Nos.	
Rawsilk:MT,			
	Muga		
Year	**Dfls**	**Reeling Cocoons**	**Raw Silk**
2002-03	96.38	5023	102
2003-04	101.17	4866	105
2004-05	105.45	5198	110
2005-06	105.69	5228	110
2006-07	105.49	5159	115
2007-08	98.84	4933	117
2008-09	117.24	5880	119
2009-10	99.52	5155	105

Source: as reported by DOSs of States

Mulberry silk is chiefly produced in five states viz, Karnataka, Andhra Pradesh, Tamil Nadu, West Bengal and Jammu & Kashmir. Due to heavy demands of mulberry silk through out the country for the production of saris and grey fabrics, mulberry silk production is now increasing steadily and spreading to many non-traditional states viz. Rajasthan, Gujarat, Uttarakhand, Bihar, Orissa, Assam, Kerala, Himanchal Pradesh and Madhya Pradesh. Among all these states Karnataka is now contributing over 60 per cent of the total silk produced in the country. Tasar silk is produced in Jharkhand, Bihar, Chattisgarh, Maharastra, Orissa, and Andhra Pradesh. The eri silk production is chiefly restricted to North Bihar and North-East state of India, Oak tasar sericulture is restricted to North Eastern and North-Western part of India. North-East India is unique in its large biodiversity of sericigenous insects. It contributes to approximately 14% of the total raw silk production in India, but the share of eri is 97% and muga is 99.8%. Oak tasar culture is practiced in the entire sub-Himalayan belt right from Manipur, Nagaland, Arunachal Pradesh in the North-East to Jammu & Kashmir, Himanchal Pradesh and Uttarakhand in the North-West touching the fringes of Assam and Meghalaya. Manipur, Mizoram and Nagaland are three main states of North-Eastern region which can afford great potential for exploitation of natural oak tasar silk production.

Eri silkworm rearing is being practiced traditionally in rural areas of India especially in North eastern region. Of late eri silk production is increasing steadily in Bihar, Andhara Pradesh and Orissa and the muga silk production remained stagnant during the last few decades Today muga silk production is only 0.53 per cent of the total silk produced in India. This silk is endemic to Assam and North-Eastern states since high humid climate is essential for muga food plants and muga silkworm rearing. The silk production and its export has also increased.

During 2001-02, earning from silk goods export was 2359.56 crore however during 2009-10 it has increased to 2871.79 crore. Dissemination of advanced technologies supplemented by the quality seed supply, extension programmes, training, technical assistance, etc., have resulted in improvement of production, productivity and quality of mulberry cocoons in the field. There has been considerable impact on increase in uptake of silkworm seed and enhancement of raw silk production per unit area. There has been upgradation of infrastructure of farmers, improvement of skills and knowledge of the farmers and paved way for strict discipline in sericulture practices. This has been made possible through technology demonstration programmes, adoption of new technologies in mulberry cultivation, silkworm rearing and disease management. Besides this other extension activities viz. Krishi mela/ field days/farmers days/Exhibition/Audio/visual/Vichar gosthi/Work shop/film shows are undertaken to reach more number of beneficiaries and keep abreast of the latest methodologies among the farmers. The growth and development of sericulture depends on the rapidity with which the new technologies and practices are developed. The success of Brazil, South Korea and China in improving the productivity is mainly attributable to the technological advancements, systematic adoption and meticulous practices. Indian scientists have also significantly contributed to the development of technologies for silkworm rearing practices.

In order to increase the silk production, productivity and quality of silk, it is better to minimize the incidence of pests on silk host plants and attack of parasites and predators on silkworms. Insect inflicts considerable damage to almost every part of the silk host plants and the seriousness of the attack is derived by the nature of the injuries inflicted and the susceptibility of the plant concern to insect attack. The biting and the sucking types of insects generally attack silk host plants. It causes direct loss by feeding and indirect loss through helping in the transmission of diseases – bacterial, fungal and viral. Defoliators cause maximum injury by feeding on the growing points of the plants. They make holes by feeding or roll up the leaves and feed from the inside or only feed on the layer of surface tissue. Sometimes they may live concealed under loose bark of the plants or cut the tender stems of the plants.

Insects with sucking habits cause general chlorosis of the leaves or a silver whitening of the leaf surface or yellow specking or brownish necrotic lesion. Crinkling and curling of the leaves is a common effect. Injuries are also caused by internal feeders, feeding from within the plant tissues, some called the borers, boring the internal tissues and other called the minors, mining the leaf tissue. Many plant tissues react to insect salivary toxins during feeding, by the formation of galls or abnormal out growths and due to this toxaemia the growth of the plants may be impaired. Another aspect of injury is through feeding by subterranean insects living in the soil, feeding on the roots of the plants by chewing, or boring, sucking, or by formation of galls. In general the attacked parts result in stunting, discoloration, withering.

The indirect effect of feeding is evident in the loss of quality of produce through reduction in the nutritive value. Insects are indirectly responsible for the severe damage and loss caused through mechanical of biological transmission of diseases agents The natural equilibrium or stability of the insect population is disturbed resulting in very favourable opportunities for abnormal increase of populations of species making them reach pest proportions. They reach a pest status when they are responsible for 5 per cent of the loss of yield and are designated minor or major pests when the loss ranges from 5 to 10 per cent and more respectively. According to periodicity of their occurrence insect pests are said to be regular, occasional, seasonal, persistent or sporadic and sometimes severe infestations often result in epidemic. The major causes for outbreak of insect pests is through destruction of forests or by bringing them under cultivation, destruction of their natural enemies – predators and parasites, intensive cultivation of the crops and new strains, accidental introduction of foreign pests, etc. (Van Lenteran, 1987; Waage, 1983). The assessment of pest population is the most important aspect before adoption of the control measures. Applying various sampling methods usually does pest assessment. The basic objective of the insect control programme is related with avoidance, elimination or reduction of the factors, which promotes excessive multiplication of insects. For efficient functioning of indirect control measures, a through knowledge of their life cycle, their pest status, distribution, periodicity, host complex, behaviour among others, is pre-requisite, essential for timing of the controls (Wheatly *et al.*, 1989). The economics of the control measures is also very vital factors; therefore, any control measures devised should be practicable, cheap and effective. Insect control methods have been broadly classified into the natural and applied control methods. Natural control methods is related with manipulation of a biotic factors viz. temperature, humidity, rain fall, soil conditions, wind velocity, topographic factor including barriers like mountain ranges, large bodies of water, thick forests restricting the spread of insects and natural enemies like the predators and parasites which increase with the increase of populations of insects pests.

Applied control methods are mostly divided into the preventive and curative or direct methods. Preventive or prophylactic methods to ward off pest invasions include measures such as removal of weeds, grasses and dead branches form the field; through usages of good seeds, proper cultivation, maneuvering, irrigation, removal of crop remains and stubble, growing of pest resistant varieties, treatment of seeds with chemicals, rotation of crops, mixed cropping, changing of sowing, planting and harvesting times, cutting and ranking of field bunds, etc. Among the direct methods are mechanical and physical control methods. The physical methods involve the use of electricity, sound wave, infrared and X-rays. The mechanical method is concerned with the operation of machinery such as hopper dozers, fly and maggot traps, light traps, centrifugal force machines and manual operation involving the hand picking of egg masses, larvae and adults. Chemical methods include the use of insecticides, repellants and attractants, antecedents and chemosterilants. Biological control is applied to control the pest population by the introduction of parasites, predators and pathogens of the pests. Legal or legislative methods of control are also adopted to check the introduction of exotic pests, diseases, weeds, etc., with a view to help in the prevention of their spread within the country. Therefore, an integrated approach needs to be developed by understanding the bioecology and behaviour of the pests, parasites and predators existing in the sericulture ecosystem.

CHAPTER-2

Silkworm and Its Host Plants

Silk producing insects are highly specialized groups of invertebrates belonging to the family bombycidae and satuniidae in order Lepidoptera grouped under the largest of animal phyla, the Arthropoda. They are also known as sericigenous insects as the outer layer of the silk filament comprises a protein called sericine. Based on the feed consumed, they are largely divided into mulberry silk insect, *Bombyx mori* L. (Bombycidae), feeding exclusively on mulberry (*Morus spp.*) and Non-mulberry silk insects feeding on a number of plants other than mulberry. The family bombycidae is characterized by the absence of proboscis and presence of bipectinate antennae in both the sexes. The legs are hairy and without spurs. Bombycidae includes fifteen silk producing insects species of small dull moths. The ecological distribution and feeding behaviour is different and silkworms are fully domesticated. The *Bombyx mori* is the well known silkworm of this family which has been introduced into the many parts of the world for commercial purposes. It is now entirely domesticated and is not known in the wild state. The distribution of mulberry silk fauna is presented on Table 2.1. A number of local races exist, and Hutton, Cotes and others have regarded these as distinct species. They differ chiefly in the number of annual broods, which are largely dependent upon climate and completely domesticated. One important difference induced by domestication is the variation in the number of broods; this is a matter depending largely upon climate, the one brooded (univoltine) being probably the normal habit in a cold climate with short summer, the multivoltine found in warm moist localities. The natural food in all cases is the leaves of mulberry (*Morus alba*) and the silk produced is white or yellow. Seitz (1928) reported the silk producing bombycoid species of the oriental region. Mookerjee (1919) opined that *Bombyx mori* probably dispersed from the Himalayan country to conducive and favorable regions particularly the oriental countries – carried by human agencies where the species underwent local variations as time passed. Robinson (1971) after a series of kariological studies of the races of south-east Asia and north-east India was of the opinion that the region may claim to be the likely place of the origin of *B. mori*. Danilevskii (1975) referred that sericulture perhaps originated in the region of south-east Asia.

Non-mulberry silk insects are polyphagous unlike the mulberry silk insect and are largely from the family, Saturniidae. Distribution of non mulberry silk fauna have been listed on Table 2.2. Saturniidae comprises the largest group of sericigenous insects belonging to the genus *Antheraea, Philosamia*, and *Attacus*. *Antheraea* comprises of more species than any other genus of sericigenous insects. Nearly 500 varieties of silk producing caterpillars live all over the world. Their size and appearance is variable and always beautiful. Only the fiber from four wild silkworms (tasar, oak tasar, muga and eri) is used commercially. Silk produced from the remainder is unsuitable for a variety of reasons: the quality is weak and inferior, the cocoons cannot be unraveled or the wild nature of the caterpillar and moth make it nearly impossible to collect enough cocoons to make it financially feasible endeavor. Much of the attempts have been made to cultivate many types of wild silkworms under captivity, but sure success has not been achieved. India is home to all three commercial wild silkworms. Rearing these wild silkworms is an ancient and integral part of aboriginal life. Rearing wild silkworms is a risky venture, and only 20-30 per cent of hatched eggs make cocoons. Thirty-eight species have been recorded in north India.

Besides, there are over fifty forms/variants/aberrant/races. Among the one hundred and sixty species of non-mulberry silkworms so far reported (Seitz, 1933) *Antheraea mylitta* Drury, *Antheraea proylei* Jolly, *Antheraea assamensis* Ww., *Antheraea pernyi* Guer, *Antheraea yamamai* Guer and *Philosamia ricini* Hutt. are exploited for commercial silk production and only limited studies so far has been conducted to evaluate the economic importance of other sericigenous species. *Antheraea yamamai* is the Japanese oak silkworm which is reared on large scale in that country, and was introduced into Europe in 1861. *A. pernyi*, the Chinese oak silkworm, yields Shantung silk which is pale buff in colour and largely exported. *Antheraea paphia or mylitta* and *A. assama* are polyphagous and produce silk after feeding on various food plants. The Emperor moth *Saturina pavonia* and *S. pyri* also yield silk of commercial value. *Attacus atlas* and *A. edwardsi* are among the largest moths in the world, the females having a wing expanse of about 25 cm. Among giant silkmoths, *Cecropia silkmoth* (*Hyalophora cecropia*), *Columbia silkmoth* (*Hyalophora columbia*), *Ceanothus silkmoth* (*Hyalophora euryalus*), *Promethea silkmoth* (*Callosamia promethea*), *Tulip tree silkmoth* (*Callosamia angulifera*), Sweetbay silkmoth (*Callosamia securifera*) and *Calleta silkmoth* (*Eupackardia calleta*) are most common and reported from North America. These are medium to very large-sized moths, with adult wingspans ranging from 7.5 to 15 cm. They have hairy bodies and relatively small heads. Caterpillars feed on a wide range of native and ornamental trees and shrubs. Caterpillars of giant silkworms pupate in a well-built silken cocoon. Several species yield silk of commercial value. A majority of the world's wild silk is produced by the 'Tussah (*Antheraea*)' caterpillars. They produce a honey beige coloured silk and are reared extensively in China, India, Korea and Japan. Tropical Tasar silkworm (*Antheraea mylitta*) is reared on the food plants of arjun (*Terminalia arjuna*), asan (*T. tomentosa*) and sal (*Shorea robusta*). Wild cocoons of tasar are largely collected from sal from forests. Tropical Tasar silkworm (*Antheraea mylitta*) is uni/bivoltine, polyphagous on wild forest trees; never domesticated, but fertilization takes place in the captivity from reared females; found wholly free living in forests. Silk is dirty brown and reel able, cocoon with peduncle. The genus

Shorea (Dipterocarpacae) comprising of four species viz. *Shorea robusta*, *S. assamica*, *S. telura* and *S. tumbuggaia* are available in Indian sub-continent. The sal is of great significance from silviculture point of view as it provides valuable wood as a major forest produce. Wild cocoons of tasar occasionally are collected by tribal from sal (*Shorea robusta*) forests. Commercial rearing and natural collection of tasar cocoons are conducted on all these food plants in the different region of India. Due to geographical and ecological variation in the different region, various eco races have been screened and each one is acclimatized to the specific region on specific host plants. Tasar silkworm is represented by 44 eco-races of which the economically important ones are daba, raily, laria, sarihan, sukinda, bogai, modal, bhandara and Andhra local. Majority of tasar silkworm population in nature is bivoltine (July-August, September-October) while trivoltine (November-February) is also found in few ecotypes like daba and sukinda. Tropical tasar food plants are distributed through central and southeran plateau regions and in the humid and dense forests of Bihar, Jharkhand, Chattisgarh, Orissa and West Bengal and the borders of Uttar Pradesh, Maharastra and Andhara Pradesh. There are about 11.1 million hectares of tasar food plants available in the country of which only less than 5% area is utilized for food plants and remaining lies in the deep forests and is inaccessible to tribal for tasar silkworm rearing.

Oak tasar (*Antheraea proylei*, Jolly) is a synthesized hybrid from the inter-specific hybridisation between an indigenous species *Antheraea roylei* and its Chinese counterpart *Antheraea pernyi*. Silkworm rearing is practiced generally in North-eastern and northwestern region of India. The spring crop is the main seed/ commercial crop, which starts from the first week of March. However, an additional subsidiary crop is also taken during summer or autumn by breaking the pupal diapause. The rearing schedule in North-eastern and North-western region is different due to sprouting behaviour, maturity of food plants and different climatic conditions.

Muga silkworm (*Antheraea assamensis* Westwood) is multivoltine and polyphagous insect largely reared on commercial scale in the Brahamaputra River Valley in Assam and sparsely in other north- eastern states of India. It is partly domesticated but also found wild; silk white or yellow, reel able, the pupae requiring to be stifled; cultivated in Eastern Bengal and Assam. Larva of muga silkworms feed upon mejankori leaves, *Litsea citrate.* So muga silk is known as "mejankori silk", which is attractive due to its durability, luster and creamy white shaded fibre. The silk released from these cocoons is shimmery and golden. It is rarely seen outside Assam because the Assamese people cherish it as cloth for all the important occasions in their lives.

Eri (*Philosamia*) caterpillars are found mostly in north-eastern India with a few in China and Japan. It is multivoltine polyphagous but reared only on castor; partly or wholly domesticated for rearing but also found wild rarely; Eri silkworms, though spin a continuous filament, the cocoons are not reelable like others as the cocoons are loose flossy with pointed end with a very thin filament weaker at the open pointed end and hence cut and spun Eri silkworm (*Philosamia ricini*) feeds on several varieties of food plants of which castor, kesseru, tapioca and barkesseru are important.

Apart from this there are several other silkworm species which are of economic importance and commercially exploited for large scale silk production provided some technologies are developed. The most common among them are Anaphe, Cecropia, *Actias, Cricula* and several others. Anaphe (*Anaphe infracta*) is a common silkworm found in Uganda and other part of Africa. Caterpillars feed on the leaves of a species of fig tree. The nest and cocon which are formed in considerable numbers are used for waste silk. Silk is reddish brown in colour some what like tussah silk. In southern Nigeria, anaphe silk is used by the natives in conjuction with cotton for making the so called "Soyan" cloth. Anaphae (*Anaphe infracta*) caterpillars make a five inch brown communal silk pouch. Inside the pouch, each silkworm spins its own small cocoon. In some part of Africa, these pouches are collected and sold in the markets when the moths have flown away after metamorphosis. Silk is spun from the pouch, not from the individual cocoons, to make cloth. This type of silk is produced by the anaphe silkworm which has various species like *A. moloneyi* Druce, *A. panda* Boisduval, *A. reticulata* Walker, *A. ambrizia* Butter, *A. carteri* Walsingham, *A. venata* Bulter, and *A. infracta* Walsingham. The silk is distributed in southern and central Africa. The fabric is elastic and stronger than that of mulberry silk. Cecropia (*Hyalophora cecropia*) caterpillars were familiar to the native North Americans of long ago. They use the cocoons of this caterpillar to make rattles and dancing anklets for their shamans and ceremonies. *Actias selene*, Hubn., is a very striking insect, the forewing is large, the hind wing produced into a long tail; the colour is delicate pale green, the forewing having a dark pink fore edge; in each wing is buff and red spot. The larva is green with yellow tubercles bearing spines. The species is widely spread over the hill forest areas in India, at low elevations often, but typically a subtropical and occasionally a planes species in Chotanagpur.

Cricula silk *(Cricula trifenestrata)*, produces a beautiful golden cocoon and inhabits throughout Indonesia and elsewhere in South Asia. Recently Japanese scientists and technologists succeeded in developing these cocoon sheets, yarn and textiles. Because of their natural golden colour and very fine filament, all goods commanded a price more than several times higher than Bombyx silk. The price of Cricula cut cocoon varies widely depending on its use, but now is in the range US $ 50-1000 per kg. The wild cut cocoons; Eri and Chinese tasar are sold for US $ 3-8 and US $ 5-10 respectively. It is recognized that *Cricula* cocoons are very expensive, because of their characteristic shiny colour and finer porous filament. Gonometa silk is produced by various *Gonometa* species abundantly available in various part of Africa. The larvae are reared on different *Acacia* species. Apart from this 19 species in 6 genera were recorded pupating in silk cocoons. *Argema* and *Epiphora* cocoons have quite good potential for silk production in some part of Africa. Some of the wild silk moths belong to the family Lasiocampidae and Thaumetopoeidae are also commercially exploited for silk production in south Africa. Fagara silk is produced by the giant silk moth, *Attacus atlas* and few other related species or races inhabiting India, Australia, China and Sudan. The fagara cocoons are light brown in colour and of less importance, since the silk is not commercially exploitable. Coan silk fibre is secreted by the larva of *Pachypasa otus* D. These larvae are found in Mediterranean biogeographic region. This is a polyphagous insect feeding on pine, ash, cypress, juniper and oak. The cocoons are white in colour. Mussel silk belongs to the non-insect category which is obtained from a particular mollusca, *Pinna squamosa*. The fibre is called "byssus thread", brown in colour, strong in nature and helps the animal to anchor itself to a rock or any surface of the habitat. The byssus is combed and then spun into silk, popularly known as "fish wool". Its production is largely confined to Toronto and Italy.

Weaver ant silk are tiny animals living in colonies, which consist of as many as 5 million female workers, progeny of a single enormous queen. The common species of weaver ants are *Oecophylla longinoda* and *O. smaragdina*. The *O. longinoda* is the yellow tree ant of tropical Africa, whereas *O. smaragdina* is the red ant of south eastern Asia. The nests are made of leaves folded together to form tight tent-like compartments. The leaves are held in place by scams of silk spun by larva. It comes from the larval gland, opening just below their mouth (Mohanty and Mathur, 1997).

Green lacewing fly silk is an insect of the Order Neuroptera. The important species are *Chrysoperla carnea, Mallada boninensis* and *Osmylus fulvicephalus*. The larvae live in wet moss and undergo complete metamorphosis. They are predatory, feeding largely on aphids. When fully grown, each spins a cocoon of white silk, given out from a spinneret, which is at the hind end of the body and not on the head as in the silk spinning caterpillar. The cocoons are usually attached to the leaves or barks (Burton and

Burton, 1990). The common green Autralian lacewing has attracted the attention of of researchers in various part of the world. The silk produced by the lacewing is tougher and upto six times more elastic than that produced by the silkworm. Its unique strcuture which differs from that of other silk producing insects, is also easier to recreate artifically than silkworm silk. Today we take interest in some silk producing moths for their beauty and others for their devastating effect on our trees and shrubs. We are all too familiar with tent caterpillars and gypsy moths, two wild silkworms that have become a menace. The gypsy moth, a native of Europe, was brought to Massachusetts in 1869 to be cross bred with the *Bombyx mori* silkworm in the hope of making rearing easier. Some caterpillars escaped when wind destroyed their cage, eventually creating problem all across North America. In contrast we enjoy the beauty of their large, colorful, wild silk caterpillars and moths: *Cecropia* (*Hyalophora cecropia*), Luna (*Actias luna*), polyphemus (*Antheraea polyphemus*), Promethea (*Callosamia promethea*).

SPIDER SILK

Spider silk is soft and fine and is spun by arachnids, like *Nephila madagascarrensis*, *Miranda* and *Eperia*. Spider silk is a protein fibre spun by spiders. Spiders use their silk to make webs or other structures, which function as nets to catch other animals, or as nests or cocoons for protection for their offspring. They can also suspend themselves using their silk, normally for the same reasons. Spider silk is a remarkably strong material. Its tensile strength is comparable to that of high-grade steel. (1500 MPa), and about half as strong as aramid filaments, such as Twaron or Kevlar (3000 MPa). Most importantly, spider silk is extremely lightweight: a strand of spider silk long enough to circle the Earth would weigh less than 500 grams (Shao, 2002). Spider silk is also especially ductile, able to stretch up to 140% of its length without breaking. It can hold its strength below 40 °C. This gives it a very high toughness (or work to fracture), which "equals that of commercial polyaramid (aromatic nylon) filaments, which themselves are benchmarks of modern polymer fibre technology". Hayashi (2007) developed methodology for replicating the properties of spider silk. Spiders weave their webs from proteins secreted from silk glands in their abdomen. What starts as liquid becomes a solid fibre as it's extruded from spinnerets. Some spiders make only one kind of silk. But many of the 40,000 categorized species of spiders possess a toolkit of different glands to make silk for various functions — such as constructing egg sacks, trapping and wrapping prey or creating a dragline to hang from the eave of a roof. Each type of silk is composed of a unique set of proteins that when combined make spider silks some of the toughest natural fibres known.

TRANSGENIC ANIMAL

From the past decades, attempts have been made to harvest the silk protein from the goat. This is research done by a Canadian company called Nexium. There's a gene called beta casein that's one of the milk proteins. Nexium took the on-off switch for the milk protein gene and fused that with the spider silk gene. So only a female goat that is producing milk will make [the silk protein]. Basically every morning you go out there with your bucket and you milk your goat. The silk protein will come out, but so will the other milk proteins, and then they have to be separated. A product that spiders have been crafting for perhaps 400 million years is being developed to protect military and law enforcement personnel.

The material is spider-dragline filament from which arachnids spin their webs. It is one of the strongest materials in the world—many times stronger than steel. Spider silk's tensile strength is such that it can withstand weight of up to 300,000 pound per square inch. Scientists currently are developing dragline filament for use in the next generation of bulletproof vests. Currently, bulletproof vests are made of Kevlar, which provides a dependable barrier against bullets. Soldiers and police personnel, however, report that Kevlar vests are heavy; inflexible and hot to wear. Vests made of dragline filament may resolve these problems.

Table 2.1. Distribution of mulberry silk fauna

Common name	Scientific name	Distribution
Mulberry silkworm	*Bombyx mori* L	World wide domesticated
Wild mulberry silkworm	*Bombyx mandarina*	Japan, Korea
Desi Polo, Chota Polo	*Bombyx fortunatus*, Hutt.	China, India
Nistry, Madrassi	*Bombyx croesi*, Hutt.	China, India
Naya Paw	*Bombyx arracanensis*, Hutt	India, China
Boro Polo, Bara Pat	*Bombyx textor*, Hutt.	India, Korea
Sina, Cheena, Chota Pat	*Bombyx sinensis*, Hutt	China
	Andraca bipunctata	North-east
	Ocinara albilunata	
	Ocinara diaphana	
	Ocinara cyproba	
	Ocinara lactea	
	Trilocha varians	South India
	Trilocha albilunata	
	Gunda javanica	Himalayan foot hill
	Gunda sikkima	Sikkim
	Mustilia pephara	Meghalaya
	Mustilia columbaris	Himalayan foot hill

Table 2.2. Global distribution of Non-Mulberry silk fauna

Antheraea (Tasar)	**Philosamia (eri)**
A. mylitta Drury	*P. cynthia* D.
A. paphia Linn.	*P. ricini* Boisd.
A. knyvetti Hamps.	*P. lunuloides* Reb.
A. assama Westwood	*P.obscura* Btl.
A. compta R.& J	*P. canningii* Hutt.
A. frithi Mr.	*P. walkeri* Fldr.
A. helferi Mr.	*P. pryeri* Butl.
A. roylei Mr.	*P. fluva* Jord.
A. sivalica Mr.	*P. insularis* Voll.
A. andamana Mr.	*P. vaneeckei* Wts.
A. pernyi G.M.	*P. vanderberghi* Wts.
A. yamamai G.M.	*P. luzonica* Wts.
A. pasteuri Bouv.	*P. tetrica* Rbl.
A. raffray Bouv.	*P. borneensis* Rbl.
A. jana Stoll.	*P. ceramensis* Bouv.
A. semperi Fldr.	*P. mindanaensis* Rbl.
A. cordifolia Weym.	*P. advena* Pack.
A. pratti Bouv.	
A. imperator Wts.	
A. brunnea Eecke.	
A. billitonensis Mr.	**Attacus (fagara)**
A. larissa Ww.	*A. atlas* Linn.
A. ridleyi Mr.	*A. standingeri* Roth.
A. prelarissa Bouv.	*A. crameri* Fldr.
A. surakarta Mr.	*A. edwardsi* White
A. mylittoides Bouv.	*A. dohertyi* Roth.
A. delegate Swh.	*A. taprobanis* Mr.
A. fickei Weym	*A. macmulleri* Wts.
A. pristina Wkr.	*A. simalurana* Wts.
A. sciron Ww.	*A. erebus* Fruhst.
A. harti Mr.	*A. gladiator* Fruhst.
A. gephyra Niep.	*A. lorquini* Fldr.
A. rumphi Fldr.	*A. caesar* M.&W.
A. eucalypti Sott.	*A. imperator* Ky.
A. larissoides Bouv.	*A. cynthia* D.
A. polyphemus Cram.	*A. ricini* Boisd
A. fasciata Mr.	
A. sumatrana Niep.	

A. borneensis Mr.	**Anaphe**
A. korintjina Bouv.	*A. infracta* Wals.
A. perrotteti Guer.	*A.venata* Butl.
A. yongei Wts.	*A.moloney* Druce
A. insularis Wts.	*A.panda* Boisd.
A. javanensis Bouv.	*A.reticulata* Walk.
A. hazina Butt.	*A.carteri* Wals.
A. calida Butl.	
A. morose Butl.	**Nephila etc (spider)**
A. fentoni Butl.	*N. madaguscarensis*
A. sergustus Ww.	*Miranda aurentia*
A. nebulosa Hutt.	*Epeira* spp.
A. olivescens Mr.	
A. platessa Roth.	**Pachypasa (coan)**
A. versicolor Mr.	*P.otus* D.
A. pulchra Mr.	*P.lineosa* Vill
A. ochripicta Mr.	
A. fraternal Mr.	**Pinna (mussel)**
A. cingalesa Mr.	*P. squamosa*
A. celebensis W.& S.	
A. buruensis Bouv.	
A.subcaeca Aurvil.	
A. fusca Roth.	
A. minahassae Niep.	

Antheraea polyphemus and (*Philosamia*) *advena* are endemic in the U.S.A. Besides this some other species viz. Loepa newara Mo, Loepa katinka Westw., *L. sikkima* Mo., *L. Miranda* Mo., have also been reported from various geographical region. Among Saturina species *Saturnia stoliczkana* Feld., *Saturnia pyretorum* (*S. Cidosa* Mo.), *S. grotei* Mo., *S. huttoni* Mo. (Neoris), *S. simla* Westw. (Ciligula). *S. thibeta* Westw.(Caligulal), *S. zuleika* Ho. (Rinaca) are most common. Two new species of Salssa, viz. *Salassa lola* Westw. and *Salassa royi* Elw. are reported from North America. Some other sericigenous insect viz. *Brahmaea wallichii* Gr. (= *B. certhia* F.), *Theophila huttoni* Westw. (religiosa, Helf., bengalensis, Hutt.), *Ocinara varians* Wlk., *Trilocha albicollis* Wlk., *O. apicalis* Wlk.. O. lida, Mo., *O. signifera* Wlk., O. lactea Hutt., O.diaphana Mo., *Cricula trifenestrata* Helf and *Cricula drepanoides* Mo have been reported from North-eastern region of India. Tuskes *et al.* (1996) reported about 1200 species worldwide in his book; *"The wild silk moths of North America"*. This work covers the approximately 70 species found in North America north of Mexico, and also several subspecies. Detail life history of certain familiar luna and cecropia moths have been reported. Besides this, each species entry provides some general comments and discusses distribution, adult description and biology, immature stages of the moth, and rearing.

LIST OF SOME WILD SILK MOTHS AND THEIR SILK

Wild silks are often referred to in India as 'Vanya' silks. "The term 'Vanya' is of Sanskrit origin, meaning untamed, wild, or forest-based. Muga, Tasar, Oak tasar and Eri silkworms are not fully tamed and the world lovingly calls the silks they produce as Wild silks. India produces four kinds of silk: mulberry, tasar, muga and eri. The silkworm *Bombyx mori* is fed on mulberry leaves cultivated in plantations. Silkworms are also found wild on forest trees, viz. *Antheraea mylitta* which produces the tasar silk (Tasar). *Antheraea mylitta* feeds on several trees such as *Terminaila arjuna*, *Terminalia tomentosa*, *Anogeissus latifolia*, *Lagerstroemia parviflora* , and *Madhuca indica*. Wild silkworm *Antheraea assamensis* produces muga silk, and another wild silkworm *Philosamia synthia ricini* (=*Samia cynthia*) produces eri silk. The eri silk worm from India feeds on the leaves of the *castor* plant. It is the only completely domesticated silkworm other than *Bombyx mori*. The silk is extremely durable, but cannot be easily reeled off the cocoon and is thus spun like cotton or wool.

Antheraea assamensis (Helfer, 1837). is exclusively from *Assam*. Its silk has a beautiful glossy golden hue which improves with age and washing. It is never bleached or dyed and is stain resistant. Eaarlier it was reserved for the exclusive use of royal families in Assam for 600 years. Tasar silkworm *Antheraea mylitta* (Drury, 1773) is abundantly available in tropical tasar region of India. *Antheraea pernyi* (Guénerin-Méneville, 1855). is the *Chinese Tussah Moth*. The colour and quality of the silk depends on the climate and soil. *Antheraea polyphemus* is most potential wild silkmoth of North America. *Antheraea yamamai* (Guénerin-Méneville, 1861) is generally called as the "tensan" silk moth has been cultivated in Japan for more than 1000 years. It produces a naturally white silk but does not dye well, though it is very strong and elastic. It is now very rare and expensive. *Anisota senatoria* is orange-tipped oakworm moth found in North America. *Automeris io*. - (Fabricius, 1775) is also abundantly available in North America.

Bombyx mandarina (Moore) is the possible wild form of *Bombyx . mori*. *Bombyx sinensis* was reported from China. It is a prolific sericigenous insect but cocoons are small. *Callosamia promethea* was reported from North America and reared in several part. *Gonometa postica* Walker was recorded from the *Kalahari* region. *Gonometa rufobrunnae* Aurivillius. Generally feeds on the *Mopane* tree in southern Africa. *Hyalophora cecropia*. reported from North America is one of the most beautiful moth and its quality of the silk depends on food source. *Samia cynthia* (Drury, 1773) and the *Ailanthus Silkmoth* is somewhat domesticated silkworm from China. Later on it was introduced into North America. The eri silkmoth from Assam is a subspecies of this moth (*S. cynthia ricini*). It produces a white silk which resembles wool mixed with cotton, but feels like silk.

HOST PLANTS

The natural silk is broadly classified into two types: (a) silk of plant origin and (b) animal origin. The natural silk of plant origin is obtained from silk cotton tree and floss-silk tree. Silk cotton fibre is known as *Ceiba pentandra* (Bombacaceae), Sveta salmali, Safed simal, Schwetsimal, Salmali, ijavum, Buraga, Illavi, etc. It is abundantly available in different parts of South Asian countries, i.e., India, Burma, Sri Lanka, Malaya, Java and China. There is immense potential for this fibre if it is blended with all natural and man-made fibres. This fiber is obtained from the seedpod of a tree. The fibre is made up of unicellular hair occurring in the seedpods that constitute the fruit of the tree. It grows on the inner wall of the pod. The fibre is smooth, cylindrical, hollow, thin-walled and frequently bends over itself. Owing to the smoothness of the fibre, it may be mixed with cotton, wool and spun into yarn. It is resilient, light, elastic, short and silky. It has low thermal conductivity and high ability to absorb sound. It contains highly lignified cellulose. Besides this Floss silk tree is also well known. Floss silk tree, *Chorisia speciosa* is a thorny flowering tree of the family, Bombycaceae. It is native to South America, but cultivated as an ornamental in other regions. It grows to a height of about 15 m (50 feet). The unique feature of this large tree is that the entire trunk along with its stem is covered with thorns. It is a flowering tree, bearing large pink coloured flowers.

MULBERRY

The mulberry is a fast growing deciduous plant belonging to the family Moraceae and genus *Morus*. It has a deep root system. The leaves are simple, alternate, stipulate, petiolate, entire or lobed. Mulberry foliage is the only food for the silkworm (*Bombyx mori*) and is grown under varied climatic conditions ranging from temperate to tropical. Mulberry leaf is a major economic component in sericulture since the quality and quantity of leaf produced per unit area have a direct bearing on cocoon harvest. There are about 68 species of the genus *Morus*. The majority of these species occur in Asia, especially in China (24 species) and Japan (19). Continental

America is also rich in its *Morus* species. The genus is poorly represented in Africa, Europe and the Near East, and it is not present in Australia. In India, there are many species of *Morus*, of which *Morus alba, M. indica. M. serrata* and *M. laevigata* grow wild in the Himalayas. Several varieties have been introduced belonging to *M. multicaulis*, *M. nigra*, *M. sinensis* and M. *philippinensis*. Most of the Indian varieties of mulberry belong to *M. indica*. In China there are 15 species, of which four species, *Morus alba*, *M. multicaulis*, *M. atropurpurea* and *M. mizuho* are cultivated for sericulture. The details of the mulberry varieties under cultivation in different states of India are given on Table 2.3. In the former Soviet Union *M. multicaulis*, *M. alba*, *M. tartarica* and *M. nigra* are present. Though mulberry cultivation is practised in various climates, the major area is in the tropical zone covering Karnataka, Andhra Pradesh and Tamil Nadu states, with about 90 percent. In the sub-tropical zone, West Bengal, Himachal Pradesh and the north-eastern states have major areas under mulberry cultivation.

Table 2.3 : Mulberry varieties in India

Variety	Region	Developed at	Origin
1	2	3	4
Kanva-2	South India Irrigated	CSRTI, Mysore	Selection from natural variability
S-36	South India Irrigated	CSRTI, Mysore	Developed through EMS treatment of Berhampore Local
S-54	South India Irrigated	CSRTI, Mysore	Developed through EMS treatment of Berhampore Local
Victoria-1	South India Irrigated	CSRTI, Mysore	Hybrid from S30 x Berc 776
DD	South India Irrigated	KSSRDI, Thalaghattapura	Clonal selection
S-13	South India Rainfed	CSRTI, Mysore	Selection from polycross (mixed pollen) progeny
S-34	South India Rainfed	CSRTI, Mysore	Selection from polycross (mixed pollen) progeny

1	2	3	4
MR-2	South India Rainfed	CSRTI, Mysore	Selection from open pollinated hybrids.
S-1	Eastern and NE India Irrigated	CSRTI, Berhampore	Introduction from (Mandalaya) Myanmar
S-7999	Eastern and NE India Irrigated	CSRTI, Berhampore	Selection from open pollinated hybrids
S-1635	Eastern and NE India Irrigated	CSRTI, Berhampore	Triploid selection
S-146	N. India and Hills of J and K Irrigated	CSRTI, Berhampore	Selection from open pollinated hybrids
Tr-10	Hills of Eastern India	CSRTI, Berhampore	Triploid of Ber. S1
BC-259	Hills of Eastern India	CSRTI, Berhampore	Back crossing of hybrid of Matigare local x Kosen with Kosen twice
Goshoerami	Temperate	CSRTI, Pampore	Introduction from Japan.
Chak Majra	Subtemperate	RSRS, Jammu	Selection from natural variability
China White	Temperate	CSRTI, Pampore	Clonal selection

Mulberry thrives under various climatic conditions ranging from temperate to tropical located north of the equator between 28° N and 55°N latitude. The ideal range of temperature is from 24 to 28°C. Mulberry grows well in places with an annual rainfall ranging from 600 to 2 500 mm. In areas with low rainfall, growth is limited through moisture stress, resulting in low yields. On average, mulberry requires 340m^3/ha of water every ten days in case of loamy soils and 15 days in clayey soils. Atmospheric humidity in the range of 65-80 percent is ideal for mulberry growth. Sunshine is one of the important factors controlling growth and leaf quality. In the tropics, mulberry grows with a sunshine range of nine to 13 hours a day. Mulberry can be cultivated from sea level up to an elevation of 1000 m. Mulberry flourishes well in soils that are flat, deep, fertile, well drained, loamy to clayey, and porous with good moisture holding capacity. The ideal range of soil pH is 6.2 to 6.8, the optimum being 6.5 to 6.8. Soil amendments may be used to correct the soil to obtain the required pH. Kanva-2, S-13 and S-34 varieties are recommended for rainfed (rainfall: 500-800 mm) regions of South India (Karnataka, Andhra Pradesh and Tamil Nadu) where as Kanva-2, S-36, S-54, DD, MR-2 (especially in Tamil Nadu) and Victoria-1 varieties are recommended for irrigated conditions. Some of the variety viz. S-1, S-7999, S-1635, S-146, Tr-10 and BC-259 varieties are recommended for the hilly regions of north and northeastern India. The uses of the various species of the genus *Morus* are enumerated below:

Fig. 2.1: High yielding hybrid of *Morus alba*

Morus alba

This species is cultivated in the hilly and plain areas of India (Himalayan region) for silkworm rearing. It is also used as a tree in roads and in social forestry. The fruit are made into juice, liquor and stews. The wood is used in the sports goods industry. It is also used for house building; agricultural implements; furniture; for making spokes, poles, shafts and bent parts of carriages and carts. The stem bark is used for m.king paper.

Morus indica

The cultivated forms that are utilized in silkworm rearing belong to *M. indica*. There are a few profuse fruiting varieties occurring in Maharastra and Meghalaya that can be utilized as the female parent in breeding programmes. The fruit is used for jam, jelly and juice making in Maharashtra. The pruned branches are used as fuel.

Morus laevigata

The trees of this species produce sweet fruits that are used in juice and jam making in central India. In northeast India the wood is utilized as firewood; in house building and furniture making; for making stocks, spokes, poles, shafts of carriages and casts. The wood is suitable for plywood making and panelling, carving and making of toys and tea chests. It is also used for making tennis rackets. The straight log of the tree is used as a support in house-building work.

Morus serrata

The wood is used for furniture making and carving, toy making, sports goods, agricultural implements and cheap types of rifles and guns.

Medicinal uses

The various parts of the mulberry plant find use in Ayurvedic preparations. The leaves have diaphoretic and emollient effects and are used for making a decoction that can be used as a gargle that throat inflammation. The fruits are used to treat sore throat, depression, high fever and are both a coolant and laxative. The root extract has hypoglycaemic properties. The root bark is used as an anthelmintic, purgative and vermifuge. Mulberry root juice is administered to patients with high blood pressure. The Chinese use the leaf tips from young leaves to boil with tea to control blood pressure. The milky latex is used as a plaster for sores and for the preparation of dermal creams.

Non-mulberry host plant

Most of the Non-mulberry silkworms are polyphagous which feed on various types of food plants. Silkworm host plants have been categorized in three groups, viz. (1) *Terminalia* or *Shorea* based *A. mylitta*, (2) *Quercus* based *A. pernyi*, *A. roylei, A. Proylei, A. yamamai* and *A. polyphemus* (3) *Machilus* based *A. assama*. While *Machilus/Litsaea* based *A. assama* is limited only to North-east and Himalayan range of India. Nevertheless, certain plants have been observed to yield a better crop of cocoons with superior technological properties. These are referred to as primary food plant, while the others are known as secondary food plants.

Tasar food plants

Tasar silkworm (*Antheraea mylitta)* is a wild silkworm feeding on a variety of food plants. The host plant of tasar silkworm has been categorized as primary, secondary and tertiary food plants on basis of preferential feeding and or adaptability of silkworm to their food plants. There are about a dozen of primary host plants and 40 secondary host plants. In India, the tasar larvae use three principal species of tasar food plants. They are *Terminalia tomentosa* W. & A., *Terminalia arjuna* Bedd. and sal (*Shorea robusta* Roxb.). These trees are distributed in most parts of the tasar rearing belt of India. Usually Sukinda, Bogai and Daba ecoraces feed on *T. tomentosa* and *T.arjuna*, whereas Nalia and Modal ecoraces grown completely wild, feed on *S. robusta*. Some other eco races viz. Daba, Sarihan, Bhandara, Aandhara local and Sukinda are adapted to arjun (*Terminalia arjuna*), asan (*Terminalia tomentosa*) where as Laria, Barhrwa, Railey, Modal and Nalia eco races are adapted to Sal (*Shorea robusta*). Since all these eco races prefer to feed voraciously on *Terminalia arjuna* Bedd. (Arjun), *Terminalia tomentosa* (Asan), *Shorea robusta* (Sal), and yield viable commercial cocoons therefore these plants are categorised as primary host plants where as other ecoraces viz. Tira, Jiribam and Belgaum are adapted to *Lagerstroemia parviflora*, Ber (*Ziziphus jujuba*) and Anjan (*Hardwickia binata*). *Lagerstroemia parviflora* (Sidha), *L. indica* (Sravani), *Zizyphus mauritiana* (Ber) etc. using it as secondary host plants wherever primary food plants are not prevalent (Table 2.4). Genus *Terminalia* Linn. is large tree distributed throughout the humid and semi humid tropics of the whole world. Few species however have a large area of distribution. *Terminalia arjuna* and *Terminalia tomentosa* are abundantly distributed throughout the tropical tasar region of India and tasar culture is generally practiced on these plants (Table 2.4). The important characteristics of *T. arjuna* plant is as follows.

Fig. 2.2 : *Terminalia arjuna* in tasar belt

***Terminalia arjuna* Bedd.**

Leaves subopposite, oblong or elliptic nearly glabrous beneath when adult, spikes usually panicle, fruits 1-2 in., nearly glabrous ovoid or ovoid oblong, the wings not very broad their striations curving much upwards. Attains 60-80 ft. Leaves usually 4-6 inch (sometimes 10 inch.) suddenly narrowed at the base, often cordate, obtuse or very shortly acute at the apex, - petiole nearly more than half inch, often very short , with two glands near its apex. Bracteoles very small. Calyx teeth nearly glabrous both within or without. Young ovary very short, covered with crisped brown or rufous hair. Wings of their fruit usually truncate or suddenly narrowed at the top. Bark smooth, grey, flaking of in large thin layers. Common along banks of river, streams and dry watercourses.

***Terminalia tomentosa* W. & A.**

Leaves sub-opposite, or upper most alternate, elliptic or ovate glabrous or very hairy beneath when adult, spikes panicle, fruits 1-2in., glabrous or hairy abovate- oblong, the wings broad striations carried out horizontally to the edge. Bark rough and black. Attains 60-100 ft. height. Leaves usually 4-8 inch, petiole 1/2 in. Flowers often attacked by a cynips producing numerous galls, which stimulate fruit. Bracteoles very small. Calyx teeth without tomentose, villous or glabrescent plant propagation. Generally tasar food plants are propagated through seed but now day's vegetative propagation methods have also been developed. In order to raise good nursery, seeds of *T arjuna* and *T. tomentosa* are collected from healthy and middle aged plants in the forest. Seeds of *T. arjuna* and *T. tomentosa* are collected during April- May where as *Shorea robusta* seed is collected during May-June. The viability of seed in case of *Shorea robusta* is very short (about one week) and therefore, nursery is prepared quickly after collection of seeds from the forest. *T. arjuna* and *T. tomentosa* seed can be stored for one year but most of the time nursery is prepared from the freshly collected seed. The seeds collected form the forest are sorted out based on their weight (above 3 g) and soaked in plain water for 96 and 48 h, respectively to soften the pericarp. The soaked seeds are heaped on sand beds and covered with moist gunny bags. Germination starts after 7 days and continues for about 3 weeks. The germinated seeds are sown in rooting media (soil, sand and FYM in a of ratio of 2 : 1 : 3 for arjun and 3 : 1 : 1 for asan). 2-3 months old seedlings may be used for transplantation. The seedlings raised in polythene bags are maintained for two months. During this period regular watering is required. When the seedlings reached at the height of about 8-10 inches, it is transferred in the field. The plantation is conducted generally during the month of August or early part of September. Pits of 1 x 1 x 1 ft are prepared in the field where it has to be propagated. Studies on spacing reveal that 1.2 m x 1.2 m (4' x 4') are ideal for arjun (6,724 plants/hectare) and 1.8 m x 1.8 m (6' x 6') for asan (3,025 plants/hectare). Four years old plantation may be used for tasar silkworm rearing purpose.

Plantation through vegetative means is also conducted for uniform maintenance of plantation. The gestation period of the plants raised through this technique is less therefore air layering technique for vegetative propagation of desired plant type/selected variety/biotype are adopted. The ideal season for air layering was observed to be June/July (high humidity, moderate temperature). To develop air layering, leaves are removed from the healthy twig at about 50 cm from the tip. Removing bark about 2-3 cm in width makes a girdle. In the upper cut surface cotton plug, soaked in rooting hormone (IBA – 300 ppm) is wrapped. The girdle is then covered with over night soaked sawdust ball and is then wrapped with a polythene sheet of size 25 cm x 25cm (400 gauze) and both the ends is tied with jute rope or twin thread. In order to improve the quality and quantity of leaves from the economic plantation various agronomical practices are applied for better establishment and quick growth of plants. It not only increases the quantity of leaves but also reduces the gestation period of the plant. Generally the gestation period for normal forest plantation is six years that comes down to 3-4 years if proper agronomical practices are adopted. Soil moisture content is one of the most important factors for better growth of plant.

Therefore, dry farming technique is important to retain the moisture contents in the soil. The main objective is to enhance the retention of rainwater for a longer period in the fieled and also to increase its percolation into deeper layer of soil, during monsoon season. Further, for the management of soil fertility optimum fertiliser (FYM & chemical) is pre-requisite to maintain the soil fertility. In tasar plantation (asan/arjun) fertilizer application is necessary due to consistent use of plantation for rearing purpose. It is observed that FYM @ 600 cft. /hectare or 2 kg./plant every alternate year is ideal and should be applied during November – December by making 12 – 15 cm deep circular basin around the plant. The requirement of NPK depends upon the nature of soil and age of the plant. The optimum dose of NPK for 4 – 10 years old plants are 100 : 50 : 50 whereas, for more than 10 years old plants the requirement is 150 : 50 : 50. The fertilizer is applied by deep placement method i.e. by making 3 – 4 holes (20 – 30 cm deep) round the plant. The height of the food plants in the rearing field is one of the most important factors for better management of silkworms rearing. Tasar silkworm rearing is conducted out door and therefore, it is essential to maintain the height of the food plants with good quality leaves. It can be achieved by proper pruning. Pruning is affected on more than four years old plantation. Plants may be heavily pruned at 1.8 m height, once in two years, during February. Light pruning of 2.5 cm diameter shoot is affected during February when heavy pruning is not required. For chawki garden, plants (Arjun/Asan) is pruned at 1 meter height. This helps to maintain the plant height at 1.8 m to 2.4 m level, which is desirable for net rearing. Grown up virgin plants should be slowly brought to 90 cm height otherwise mortality may occur. The primary food plants of tropical tasar silkworm is *Terminalia arjuna* Bedd (arjun), *T. tomentosa* W.&A (asan), *Shorea robusta* Roxb. (sal), *Lagerstroemia parviflora* Roxb.(sidha), *L.. speciosa* Pers (jarul), *L. indica* Linn. (saoni), *Zizyphus mauritiana* Lam. (ber), *Hardwickia binata* Roxb.(anjan). Apart from this, there are more than two dozen secondary food plants are reported, of which the most important are *Terminalia chebula* Retz. (haritaki), *T. belerica* Gaertn (bahera),*T. catappa* Linn. (jangali badam), *T. paniculata* Roth. (kinjal), *Anogeissus latifolia* Wall.(dhaunta), *Syzygium cumini* Linn.(jamun), *Careya arborea* Roxb. (kumbi), *Shorea tailura* Roxb. The food plants grow luxuriantly at low altitude (600 MS2) between about 40 north and south latitudes.

Table 2.4 : Food Plants of Tropical Tasar Silkworm

Scientific Name	Common Name
Primary food plants:	Arjun
1. *Terminalia arjuna* Bedd.	Asan
2. *T. tomentosa* W.&A	Sal
3. *Shorea robusta* Roxb	Sidha
4. *Lagerstroemia parviflora* Roxb	
5. *L. speciosa* Pers	Jarul
6. *L. indica* Linn	Saoni
7. *Zizyphus mauritiana* Lam	Ber
8. *Hardwickia binata* Roxb	Anjan
Secondary food plants:	
1. *Terminalia chebula* Retz	Haritaki
2. *T. belerica* Gaertn	Bahera
3. *T. catappa* Linn	Jangali badam
4. *T. paniculata* Roth	Kinjal
5. *Anogeissus latifolia* Wall	Dhaunta
6. *Syzygium cumini* Linn	Jamun
7. *Careya arborea* Roxb	Kumbi
8. *Shorea tailura* Roxb.	

Oak tasar food plants

The food plants of temperate tasar silkworm are mainly of the genus *Quercus* Linn (Oak). It is a very large genus of which nearly 300 species are distributed throughout the temperate regions of the northern hemisphere and extends into the tropics and subtropics of South America, India, Malaysia, and Australia often above 600 m above sea level. In India about 58 better-known species are distributed all along the western Sub Himalayan range at 1200 to 2100 m and in eastern tracks 700 to 1500 m above sea level. The important Quercus sp. available in western sector are *Quercus incana* Roxb (banj), *Q. ilex Linn* (phanat), *Q. glauca* Thunb. *Q. semicarpifolia* Smith (kharshu), *Q. himalayana* Bahadur (Moru). In the north-eastern part of the India *Q. serrata* Thunb (uyung), *Q. semiserrata* Roxb. *Q. dealbata* Hook f. & Thomas are abundantly available (Table 2.7).

Table 2.5 : Food Plants of Oak Tasar Silkworm

Scientific Name	Common Name	Distribution
Primary		
1.*Quercus incana* Roxb	Banj	North-West
2. *Q. ilex Linn*	Phanat	North-West
3. *Q. glauca* Thunb.		North-West
4.*Q. semicarpifolia* Smith	Kharshu	North-West
5. *Q. himalayana* Bahadur	Moru	North-West
6. *Q. serrata* Thunb	Uyung	North-East
7. *Q. semiserrata* Roxb.		North-East
8. *Q. dealbata* Hook f. & Thomas		North-East

In India about 58 known species of *Quercus* are distributed all along the North-Western range at 1200 m-3500 m ASL and in North-Eastern region at 600 m-1800 m ASL. Of these species, seven species viz. *Quercus serrata, Quercus semiserrata, Quercus dealbata, Quercus grifithi, Q. incana, Q. semicarpifolia and Q. himalayana* are exploited for Oak Tasar culture. These species are naturally grown in about 18,41,500 hectares of land and out about 46,500 hectares can be utilized for this culture. Manipur alone shares about 20,000 hec. of naturally grown Oak plantation. In the North-Eastern region the availability of Oak species ranges from 1,500 AMSL to 6, 000 AMSL. Out of seven species of Oak on which tasar speices can be reared, 4 species are available in North-east. They are *Q. serrata, Q. semi serrata, Q. dealbata* and *Q. frithi*. Oak plant distribution is scattered in entire North Eastern region. It has been estimated that about 40,000 hectares of Oak flora is available in Manipur out of which 20,000 hectares is within easy rich of the inhabitants and can be exploited for raising Oak tasar crop. The entire Oak flora is mainly confined to the hills where it has attained the form of small bushes and dwarf trees because of regular cutting for fire wood. *Quercus serrata, Quercus dealbata* and *Quercus grifithii* are predominantly found in Manipur. *Quercus serrata* and *Quercus dealbata* are distributed in lower altitude (2000-3000' ASL) regions as mixed vegetation while *Quercus grifithii* is mainly available in the upper ridges. In Nagaland Oak flroa is mainly available in Kohima, Phek and Tuensang districts at an altitude of 3000- 6000 ASL.

The total area under Oak plantations is roughly estimated to be 15000-20, 0000 hectares. The species available are *Quercus serrata, Quercus dealbata* , *Quercus grifithii, Quercus lineata, Quercus lamellosa* out of which only *Quercus serrata, Quercus dealbata* and *Quercus grifithii* are being utilized for feeding of oak tasar silkworms. All the oak growing areas are easily accessible in the state. In Assam North Cachar Hills of Assam, Garampani area situated at an altitude of 24000 ASL abounds in Oak tasar plants viz. *Quercus serrata and Quercus dealbata*. Total area under Oak plantation has been reported to be about 24,000 hectares out of which only 1000 hectares are utilized for commercial rearing of silkworm. The Oak tasar plantation in Meghalaya is confined to Bhoi area east of Khasi Hills and Garampani area of Jaintia Hills where nearly 23,000 hectares of *Quercus serrata* and *Quercus dealbata* plantations are available. Most of the these plants are in the form of tall tree and widely dispersed over steep hills and deep gorges. The Oak plantation in Mizoram is confined to Champai and Lunglei area where 15000 hectares are covered. *Quercus serrata, Quercus dealbata* and *Quercus grifithii* are existing tree species. Further, a total area of 1,25,000 hectares are under Oak plantation in Arunachal Pradesh. However, the species suitable for Oak rearing are *Quercus grifithii, Quercus semicarpifolia* and *Quercus dealbata*. These species cover an area of 1,70,000 hectares. Though the climate is quite ideal and suitable Oak species are available in plenty, yet population density which is very thin (7 per sq. km.) poses a great constraints for its commercial exploitation.

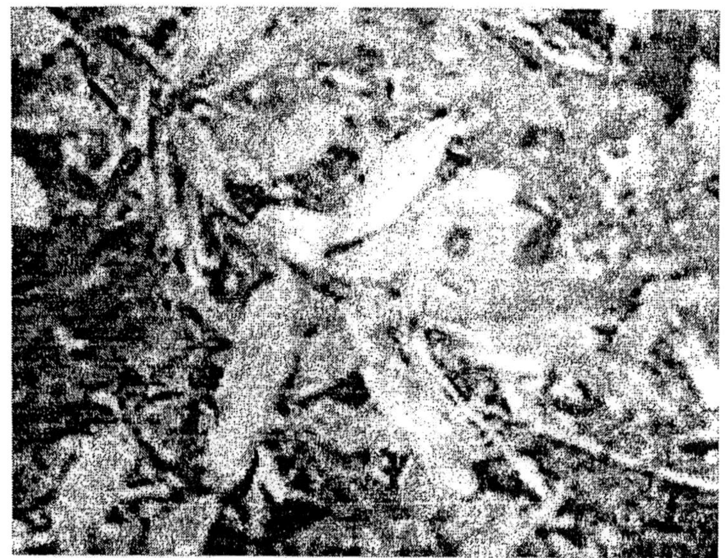

Fig. 2.3 : *Quercus incana* in temperate tasar region

In North-Western region particularly in Jammu & Kashmir, only eight districts namely, Doda, Udhampur, Kathua, Punch, Rajouri, Anant Nag, Baramula and Srinagar are having concentrated pockets of Oak flroa. The various species available in the state are *Q. incana, Q. himalayana, Q. semicarpifolia, Q. gluca, Q. ilex* and *Q. robut* etc. In Uttar Pradesh about 3.05 lakh hectares of forest is distributed in the eight districts of Kumaon and Garhwal divisions but these forests are quite dense and inaccessible. However, an area of about 16,495 hectares can be exploited for Oak tasar culture. The socio-economic conditions of hill districts namely Chamoli, Pithoragarh, Uttarkashi, Tehri, Nainital and Almora are well suited for out rearing. Main Oak species available are *Q. incana, Q. semicarpifolia,* and *Q. himalayana,* etc. The Oak plantation in Himanchal Pradesh is estimated that nearly 1,39,593 hectares of Oak tasar food plants are available in the state. *Quercus incana* covers 68,698 hectares, *Q. semicarpifolia* 69,210 hectares and *Q. dealbata* only 2015 hectares. India has twenty million acres of Oak plantation but because of various constraints potential of Oak feeding silkworm could not be exploited for silk production. Depletion of forest resources in post-independence period has posed serious challenges for retaining seri diversity in India.

During the recent years, the plantation of Non-mulberry silkworm host plants suffered a serious set back due to the damage caused by the stem borers. Further, there is large diversity in the sprouting and maturity behaviour of various species of Oak plants. The same species sprouts in different times at different altitudes. *Q. semicarpifolia* which is a slow maturing variety is abundant at higher altitudes (7000-9000 above MSL) and *Q. himalayana* is found at altitude of 5000-7000 feet above MSL. For the successful second crop rearing a concept of low altitude, middle altitude and high altitude rearing has been adopted and it has been found to be economically viable. The first crop rearing is proponed and done on *Q.. incana* at low altitudes in the month of March. A successful second crop is then raised at higher altitudes on *Q. semicarpifolia* and *Q. himalayana* where the sprouting is delayed (May-June) in comparison to plants at lower altitudes (March). In the middle altitudes rearing, the season is April/May. The cocoon harvest of the middle altitude rearing can be used for reeling purposes.

AUGMENTATION

Seeds of all oaks are generally collected during Oct.- Nov. and nursery is raised during Feb.- May. Mechanically scarified seeds yield more than 90% germination. They require soaking in water for 2-4 days. Subsequently, they are heaped and covered by moist gunny bag or thatch grass and when their germination starts, they are dibbled horizontally in rooting mixture of soil, sand and FYM of 1:1:1 ratio contained in perforated polythene tubes and irrigated regularly. Seedlings are allowed to grow for about three months in nursery until on set of monsoon. *Quercus serrata*, *Q grifthi*, and *Q incana* have been propagated through air layers and more than 75 % rooting have been reported in case of *Q. serrata* and *Q. grifthi* with IBA 250 and IBA 500 ppm respectively while *Q. incana* has been reported to yield 28% rooting. For maintenance of economic plantation, foliar application of urea (1.0- 1.5%) has been practiced in both tropical and temperate tasar culture for improvement of economic characters. Fortification of *Q. serrata* leaves with minerals (MgSO4, KH_2PO_4, NaCl, NaOH, CaCl2) dextrose, vitamins (ascorbic acid, folic acid), amino acid (glycine) and urea revealed that all the minerals decreased larval period from 1-2.5 days while vitamins have no effects on larval periods. Application of fertilizers for the maintenance of food plants is almost similar to the tasar plants. NPK is applied @ 75:25:25, 100:50:50 and 150:50:50 kg/ha in two split doses to 2-3, 4-10 and above 10 years old plants respectively. Besides NPK , FYM is also applied @1000, 2500, 900 and 1250 cft/ ha/year to *Q. serrata* Further, N75 P25 K38, Azobacter 10, Azospirilium10, Phosphatica20 Kg/ha dose may be applied instead of N150 P50 K50 Kg dose /ha/year for two years in single dose followed by Azobacter10, Azospirilium10, Phosphatica20 Kg/ha during third year. Fertilizer and manure application have been found to increase leaf yield in Oak tasar plants by approximately 50-60 per cent.

Oak plants also play an important role in the maintenance of ecological balance in entire north-east region. It has multipurpose use and highly economical for the welfare of the people of the region. There is vast flora of Oak plants and some of which are used for sericulture and large part of it are still unutilized. The pruning of the Oak plants has to be adjusted to provide silkworm quality foliage during suitable rearing season when the temperature and humidity condition are ideal for silkworms. It has been observed that pruning done during September ensures early leaf sprouting by the end of February or first week of March at low altitude (750 m) and ten days later at middle and higher altitude (1200 m and above) as against sprouting in the middle of March and early April respectively. For second crop rearing pruning may be given one month earlier to onset of rearing. It has tremendous impact in increasing the Oak tasar silk production in North-Eastern region. Due to different climatic conditions, the crop pattern in the north-eastern and north-western region is quite different and therefore the crop cycle is also different in both the region. In

North-Eastern region, the spring crop is the main seed/commercial crop which starts from the first week of March when the natural sprouting takes place in *Quercus serrata* plants. However, an additional subsidiary crop can also be taken up during summer or autumn by breaking the pupal diapause subjecting seed cocoons to photoperiodic treatment and by giving light pruning to plants well in advance. Pre-poned rearing in spring by brushing during first week of March or earlier at low altitude and ten days later at middle/high altitude instead of conventional brushing in late March or April in Manipur States ensures successful cocoon harvest. Similarly in other Eastern States it is recommended that brushing to be synchronized with the early leaf sprouting and its development to avoid high temperature during late stage silkworm rearing. In North-western region, the sprouting behaviour and the maturity of the various species is different. However, even the same species sprouts in different times at different altitudes. *Q. semicarpifolia* which is slow maturing variety is abundant at higher altitudes (7000-9000 ASL) and *Q. himalayana* is found at an altitude of 5000-7000 ASL. To raise a successful second crop rearing a concept of low altitude, middle altitude and high altitude rearing has been adopted and it is found to be economically viable. The first crop rearing is preponed and done on *Q. incana* at low altitude in month of March. A successful second crop is then raised at higher altitudes of *Q. semicarpifolia* and *Q. himalayana* where the sprouting is somewhat late (May-June) in comparison to plants at lower altitudes (March). In the middle altitude rearing, the season is April/May. The cocoon harvest of the middle altitude rearing can be used for reeling purposes. In the North-west the main problem had been the failure of the second crop and consequently dependence for seed stock on the states. Preservation of seed cocoons from spring to spring affects moth emergence, egg laying and hatching adversely. A break through in stabilization of seed crops has recently been achieved by pre-poning the first crop on *Q. incana* at lower altitude (2000'-2550') followed by and early second rearing on Q.*semicarpifolia* at high altitude places (7000' above). This finding has not only ensured successful harvest of 30 cocoons as against complete failure earlier but also gives better quality cocoons with improved fecundity capacity.

MUGA FOOD PLANTS

Muga silk production being an out door activity, requires adequate food plants. Som tree is the most preferred food plant providing food material for muga silkworm rearing in the north-eastern region and soalu is used as secondary plant. The polyphagous muga silkworm feeds primarily on the leaves of *Persea bombycina* King (Som), *Litsaea polyantha* Roxb. (Soalu). The important secondary food plants are *Litsaea citrata* Blume (Mejankori), *L. monopetala* Pers(Scalu), *L. salicifolia* Roxb ex. Wall (Digloti), *Michelia oblonga* Wall (Chapa), *Cinnamomum obtusifolium* Nees (Patihonda), *Symplocos grandiflora* Wall (Bhomlati), *Magnolia sphenocarpa* (Chapa/Panchapa), *Celasstrus monospermal* (Bhomloti) and *Zizyphus zuzuba* (Bagori) (Table 2.6).

Table 2.6 : Food plants of muga silkworm

Scientific Name	Common Name	Distribution
Primary food plants		
1. *Persea bombycina* Kost	Som	North East/West
2. *Litsaea polyantha* Roxb	Soalu	North East/West
Secondary food plants		
3. *Litsaea citrata* Blume		
4. *Litsaea monopetala* Pers	Mejankori	North East
5. *Litsaea salicifolia* Roxbex.	Scalu	North East
Wall	Digloti	
6. *Michelia oblonga* Wall		
7. *Cinnamomum obtusifolium*	Chapa	North East
Nees	Patihonda	
8. *Symplocos grandiflora* Wall		North East
9. *Magnolia sphenocarpa*	Bhomlati	North East
10. *Celasstrus monospermal*	Chapa/Panchapa	North East
11. *Zizyphus zuzuba*	Bhomloti	North East
	Bagori	North East

Fig. 2.4 : Persea *bombycina*

Fig: 2.5 : *Litsaea polyantha*

Persea bombycina King (som) is classified on the basis of leaf shape into sixteen morphotypes, of which the three more significant are Naharpatiya, Azarpatiya and Jampatiya. Naharpatiya is considered as one of the best food plant for muga culture. P. *bombycina* and P. *monopetala* are distributed in India, Nepal, Burma, Khmer, Malaysia and Indonesia. In India these plants are common in Assam, Meghalaya, Arunachal pradesh, Tripura, Nagaland and Manipur and West Bengal. *L.. citrata* is found throughout the sub -Himalayan belt. Though the distribution of plant species is wide spread yet, its use in muga culture is restricted to Assam valley. Assam has some other important species of *Litsaea* like *Liglutinosa* C.B. Robin, *L. sabifera* Pers., *L. umbrosa* etc. but none of those have been used as a food plant for muga silkworms. During early 70s muga plantation was more than 9200 hectares but there has been a substantial decline and currently it reached at around 4730 hectares. The existing area under muga food plants is estimated at about 3100 ha with Assam accounting over 75%, Meghalaya around 15 % and Manipur about 7 % of the total area. There has been gradual increase in the area under food plants since the Muga host plant augmentation programme was implemented by the Central Silk Board.

ERI FOOD PLANTS

Ericulture primarily comprises two activities, viz., food plant cultivation and silkworm rearing as on-farm activities followed by cocoon processing activities, spinning, dyeing, weaving, etc. The eri silkworm, *Samia ricini*, feeds on a large number of food plants (Table 2.7); most preferred being castor (*Ricinus communis*), followed by kesseru (*Heteropanax fragrans*), cassava (*Manihot utilissima*) and barkesseru (*Ailenthus* spp).

Table 2.7 : Host plants of eri silkworm

Scientific Name	Common Name	Distribution
1.*Ricinus communis* Linn.	Castor or era	Northern /
2. *Heteropanax fragrans* Seem	Kesseru	North-East
3. *Evodia fraxinifolia* Hook	Payam	North-
4. *Manihot utilissima* Phol	Tapioca	East/South
5. Plumeria acutifolia	Gulancha Phool/Champa	India
6. *Jatropa curcas* Linn.	Bhotera	
7. *Ailanthus excelsa* Roxb.	Barkesseru	
8. *Sapium ellgeniafolium*	Korha	
9. Zanthorylum yhesta	Bazraman	
10. Hodgsonia heteroclita	Thebow	
11. *Carica papaya* Linn.	Papaya	
12. *Cinnamomum glanduliferum* Messin	Cinnamomum	
13. *Micromulum minutum*	Phootkho	

Food plants of world saturniids are presented on Table 2.8. Majority of these plants belong to family Euphorbiaceae. Of these, castor (*Ricinus communis*), Kesseru (*Heteropanax fragrans*), Payam (*Evodia flaxinafolia*), Tapioca (*Manihot utilissima)* and barkesseru (*Ailanthus* spp) are largely used and castor has been so far regarded as the best followed by kesseru.

Table 2.8 : Food plants of World Saturniidae

Scientific Name	Common name
1	2
Ailanthus altissima	Chinese Tree of Heaven
Althaea rosa	Hollyhock
Apium graveolens	Celery
Azadirachta indica	Persian lilac
Berberis	Barberry
Carica papaya	Papaya
Carpinus betulus	European hornbeam
Cassia fistula	Indian laburnum
Coriaria nepalensis	Tanner's tree
Cornus.	Dogwood
Euodia fraxinifolia	Poyam
Firmiana simplex	Chinese parasol tree
Forsythia	Golden-bells
Fraxinus	Ash
Gmelina arborea	Gomari

1	2
Heteropanax fragrans	Keseru
Ilex chinensis	Kashi holly
Jatropha curcas	Barbados nut
Juglans cinerea	Butternut
Juglans regia	English walnut
Lagerstroemia indica	Crapemyrtle
Lawsonia alba	Indian privet
Ligustrum	Privet
Lindera benzoin	Spicebush
Liquidambar styraciflua	Sweetgum
Liriodendron tulipifera	Tuliptree/White poplar
Manihot ultissima	Cassava/Tapioca plant
Michelia figo	Banana shrub
Plantanus	Sycamore
Plumaria acuminata	Temple tree
Prunus serotina	Wild black cherry
Pyrus malus	Apple
Rhus typhina	Staghorn sumac
Ricinus communis	Castor-oil bean
Salix	Willow

1	2
Sassafras albidum	Sassafra
Syringa vulgaris	Common lilac
Tilia	Basswood/Lime/Linden
Zanthoxylum acanthopodium.....	Ash
Zanthoxylum hostile	Zanthoxylum
Zanthopodium alatum	Zanthopodium
Zizaphus jujuba	Zizaphus jujuba

Food plants preference of almost all commercially exploited silkoworms have been tested. Earlier workers have reported on the rearing performance of eri silkworm on castor (Dookia, 1980; and Sachan and Bajpai, 1972). Saratchandra and Joshi (1988) tested the rearing efficacy of different varieties/cultivars of tapioca and reported that some improved tapioca varieties are as good as castor. H-1423, H-97, H-972, H-648 and H-2304. H-648 gave 100% ERR followed by H-97 (96.67%), H1423 & H-972 (93.33%) and H-2304 (88.9%). Feeding with H-3641, H-1423, H-648, H-1253, H-2304 and H-972 varieties yielded male cocoons weighing around 2.0 g. H-1423 yielded the highest female cocoon weight of 2.54 g. Highest shell weight was reported to be 0.26 g and 0.283 in male and female respectively.

CULTIVATION OF FOOD PLANTS

In Assam and other North-Eastern States, there is no systematic cultivation of food plants for feeding eri silkworms. Castor grows abundantly in and around residential areas, the leaves of which are harvested and used to feed silkworms, but no seed is harvested. However in the East and West Garo Hills districts of Meghalaya tapioca is cultivated for the tubers. However, the seed farms largely depend on castor and kesseru. Kesseru being a perennial plant, it is also raised in village grazing reserves, for common use by the farmers. Use of payam or barkesseru is not a regular feature, but is practiced in forest areas. Thus, the rearer often collects leaf from scattered castor or Kesseru plants. However, in other States of India, castor is cultivated for the production of oil seeds and tapioca for its tubers, which is used in sago industry. Tapioca is extensively cultivated in Tamil Nadu and Kerala, while castor is cultivated in Andhra Pradesh, Maharashtra and Gujarat. The available information on the cultivation of eri food plants is presented in the following paragraphs.

CASTOR

Castor (*Ricinus communis* L.) is the most preferred food plant for eri silkworm rearing. It belongs to family Euphorbiaceae. Dioscorides named castor as *kroton* or *kiki* in Greek and the castor seed and the dog tick are known in Latin as *Ricinus*. The genus is represented by a single species. English traders who confused the oil with that of *Vitex agnus-cactus*, known to Spanish and Portuguese in Jamaica as *agno-casto*, called it as 'castor'. In Sanskrit, it is known as *rivuka* or *eranda*. Castor is widely distributed throughout the world particularly in the warmer regions although the origin is obscure. Weiss (1971) mentioned that its origin was probably Ethiopia. Russian plant breeders considered that there were four large centres of variability, each with its own specific plant characters and most important are Iran-Afghanisthan region, Palestine-SW Asia, India-China and the Arabian Peninsula (Moshkin, 1986). All types of castor plants have the same chromosome number $x = 10$, $2x = 20$. Natural polyploidy is probably absent. Incompatibility has been reported between strains from extremely different environments. Though the agricultural scientists have carried out considerable work, their focus has been totally on the seed yield and the oil content. Very little has been done on the production and quality of the leaves and their utility to eri silkworm. For oil seed production, dwarf and short duration crops are preferred where as long duration or perennial plants are advantageous to leaf production at lower costs.

Fig. 2.6 : Ricinus communis (Castor)

In Assam and other north-eastern states of India where ericulture is practiced, castor cultivars are identified by the leaf colour and coating. They are green non-bloomy, red non-bloomy, green bloomy and red bloomy. Of them green non-bloomy was preferred as it ensures a better survival (87%) of the silkworms (Ann. Rep., 1974-75). Castor may be grown as annual or perennial and is commonly seen-one having pale green and the other with violet leaves. Both are equally suitable for feeding eri silkworms. September-October and March-April are the sowing seasons in Assam, but April is the most ideal season (Suryanarayana *et al*, 2003). After ploughing the land 2-3 times to a depth of 20-25 cm, pits of the size of 20 x 25 x 25 cm (L x B x D) are made at a spacing of 1 mtr apart followed by application of 1 kg farmyard manure (FYM) per pit. Castor beans are coated with Bavistin at the rate of 2 gm per kg of seed to prevent fungal infection and sown @ 2 beans per pit at a depth of 2.5-3.0 cm and irrigated immediately. Germination takes place in about 10 days. A basal dose of chemical fertiliser, NPK at the rate of 24:16:8 per acre is applied one month after germination. Only one seedling per pit is retained and the other removed. When the seedlings are three month old, 12 kg of nitrogen is applied per acre. Hoeing and weeding has to be done before application of fertiliser. 4-5 leaf harvests can be made the 1st crop during May-June yielding 2,000 kg, 2nd during July-August with an yield of 1,800 kg, 3rd crop during September-October with 1,200 kg, 4th during November-January with 1,000 kg and the last one during February-April with an yield of 500 kg of leaf per acre. Considering the poor yield, the 5th crop and the sowing season starts in March, the last crop has to be ignored. Experiments conducted at Titabar (Assam, India) indicated that 50% harvest of castor leaves which amounts to 7,430 kg resulted in reduction of seed yield by 13.69% (157 kg) and (Saratchandra and Joshi, 1986). Seed loss was negligible with harvest of 25% of the leaf. However, it may vary slightly in different states depending on the agro-climatic situation prevailing there. In Andhra Pradesh where castor is grown for its seed, Khariff crop is sown in June-July lasting about 150-180 days and Rabi crop in October-November lasting for 210 days (Jayaprakash, et al, 2003). In Uttar Pradesh caster is grown on bunds or as an intercrop with chilli and is sown between July to October (Pandey, 2003). Raghavaiah (2003) reported that a hybrid variety of castor, DCH-177 yielded 15.4 mt of leaf per hectare per year in Bangalore good enough to rear 1196 dfls which can yield 1101 kg green cocoon equivalent to 127.5 kg shells besides castor yield of 542 kg/ha. The work of ICAR complex of Imphal indicated a higher consumption index for eri silkworm in the local pink varieties followed by DCS-9 and DCH-177. He also mentioned that 25-30% defoliation did not result in significant yield reduction. If, due to some reason, the germination is not up to the mark or, the new plants appear weak, they should be uprooted and, in the gaps fresh, healthy seeds should be sown. To get more branches and leaves, the perennial variety requires light pruning when it attains a height of about 5'. The object of thinning is to cultivate the healthiest plants and give proper spacing plant to plant. Even a stick, about 1 ¼ inch thick and 8 inches long, may be used to drill holes and sow seeds. To feed the worms, leaves should be harvested from the healthy plants. Dry leavers and twigs must be removed from the plants and the plots should be weeded frequently to keep them clean. A full-grown perennial caster are planted at a spacing of 12' x 12' apart and it is estimated that one acre of this may yield about 20,000 kg of leaf (Sarcar, 1980). For successful cultivation of castor trees, light clay soil is required. However, light sandy soil is always preferable to clay soil. Though potash is necessary in the soil, it is replenished by application of ash of the burnt weeds, leaves, stalk and stems. Cultivation of leguminous cover crops like sun hemp, daincha, cowpea, mung, matkalai etc., in castor plantation adds nitrogen and humus in the soil. Castor and mustard oil cakes may be used as manure. A dose of lime is necessary, every two or three years, when oil cakes are used. Inorganic manures may be applied as top dressing either at the beginning or at the middle of the crop. The castor seeds do not mature simultaneously and mature capsules will shed the seeds if the harvest is delayed until all the fruits of the spike dry on the plants. Therefore, the capsules are harvested every 3-4 days. To reduce the manpower requirement in harvesting the capsules, dwarf varieties and short duration crops the capsules of which do not open automatically on maturity are preferred where the crop can be harvested mechanically or whole spikes can be harvested at a time.

KESSERU

Kesseru (*Heteropanax fragrans*) is a perennial tree found in wild in north-eastern region of India and is also cultivated for eri silkworm rearing. It is extensively grown on embankments, in village grazing reserves and around homestead lands in the region. Kesseru can grow in acidic soils. High sloppy lands are well suited for kesseru. Plants are first raised in nursery and later transplanted. Flat and well-drained land is essential for raising nursery. February is an ideal month for raising nursery. Land is ploughed 2-3 times thoroughly and then levelled. 6 x 2 m size beds are raised to about 15 cm height and 6 cft of farmyard manure is applied and thoroughly mixed along with soil before raising the beds. Kesseru fruits are harvested in February, which are soaked overnight and the seed separated from the pulp by rubbing against gunny cloth. Placing them in water as they are easily sunk while non-viable ones float can identify viable seeds. Treat the seeds with Endophil-M 45@ 2-3 gm per kg of seeds to prevent fungal infection. Around 800 seeds are sown at a spacing of 15 x 10 cm and covered with a thin layer of straw to retain moisture. Sprinkle water regularly in the morning and afternoon in dry season. Remove the straw after germination of seeds. Light weeding is essential at an interval of 20-30 days till they attain a height of 20-25 cm, which takes about 6 months. Post-monsoon period is good for transplanting kesseru seedlings. Land should be ploughed to a depth of 20-30 cm, levelled and 30 x 30 x 30 cm pits made at a spacing of 2 m x 2m. On hilly areas a spacing of 3 x 3 m spacing is followed. About 5 kg FYM is added to each pit and mixed thoroughly with the soil. Six-month-old seedlings are transplanted preferably on a rainy day. Light hoeing and weeding is carried as and when required. About 5 kg FYM is applied to each plant in April and NPK @ 50:30:10 kg per acre per year in two split doses, i.e., in April and September. Pruning is done at a height of 1.75 m after attaining the age of 3 years. Pruning can also be done at slightly higher levels to avoid cattle grazing when planted in open fields or village grazing reserves. Plants can also be raised from cuttings of about 2 ft. long, taken from one-year-old plants. Three crops can be harvested in a year the first two crops yielding 2,700 kg each during April-May and August-September and 1,350 kg in the third crop during December-February. Pruning is essential to induce branching. Annual pruning or pollarding is done in the month of February. Earlier literature indicates a slightly different schedule of pruning and harvest which we may ignore for a better practice. A full-grown tree in isolation can yield up to 200 kg leaf in a year.

TAPIOCA

Welzen (1997) revised the introduced species of *Manihot* (Family: Euphorbiaceae) Miller in Malesia. He observed that *M. utilissima* Phol, 1827 and *M. aipi* Phol, 1827 are junior synonyms of *M. esculenta* Cranz, 1766. It is also known as cassava or arrowroot and is cultivated for its tubers. The leaves are also used as forage for eri silkworm. Though considerable work has been done considering it as a tuber crop, very little information is available on the productivity and quality of leaves as fodder to eri silkworm. Plantation is done in June-July and tubers are harvested in December end to January end (Prakash *et al*; 2003). Well-drained soils are good for tapioca cultivation. Cuttings of around 25 cm are planted directly in the soil at a spacing of 90 x 90 cm. One acre of plantation yields 6-7 mt of tubers. Experiments conducted at RSRS, Titabar, gave highest yield of tubers, i.e., 8057 kg/ha (Ann. Rep., 1983-84)

GENERAL SERICULTURE

MULBERRY CULTURE

The mulberry silkworm, *Bombyx mori* (Bombycidae : Lepidoptera) is domesticated spread from China to India and Europe. Its domestication dates back too far to be recorded. As a result of domestication in various climates, there is marked variation in the morphology, physiology and voltinism of the silkworm. One of the most important differences induced by the domestication is the variation in the number of broods. It largely depends upon temperature and other abiotic factors. Some silkworms breeds are univoltine, bivoltine and multivoltine based on the number of generations in a year. The distributing of mulberry silk fauna is presented on Table 2.6. Univoltine breeds have only one life cycle in a year and are generally reared in colder regions, due to the hibernating characters of the eggs. The embryo continues to grow halfway and enters a period of diapause. The embryo resumes its growth only in the early part of the spring and hatches out as larva. Bivoltine silkworms have an alternating life cycle of diapausing and non diapausing eggs. It has non-dormant eggs in the first generation and dormant eggs in second generation of the year. They hatch twice in a year and are suited to warmer regions. The bivoltine varieties are healthier and display a high cocoon producing capacity and the larval period is shorter as compared to univoltine varieties. Multivoltine varieties produce non-dormant eggs. They lay eggs several times in a year and produce new generations. Normally the embryo starts developing and after 10 days of embryonic growth, hatches out as larva. Such egg which completes their embryonic period in 10 days is called non-hibernating eggs. This character enables the breed to complete more than five life cycles in a year and as such the breeds are called multivoltines. They are suitable for high temperature region. Under tropical condition, these varieties have a shorter larval period and are resistant to diseases. But their cocoon quality and filament productivity are clearly lower than those of the bivoltine hybrid varieties. Therefore, among the silkworm varieties bivoltine silkworms are superior both in the quality of the larvae and the cocoons. Their eggs can be hatched anywhere in any season, if cold storage facilities are available and artificial hatching techniques are used. Thus, if efficient modern sericulture is to be adopted in the tropical region, it is important to use bivoltine varieties.

The growth of the silkworm is markedly affected by the climatic conditions of the rearing site, i.e., temperature, relative humidity, air quality, airflow, light, etc. It is important to adjust the microclimate of the rearing site to make it suitable for growth of respective instars so that silkworms can produce cocoons to the maximum possible extent. The weather conditions outside the rearing house also exert a significant effect not only on the microclimate but also on the nutritive value of the mulberry leaves. Mulberry silkworm is commercially exploited for silk production. The silkworm rearing is conducted in various climatic conditions and the silkworm after 10 days of pupation in the cocoon emerges out as moth. Reproduction is the main activity of the moth. After copulation, the female moth lays 400 to 500 eggs. There are five larval instars. The first instar period varies from 4-5 days. The newborn silkworm is black, hairy and about the size of 3 mm. It eats continuously mulberry leaves for 4-5 days, and then rests 20-24 hours before shedding its skin. This process of eating, resting and shedding is called an instar. The second instar is the shortest among all the instars. The larval duration is of 3-4 days. The silkworm eats so much and grows so fast that its skin cannot keep pace. After the rest period the skin breaks around the head and larva wriggles out. Third instar period is of 4-5 days. During each rest period, the silkworms are very sensitive and must not be disturbed. After each instar, the silkworm's skin is lighter in colour. After entering fourth instar, the larva starts voracious feeding. Total duration is of 5-6 days. The silkworm's chalky white body has nine segments and a large head. The skin feels cool and smooth like velvet. Spiracles on the sides of each segment are used for breathing. Fifth instar larval period is of 7-8 days. The last days of silkworm's life are spent transforming mulberry leaves into liquid silk. The silk is stored inside the body in two pouches. The final instar larvae make cocoons by the process of spinning. The silkworms grow 10,000 times in 23-25 days to spin a protective covering called a cocoon. It makes a strand of silk as long as 1200m. The process of metamorphosis is a complicated process. The larvae change in to pupa inside its cocoon, and the pupa is safe from predators. It goes into a deep sleep for 11-12 days, while its body goes through a change, or metamorphosis, from pupa to moth. When the pupa becomes a fully grown moth with wings and antennae, it spits brown juice, dissolving a hole in the cocoon. The flightless moth wriggles out and begins searching for a mate. It has a life span of 4-7 days. The female moth produces a special scent called a pheromone to attract the male moth. One or two days after the male fertilizes the eggs inside the female's body, she lays 300-500 eggs. The eggs hibernate until the next spring, then they hatch and the cycle begins again.

On the basis of aboriginal land, *B. mori* is classified into Japanese, Chinese, European, Korean and Tropical races/ breeds. These are further grouped according to their voltinism and moultinism. All these breeds have different characteristics and they are easily differentiated on the basis of their morphological characters over a long period of time. Japanese and Chinese varieties are commercially exploited in the temperate region whereas in tropical zone, multivoltine tropical breeds are commercially exploited due to its temperature tolerance capacity. Large number of improved silkworm breeds has been evolved for commercial exploitation and much of the research and development projects have been launched for the improvement of silkworm breeds in different regions and seasons, mainly by pure line selection. Earlier only pure breeds were utilized for reeling cocoons but later on after startling findings of Toyama (1906), cross breeds were utilized for mass cocoon production.

For the production of commercial silk, generally a breed with peanut shaped cocoon (Japanese or European breed) is crossed with a breed that produces oval cocoons (Chinese breed) in order to obtain high hybrid vigour in F1 hybrid. There was a similar trend in other silk producing countries like China and Korea. Moreover, for enhancing the quantitative characters and vigour of both the peanut and oval cocoon breeds, the characters of the European monovoltine, Japanese mono and bivolltine, Chinese mono and bivoltine, and multivoltine breeds were also introduced. Therefore, the majority of today's commercial silkworm races, whether they were peanut or oval shape cocoons, they are semi bivoltine. Tri-moulter varieties display a low productivity but thin denier and better silk quality, whereas five moulter and six moulter breeds are not completely fixed. Therefore, presently 4-moulters, in which the moultinism is stable and which consist of many breeds because of advanced breeding, are commonly used. Several attempts have been made to evolve superior hybrids in tropical regions. For obtaining better hybrids, eggs of superior breeds are essential. The cocoons of superior improved breed is separated and selected by line selection in peanut shape cocoon breed and oval shape cocoon breed separately and these breeds are crossed. The characters become distinct in F2 generation onwards. In the second stage, attempt is made to improve the quantitative and qualitative characters of two selected lines and to under take cross breeding to develop a new line from the progeny obtained by crossing a peanut shaped cocoon line with another peanut shaped cocoon line and between two lines with oval cocoon shape. This method is mainly used for improving silkworm breeds in major sericultural countries. Presently, in order to develop a peanut shaped breed, another breed with peanut shaped cocoons are used. Similarly, for the development of oval type, breeding resource material with oval shaped cocoons is used in order to reduce the fixation period of cocoon shape. Several techniques viz. single cross, utilization of three parents, F1 hybrids for four way crosses and 3-way crosses, use of mutant breeds, use of multivoltine silkworms, cross breeding, region and season specific races have been developed to increase the silk productivity. After emergence, the moths are allowed to copulate only when their wings are fully developed. Copulation is allowed for 2-3 hours and then the moths are separated and the female moths are placed on egg cards on which they are allowed to deposit eggs under protection. The male moth can be used again if placed in a cool dark area at about 5-10°C in container. Loose eggs are also prepared by allowing the moth to deposit the eggs on single moth cards or individual moth may be placed in a cell and inspected individually to minimize the wastage of eggs because of spread of the diseases.

Normally the eggs are produced during May to October for rearing in the following spring. Therefore, preservation of eggs from the time of their collection up to start of the incubation is important part for sericulture practices. In the tropical region, however, since there is no winter, for breaking the dormancy of the embryo, the hibernated eggs have to be stored at cold temperature. The eggs are collected in the dormant stage and are hatched artificially, except for the eggs that will be used in next year. Because the minimum temperature is very high in the tropics, if the eggs are allowed to hibernate until the next year, the embryo may be unable to break its dormancy and it may die. Therefore, if bivoltine breeds are used in the tropical sericulture, a combination of artificial hatching and artificial over wintering should be applied. Parent silkworm rearing is avoided and the rearing is conducted in planned manner. For artificial hatching of silkworm eggs, acid treatment is given at appropriate time for uniform hatching. After oviposition, the eggs are maintained at 25°C and after 20 hours they are dipped in acid.

Depending on the temperature of the HCl solution, the method is classified into two types viz. hot acid treatment and cold acid treatment without heating. In hot acid treatment, Hydrochloric acid, with a specific gravity of 1.075 at 15°C is heated to 46°C. The specific gravity of the HCl at 46°C becomes 1.075. The suitable acid treatment time is 4–6 minutes, depending upon the breeds. The normal temperature acid treatment is conducted without heating the acid. HCl with specific gravity of 1.10 at 15°C is used. The suitable temperature for the treatment is 25-27°C. at 25°C specific gravity of the acid becomes 1.105. The suitable immersion time is 60-90 minutes depending upon the silkworm breeds. Apart from this there are several ways of acid treatment. Acid treatment after chilling is conducted for the eggs, which have been kept in the cold storage for more than 60 days. Hot acid treatment, HCl treatment without heating, cold storage system after acid treatment is applied to schedule the rearing of mulberry silkworms.

Fig. 2.7 : Bombyx mori feeding on mulberry leaves

Silkworm rearing

According to the number of generations that can be raised under natural condition in a year, silkworm varieties are classified into univoltine, bivoltine and multivoltine varieties. The univoltine varieties always lay dormant eggs which over winters (hibernate) and hatch in the following year. Thus these varieties are adapted to colder regions. The bivoltine varieties lay non-dormant eggs in first generation and dormant eggs in second generation of the year. They hatch twice in a year and are suited to the warmer regions. Compared to univoltine varieties, the bivoltine varieties are characterized by a shorter larval period, are healthier, and display a high cocoon producing capacity. Multovoltine varieties always produce non-dormant eggs. They lay eggs many times in a year and produce new generations. They are suitable for high temperature regions like the tropics. Under the tropical environment, these varieties generally have shorter larval period and are resistant to diseases. But their cocoon quality and filament productivity are clearly lower than bivoltine hybrid varieties.

Therefore, among the silkworm varieties, bivoltine silkworms are superior both in the quality of larvae and cocoons. Their eggs can be hatch anywhere in any season, if cold storage faculties are available and artificial hatching techniques are used. Thus, if efficient modern sericulture is adopted in the tropical region, it is important to use bivoltine varieties. The growth of the silkworm is markedly affected by the climatic conditions of the rearing site, i.e., the temperature, relative humidity, air quality, air flow, light, etc. It is important to adjust the microclimate of the rearing site to make it suitable for growth of the respective instars so that silkworms can produce cocoons to the maximum possible extent. The optimum rearing temperature and relative humidity for bivoltine silkworm are 27-29°C and 90 % for Ist instar, 26°C and 85% for the 2nd instar, 25°C and 80 % for 3rd instar, 24°C and 70–75 % for 4th instar, and 22–23°C and 60–65 % for 5th instar. In the tropics, however, such temperature and relative humidity ranges can not be easily maintained because of the effect of the high ambient temperature, etc. on the microclimate in the rearing houses. It was observed that the silkworms in all instars can grow more or less normally if the maximum and minimum temperature are about 30 and 20°C respectively, and that silkworm can tolerate even a high temperature of 33–35° C without a considerable adverse effect if the exposure is short.

Apart from temperature and relative humidity, the suitability of microclimate of the silkworm rearing beds is determined by the freshness of the air and the rate of the air changes. Under good ventilation, the tolerable ranges of the temperature and relative humidity become wider. For example, even when the weather is hot and humid, if the silkworm's bed is well-ventilated, the stuffiness can be removed and the evaporation of water from the body of the silkworms can be enhanced and they will be cooled. In such cases, even a temperature of 30°C within the rearing room may not be too high. If the weather is cool and moist, by heating and ventilation, temperature increase and decrease of the relative humidity can be achieved simultaneously. The water content of the body of the silkworm changes with growth stages. The water content is about 76 % immediately after hatching and continues to increase up to the first moulting. It is about 87 %, with slight fluctuations, during the 2nd to 4th stage and decreases day by day during the 5th instar, reaching a value of about 75 % when the silkworms matures. The first instar is described as the water accumulating stage, the second to fourth instar as the water releasing stage. Therefore, the rearing condition should match the physiology of the silkworms. The first instar should be reared under a moist environment and fed with mulberry leaves with high water content, and for the fifth instar larvae the environment should be relatively dry and well-ventilated. Feeding of silkworms is one of the most important aspects for successful rearing. The younger the silkworms, the higher the digestibility, i.e., the amount of mulberry feed provided is larger than the amount eaten, because, when, the silkworms are young in particular, chopped tender leaves are given which dry up very quickly. Therefore, an excess amount is provided so that there is no shortage of feed. Moreover, since the absolute amount of feed required for young silkworms is very small, the wastage is also very negligible. On contrary, in the 5th instar, the food consumed amount 85 % of the total food requirement for all the stages.

Therefore, feed should be supplied rationally without much wastage not only form point of view of the utilization of the mulberry leaves, but also considering the labour requirement for harvest, feeding, etc. The major difference between the 5th instar and the other stages is that at this stage silk material is produced. The proportion of the weight of the silk gland, compared to the total body weight, is less than 5% in the first to the 4th instar, while in the 5th instar the proportion sharply increases, reaching a maximum of more than 40%. This stage is the most important stage with regard to the production of cocoon filament. The amount of mulberry fed and the rearing density influence the cocoon quality and cocoon ability. The larvae grow large if the rearing beds are large and large amounts of mulberry leaves are fed. The larval density in the rearing and amount of feed should be determined in taking into account the utilization efficiency of the rearing facilities, labour required for harvesting and feeding mulberry, and economical use of mulberry leaves. Keeping in view the above points, the rearing and handling techniques of silkworms can be broadly divided into two categories, i.e., first to 4th instars and 5th instar. Up to 4th instar emphasis should be placed on making the silkworms healthy by rearing them in a hygienic environment.

During the 5th instar the improvement of the cocoon quality and cocooning efficiency and economical use of labour should be given priority. The key point in rearing silkworms rationally is to handle moulting in such a manner to improve synchronisation, in each instar. In short, the last feeding before entering moulting should be late. Thus the major strategy is to give the first feed only after all the silkworms have come out of the moult. This is important from point of view of labour saving in operations like moulting and for making the cocoon quality uniform. The fourth moult is the longest compared to other moults. If the fourth moulting is yet synchronized well, the first feeding should be given on a net during the moulting. The silkworm that have moulted first and climbed on the net should be separated from the silkworms, which moulted late. Therefore, when the silkworms are healthy and reach the mature stage the groundwork for production of superior cocoons has been laid. The operations of mounting onwards determine the production of good quality cocoons. The mature silkworms are carefully transferred onto cocooning frame and maintained under an environment suitable for spinning, cocooning and achieving high cocoon quality. During spinning and cocooning, dehumidification and aeration are important, in order to improve the quality of the cocoon filaments, particularly the reelability. Therefore, the basic points in silkworm raring technology are to promote the growth and development of silkworms and to improve the quality of cocoons produced. The type of mulberry leaves fed, the microclimate, and the pathogens present are the factors that affect the growth and development of the silkworms. The major objective in rearing of silkworms is to obtain healthy silkworms. For this purpose they should be reared under hygienic environment where the temperature and relative humidity are maintained within suitable ranges and a sufficient amount of nutritions mulberry leaves of good quality is supplied.

Young silkworm rearing

The rearing methods for young silkworms can be classified into box rearing and covered rearing with damp-proof paper, foam pads, depending on the method of covering of the rearing beds; into chopped leaf rearing, chopped bud rearing, etc., depending on preparation of the feed; shelf rearing, suspended bed rearing and mechanised rearing, etc., depending on types of rearing bed used; and also into individual farmer rearing and cooperative rearing. In both the box and paper cover rearing, the rearing bed is kept moist taking advantage of resistance to high humidity of the silkworms. In this way the mulberry leaves can remain fresh for a longer period of time. The silkworms are well-fed and mulberry leaves consumption and labour utilization become economical. For box rearing, either a wooden box or a galvanised iron box, etc. (70 x 85 x 15 cm) can be used. Particularly during the first two days from start of rearing, wet news papers of other materials are placed around the rearing bed to keep the interior of the box humid. Thirty minutes before the feeding the upper lid is removed to allow fresh air to penetrate and also to allow light into the box so that the silkworms can climb up. During moulting slaked lime of carbonised rice husks may be sprinkled on the rearing bed for drying. In paper covered rearing, water repellent paraffin paper is placed below the rearing bed and use for covering the bed from above, during 1st and 2nd instars to prevent the drying of rearing beds. In the 3rd instar paraffin paper covers only the above part. While placing the covers, care should be taken so that the covering papers rests lightly on the rearing bed and does not become attached to it tightly. The edge of the covering paper may be pressed down with the stick to reduce the airflow. In co-operative rearing of young silkworms, since the silkworms are reared in an exclusive and well-equipped facility apart from the rearing centres for grownup silkworms, under the supervision of experienced technologists who apply the most appropriate disease control measures, the silkworms grow well.

There is a great deal of economy of labour also because a large number of young silkworms are reared together. Such cooperative rearing contributes significantly to the stabilization of the cocoon crop because the young stage, when the silkworms are susceptible to may diseases, they are reared under hygienic conditions. However, total crop loss may occur if cooperative rearing fails. Cooperative rearing of young silkworm also contributes to obtain uniform shaped cocoons in an area. Compared to cooperative rearing up to 2nd instar, such rearing up to 3rd instar requires about twice as many facilities and a three times larger common mulberry farm. However, cooperative rearing is preferable for stabilizing the crop and to avoid separate of the 3rd instar after the silkworms are distributed. The indoor climate is maintained by regulating the temperature and humidity in the rearing rooms. During 1st instar rearing temperature in the range of 27-29° C and 90 % of relative humidity is maintained before start of the rearing. During second instar rearing the temperature and relative humidity should be maintained, respectively, at 26-28° C and 85 % and 25°C and 80 % during 3rd instar. However, in each of these stages, during the period of moulting, the relative humidity should be brought down to about 70% to dry the rearing beds. During this period quality mulberry leaves are provided for proper growth and development of the silkworms. The suitability of the mulberry leaves for feeding young silkworms can be determined by the position of the leaves on the shoot. Taking the largest glossy leaf (the largest leaf near the shoot apex, among all the young glossy leaves) as the reference point, up to two leaves below this level are suitable for the 1st instar larvae, up to the third and 4th leaf below the reference leaf is suitable for the second instar and up to 5th to the 6th lower leaf for third instar. In the case of bud nipping, the suitable range correspond to the shoot lets up to the 5th to 6th open leaf for the 1st instar, up to the 6th to 7th leaf for the second instar and up to the 7th to 8th open leaves for the 3rd instar. Harvesting of leaves should preferably be performed early in the morning and the harvested leaves must be kept in a cool area moistened with water. Before feeding, the mulberry leaves are chopped finely to sprinkle them uniformly on to the rearing beds and also to enable to silkworms to nibble the leaf blades.

Characteristics of young silkworms are different from grown silkworms like higher growth rate, low ingestion, and higher percentage of digestion. The rearing conditions and feed requirements have been determined on the basis of characteristics of young silkworms. Rearing of first and second instar is considered as young silkworm rearing. Rearing temperature and relative humidity are maintained at 28–29°C and 85–90% respectively. Young silkworms are fed with nutritious mulberry leaves containing water (80%), protein (22%), carbohydrate (11%), minerals and vitamins. Success of cocoon crop depends upon proper rearing of young silkworms.

Late stage silkworm reaing

The grown silkworm stages include the 4th and 5th instars. However, these two instars larvae are very different physiologically. Since the 4th instar is closer to young stages, emphasis in rearing should be placed on maintaining a germ free environment with suitable temperature and relative humidity and feeding sufficient amount of nutritive leaves so that the silkworm will grow and remain healthy. In the 5th instar the weight of the silk gland shows a rapid increase, accounting for as much as 40% of even higher of the total weight. This is an important stage for the production of silk. The feed requirement of this stage is nearly 85% of the total requirement of all the larval stages. This is the stage at which mulberry leaves should be most efficiently used and labour should be saved for the harvest and feeding operations. Emphasis should be placed on cocoon production and productivity of labour during 5th instar. Larvae are susceptible to high temperature and high relative humidity as well as poor ventilation. This is a stage where they have a voracious appetite and at the same time their body water is released. Therefore, good ventilation should be given to bring down their body temperature and to remove vapours, harmful gases, etc., that arise from the large quantities of excreta generated due to the large amount of food consumed and other litter. In seasons like dry seasons in the tropics, when the relative humidity becomes as low as 50%, freshly harvested mulberry leaves should be given. Because of their good appetite, 5th instar larvae feed well even if the quality of the leaf is not very high. However, in long run poor quality cocoons will be produced because of insufficient nutrition and moreover, the larval period will increase if the larvae are fed with insufficient quantity of mulberry leaves.

Therefore, suitable environmental conditions should be made to produce good quality cocoons economically. For this purpose, mulberry leaves, labour, silkworms rearing rooms, equipment, etc., should be used efficiently to carry out rational rearing. Grown silkworms have low resistance to high temperature, high humidity and poor ventilation. Grown silkworms are reared at 23 - 25°C room temperatures with proper ventilation and 65 – 70% relative humidity. Care is taken to feed the silkworms with adequate quantity of mulberry leaves, as maximum amount of leaves are required during rearing of grown silkworms. During spring season when the temperature is low, more quantity of mulberry leaves are fed during day time whereas in other seasons when temperature is higher, more feeding is given during the night. The rearing methods for grown up silkworms are classified into shoot rearing, un-chopped leaf rearing, cut shoot rearing, etc., depending on the type of feed given; shelf rearing (tray rearing), flat rearing, mechanised rearing, etc., depending on the type of rearing bed used; and 3-tier shoot rearing is recommended.

Shoot rearing

In shoot rearing method, providing whole mulberry shoots instead of leaves carries out rearing of last three instars. Rearing is conducted on shoot racks. In this method, only one bed cleaning is done after fourth moult, handling of silkworms is minimized and cost of cocoon production is reduced. Shoots are placed alternately in order to insure equal mixing of shoots. Shoot rearing is the method in which the mulberry shoots are harvested and fed as such to the silkworms. Compared to the un-chopped leaf method of rearing, the shoot rearing method requires less labour for harvesting the mulberry plants, feeding and also for collecting mature silkworms. Moreover, since the rearing bed is arranged in three-dimensional manner the silkworm does not press the mulberry leaves down and they do not wilt quickly; they are utilized more efficiently. The rearing bed required for shoot rearing is 350 and 700 square feet for 100 dfls (40,000 larvae) in the 4th instar and 5th instar respectively. In the tropical regions where silkworm-rearing rooms are essential for protection from parsitic insects and ants, the room must be used efficiently. An effective method of utilizing the space in shoot rearing method is to rear the silkworms either on multi-tiered boards, or suspended boards. In un-chopped leaf rearing method either the leaves are plucked in the filed or are stripped from the harvested branches.

Thus this method requires more labour than shoot rearing. It is usually adopted for self-rearing and mechanised rearing of silkworms. In the shoot rearing, larval density is about 140 individuals in the 4th instar and about 70 individuals in the 5th instar per square feet of rearing bed. In cut shoot rearing, the shoots, are cut into 10–20 long pieces for feeding. The method is used mainly in mechanised rearing of grown up silkworms. The temperature and relative humidity must be regulated to reach 23–24° C and 75% RH during the 4th instar and 23° C and 65 % RH during the 5th instar. Special care should be taken so that the temperature does not drop below 20°C during the 4th instar and does not rise above 30° C during the 5th instar by allowing cool air to penetrate into the room, etc. During late age rearing, 23–25°C room temperature and 65–70% relative humidity are ideal. Rearing can also be conducted at higher temperature by proper manipulation through cooling devices, regulation of humidity, and frequency of feeding and proper ventilation. Grown silkworms are fed with mulberry leaves with less moisture content than the young silkworms. Harvesting of leaves is done during cooler hours and leaves should be plucked from the garden, which is 45–55 days mature. Leaves are preserved properly in leaf chamber covered with wet gunny clothes and shoots are preserved vertically.

Mounting and harvesting of cocoons

During spinning time, 23–24°C room temperature, 65–70% relative humidity and good ventilation are ideal. Climatic conditions during 25–30 hours after mounting have close relation so far the cocoon quality is concerned. Temperature more than 26°C and humidity more than 70% during spinning adversely affect the reeling quality of cocoons. Mature silkworms are collected in several ways like Jobrai method, net method, hand picking, self-mounting, etc. Jobrai method is most widely used for shoot rearing. In this method, when about 30% larvae start spinning, nets or ropes are kept over the rearing bed. Mulberry shoots along with mature larvae are shifted to other place. 2-3 shoots are shacked manually. Larvae are separated and mounted. In net method, a net of 1.5 mesh size is spread over the rearing bed. Larvae come up on the net. Mature larvae are separated and mounted. Hand picking method is laborious. Larvae are picked up by hand from the rearing bed and mounted on mounting frames. In self-mounting, cocooning frames are kept on the rearing bed at the time of spinning. This method is used for self and floor rearing. Several types of mountages are used for mounting namely, chandrikes, collapsible plastic mountages, bottle brush montages, rotary mountages etc. Cocoons obtained from rotary mountages are better compared to other mountages.

Sorting of cocoons

Cocoon sorting is done after removal of cocoon floss. Bad cocoons, irregular shaped, stained, double, open end, thin end, mountage pressed cocoons are rejected. After sorting of cocoons, good cocoons are kept in single layer. For marketing, cocoons are shifted to the market during cooler hours in a bag. In India, there are lot of variation in the availability of cocoons in the market. In tropical region of Karnataka and West Bengal hybrids of multivoltine and bivoltine breeds are popular. "Pure Mysore"is the most popular multivoltine breed in Karnataka, Tamil Nadu and Andhra Pradesh. These are flossy greenish in colour. In West Bengal, "Nistari" race is reared. Cocoons are golden yellow coloured and flossy. The silk content in multivoltine is poor as compared to bivoltine. Earlier, bivoltine silkworms reared in Karnataka were Kalimpong A, NB7, NB4D2 and NB18. The former two races spin oval shaped white cocoons, while the latter spin peanut shaped white cocoons. Presently, CSR2 and CSR4 are popular bivoltines in Karnataka as well as in other parts of the country. Hybrids of multivoltine and bivoltine are hardy and have more silk as compared to multivoltine. Hence, hybrids are reared for commercial silk production. In West Bengal, multivoltine (Nistari) are popular in two seasons, namely, Jesta (May) and Bhaduri (August) and multivoltine hybrids in Agrahani (October – November) and Chitra (January – February) seasons. In Jammu and Kashmir, hybrids of bivoltine races SF19, YS3, B40 (Japanese) and Yawki, Haulak, Chang nunj (Chinese) are reared.

TASAR CULTURE

Tasar culture in India is an age-old tradition but mostly restricted within the tribal communities/ aboriginals inhabiting the central plateau covering the states of Jharkhand, Bihar, Chattisgarh,

Orissa, West Bengal, Maharashtra, Rajasthan and to little extent in eastern and western hilly tracts, covering the states of Andhra Pradesh. and Karnataka. Tasar silk is produced by *Antheraea mylitta* Drury. (Saturniidae : Lepidoptera). It is holometabolus, multivoltine , semi domesticated, polyphagous and distributed mainly in the tropical belt of India. *Antheraea mylitta* is characterized by prominent bipectinate antennae in both sexes, minute labial palpi, short maxillae and absence of franulae. *Antheraea* sp. includes a number of tropical and a few temperate sericigenous insects with a transparent eyespot at the centre of each wing, the larvae are stout and smooth and have a rudiment scoli. The wild type *Antheraea paphia* Linnaeus is abundantly available in Orissa. Most of the moths are large in size under the genus *Antheraea* which is abundantly available in several part of the world. *A. mylitta* in India, *A. pernyi* in China and *A. yamamai* in Japan are commercially exploited. Besides, there are smaller species in various part of India. These are *frithii, roylei, knyvetti, sivalica, proylei,* etc. Commercial exploitation of these species except *proylei* (*pernyi* x *roylei*) is not conducted. Most of them exist in the sub-tropical Himalayan belt and the eastern hills. They are polyphagous and thrive on leaves of *Terminalia arjuna* Bedd., *T. tomentosa* W.&A., *Shorea robusta* Roxb. and secondarily on variety of plants. *A. mylitta* is also referred as *A. phaphia*. Seitz (1933) refers that *paphia* as an extremely variable common species. It is similar to *pernyi* and *mylitta*. *Paphia* moths invariably exhibit a distinct inside red and outside white post-median line. The orange yellow species is taken as genus *mylitta*. According to some, *paphia* is synonym of *mylitta*. Arora and Gupta(1979) grouped the *mylitta* species into three sub-species viz. *Paphia Paphia, Paphia-mylitta* and *Paphia cingalese*. The second one *paphia – mylitta* is cultivated form. *Frithi* differs from *mylitta* in having large hyline spots on both wings. A. r*oylei* on the other hand, resembles in general colour orientation in moths. The species can be distinguished in having smaller hyaline spots on both wings. *Helferi* is akin to *knyvetti-helferi* differs from *knyvetti* in having well developed black blotch above the ocellus.

Antheraea mylitta is represented by 44 eco races, of which the economically important ones are Daba, Raily, Laria, Sarihan, Sukinda, Bogai, Modal and Bhandara. Bivoltine daba and trivoltine sukinda is commercially reared in Orissa and Jharkhand. *A. paphia* includes three races, i.e., the univoltine modal, the bivoltine nalia and the trivoltine bogai. The distribution of the stated eco-races is specific to eco-geographical and eco-climatological status. The wild tasar, *A. paphia* is found in the forest of Simlipal Biosphere Reserve in the state of Orissa, which is hilly and undulating with diverse meteorology and ecology. At the highest altitude (601-1000m ASL), voltinism is unique and with the decrease of altitude (301-600m ASL) and (50-300m ASL), voltinism increases. Thus, the species is distributed to a wide range of geographical area between 12 to 30° North latitude and 72 to 96° East longitudes, having diversified eco-meteorological condition. The species is an insect herbivore and produced 2-3 crops in a year (July–August, September–October and November–February). Central India is the center of diversity of the tropical tasar silkworm *Antheraea mylitta*. Among various wild silk producing insects of India, the tasar silk insect, *Antheraea mylitta* Drury is the most potential one and exploited commercially at large scale.

The egg is oval, dorsoventrally flattened and bilaterally symmetrical along the anteroposterior axis. It is about 3 mm in length and 2.5 mm in diameter and weighs approximately 10 mg. Freshly laid eggs are dark brown in colour due to the deposition of muconium. After washing, it becomes white, light yellow or creamy. The larva is typically eruciform and has a typical hypognathous head with biting and chewing mouthparts. The newly hatched larva is dull brownish yellow with black head, measuring about 7mm long and 1 mm diameter and weighing about 8 mg. The body normally turns green and the head brown after about 48 hours, but also yellow, blue and almond colour larvae are seen occasionally. The larva at maturity measures about 13 cm long and 2.1 cm diameter and weighs about 50 g. The prothoracic hood of the first instar larva bears a dorsal oval black spot, which becomes M-shaped in the early second instar and later to V-shape with two dots. These marks are absent in the third instar, but reappear in the fourth and fifth instars as two semi lunar red markings corresponding to epicranial plates. The anal flaps bears a triangular black mark early in the first instar, which becomes V- shaped and brownish from the second instars onwards. The black triangular marks present on each of the claspers during the first two instars turn brown in the successive instars. The early first instar larva has a black mid dorsal line extending from the first to the seventh abdominal segment and a dumbbell shaped black mark on the eighth segment. The first seven abdominal segments have a pair of ventrolateral lines, and on eighth abdominal segment there is a single oblique line. The ventral lines are replaced by V- shaped marks late in the first instar. Each of the abdominal feet bears a horizontal black mark. All these markings disappear in the later stages of development. First to fifth instar larvae have a red mid ventral line extending from the second to the tenth of the abdominal segment. A brown upper line in fourth and fifth instar larvae borders this line. The sexual marking appear late in the fifth instar as milky white spots on the ventral surface of the eighth and ninth abdominal segments. In all larval stages various types of tubercles are present on the body. They are orange red in the second and violet in third to fifth instars. There are several hairs and setae, which are white, minute and irregularly distributed over the body. Silvery white shining spots appear during the third instar at the base of the upper lateral tubercles of the second to seventh abdominal segment. Six regular and thirty-eight irregular patterns of spots have been recorded. Shining spots are also present at the base of dorsal tubercles; these are brick red in green and yellow larvae and white in blue and almond coloured larvae. The tasar silkworms moult four times and pass through five instars. Total length of the larval period is an interaction between the genetic character and the environment. Depending up on the climatic conditions it ranges from 30 to 70 days during different seasons. In *A. mylitta* the larval period ranges from 30-35 days during July-August, 40-45 days during September-October and 60-70 days during November-January crops. The larva undergoes major changes during this stage besides increasing in size and weight and finally metamorphoses into pupa after making a cocoon of silk around it. The pupa or chrysalis is the quiescent or resting stage between the actively feeding larval and reproductive adult stage characteristic of all holometabolous insects. In insects, three types of pupae are generally found viz. pupa exarata, pupa obtecta and pupa coarctata. In pupa exarata the legs, antennae and wing rudiments lie free and are not soldered to the body. This type is characteristic of all hymenoptera, neuroptera, coleoptera and trichoptera. In pupa obtecta, the legs, antennae and wings are soldered or fastened to the body of pupa along its length. This type is characteristic of lepidopteran and dipteran insects. In pupa coarctata, the last larval moult skin is retained and hardened to form a protective case for the pupa, which is developed inside it. This case is referred to as the puparium, which is almost barrel shaped and is segmented, but without any outward sign of appendages as seen in cyclorrhaphous Diptera. The pupa of silkworm falls under the second category with well-defined segmented body. It measures about 4.5 x 2.3 cm. In size and weighs about 10.3 g. It is dark-brown in colour. The genital markings are present on the ventral side of the eighth and ninth abdominal segments. Diapause is a period of apparent cessation of development in the life cycle of insects.

This is a mechanism to survive through adverse climate and is often observed in forms living in temperate region. In temperate region diapause is an over wintering device. It is understood that the insect senses the approaching winter from the changes occurring in the host plant and prepares for diapause. Diapause manifests only one stage of the life history. *Bombyx mori* shows egg diapause; species of *Antheraea* and *Samia* show pupal diapause and several other insects show larval and adult diapause. The number of generations an insect completes in a year is referred to as voltinism. When one completes only one generation in a year, it is referred to as univoltine as in case of oak tasar silkmwrm *Antheraea proylei*. When there are two generations, it is known as bivoltine, if many, they are known as multivoltine. Diapause can either be induced or broken artificially. In mulberry silkworm diapause is broken by either acid treatment or chilling of eggs and induced by incubating eggs at higher temperature and rearing at lower temperature. In tasar silkworm, pupal diapause is terminated by light treatment as photoperiod plays a deciding factor in inducing diapause. Most of the temperate insects develop without diapause in prolonged illumination and these are known as long day insects. Eggs exposed to light for 14 hours or less will result in the occurrence of non-diapausing eggs in the following generation. After voracious feeding through the five instars and attaining the peak weight and size, the larvae attain maturity, when the silkworms stops feeding and empty the alimentary canal. The last excreta come out as green semi-solid mass followed by colourless slimy substance. The larvae after taking rest for 10-25 min. crawl for 1-1½ hr. in order to find a safe place for cocooning. Generally it selects a place having a group of at least three leaves. The process starts with forming a hammock followed by formation of ring, peduncle and cocoon. The larva throws out silk thread through its spinneret, tags the leaves and pulls them together to give it a shape of small nest. The larva then goes inside the space to spin the hammock. During hammock formation the secretion of silk is irregular. The hammock is usually a cone shaped scaffolding of loose silk threads with an opening on the top. It takes about 6 hr. to complete hammock formation. The larva then crawls out of the opening to the respective twig to form the ring. The process starts with the peeling of bark with its mandibles and completes in 24-30 min. Soon after, the larva starts throwing silk for ring formation, the head moves forward and backward during peeling and ring formation. This is followed by peduncle formation. During ring formation itself a foundation is laid by connecting the ring and the top of the hammock by 2-3 strands, but later it is strengthened by adding more silk by to and fro movements of the head in a straight line. To carry out this operation, the larva divides the entire length of the peduncle into 2-3 portions for convenience. The average time taken by the larva for the construction of ring and peduncle is 10 hr. Immediately after completing the ring and peduncle, the larva enters the hammock and spins the cocoon. Initially it throws silk to strengthen the hammock, and then actively fills up the gaps between the fibres making it thick to such an extent the larva become invisible. This is the other most layer of the cocoon formed in the first 24 hr. The inner layers continue to be formed for another 2-3 days without breaking the continuity of the thread. During cocoon formation the larva moves its head in the form of horizontal eight (¥). Though at the time of cocoon formation the larva moves its body in different directions, on completion it rests in the cocoon with its head towards the peduncle. After making the cocoon the larva discharges a light milky or milky yellowish excreta, which continues for 2-3 hrs, moistening the whole cocoon. The discharge on drying helps the cocoon to harden and imparts colour to it. This excreta contains acid urate of ammonia and its consistency depends upon the amount of lime, carbon and ammonia present in it. Spinning of cocoons complete in 4-6 days depending upon the seasons and pupation takes another 4-6 days. The cocoon is single –shelled, pendent, oval, and closed having a hard non-flossy shell with fine grains and is reelable. At the anterior end there is a well-formed dark brown peduncle with a ring at its distal end. The cocoons are generally yellow or grey. Females spin larger cocoons than the males. The moths exhibit distinct sexual dimorphism. Females are bigger (4.5 cm) with a distended abdomen and narrow bipectinate antennae of 1.5 cm long. Males are smaller (4.0cm) with a narrow abdomen and broad antennae. The females are polymorphic in colour, being grey and yellow, whereas males are brown Yellow. Brown females are rare. Mouthparts are reduced, as these are non-feeding stages. The mandible and labial palps are modified into vestigial proboscis. In male the wing span is about 16 cm. whereas in female it is about 18 cm. Post-median line (PM) is red with a white line on the border. Ante median line (AM) is usually black or dark brown in colour. The ocellus with a hyaline area is prominently positioned at the center of the wing.

Feeding behaviour

The eggs hatch at early morning and the newly borne larvae are extremely active and soon start feeding from the margin of the leaf proceeding towards the midrib. The gap between hatching and feeding remains 1-1½ hr. in *A. mylitta*. While feeding, the younger larvae sit directly on the leaf margin while the older ones take the support of twigs or petioles. The claspers and the abdominal feet are used for holding the leaf and movement, while the thoracic legs are used to bring the leaf margin between the mandibles. The larvae rest after every stretch of feeding. The resting larva raises its head upwards and sits in a typical posture. The duration of each rest does not normally exceed 5 min. The resting period decreases with the increase in the feeding period, i.e., with the advancement in age. The duration of continuous feeding at a stretch increases with the increase in the age and size of the larva, but does not exceed 1hr, the average being 30 min. The younger larvae prefer soft and tender leaves while the mature ones prefer mature leaves. The consumption of leaf during the first four instars accounts for only 20% of the total consumption through out the larval period, the rest 80% consumed during the fifth instar alone. The larvae also rest for brief periods during cocoon formation and moulting (also known as ecdysis). Prior to moulting, the larvae stop feeding; take a firm grip on the underside of the leaves or the twigs with the help of their abdominal feet and claspers. The anterior half of the body remains suspended obliquely with the dark head protruded and the prothoracic hood fully extended, which is bent ventrally in a typical posture and remain motionless. During ecdysis, the larva first sheds its head capsule and then pushes itself out anteriorly by splitting the old skin along the mid-dorsal line. The fresh moult larva has tender and pale skin. After moulting it rests about 5 min. and eats its cast skin. It then takes rest for ½ - 1 hr. and slowly resumes feeding. The body of the larva just before moult attains maximum size accommodated in the congested integument; and as a result the head appears smaller. As it comes out of moult the head looks bigger. The durations of the successive larval instars vary from 3-4, 4-5, 5-6, 7-9 and 10-15 days in *A. mylitta*. Mounting heavy larval population on the trees results in over crowding leading to malnourishment resulting in stunted growth, easy spread of diseases followed by heavy mortality and poor yield and quality besides requiring frequent transfer of the larvae. Therefore, it is suggested to mount only optimum number of larvae on each tree after assessing the availability of leaf. The growth rate is highest during the first instar, but slowly decreases towards maturity as observed in *A. mylitta*. The behaviour of the growth rate deviates slightly from this general rule. It goes on decreasing from first to fourth instar, but again shows increase during 5th instar. The larva increases in length by about 20 times and in weight by about 6,250 times from hatching to maturity.

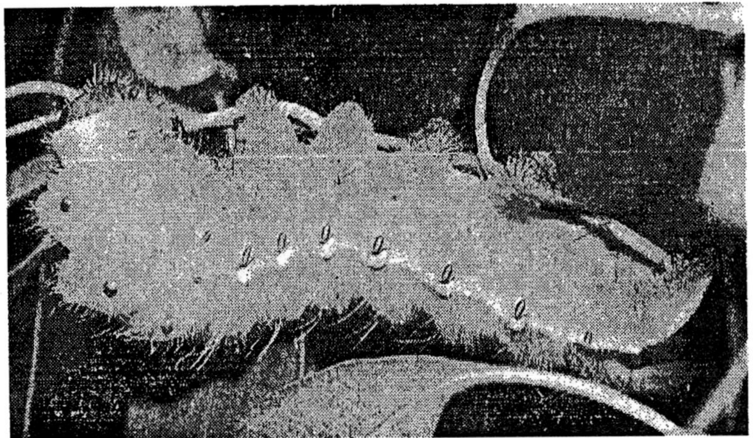

Fig. 2.8 :*Antheraea* mylitta larvae feeding on *T. arjuna* branches

Silkworm rearing

Silkworms are reared out door and therefore, the growth and development of silkworm larvae is highly affected by various biotic and a biotic factors. Among a biotic factors the environmental conditions like temperature, humidity, light, wind, rain, hailstorm, etc., play important roles in determining the different larval activities. Among all environment factors temperature plays a very important role in growth and development of larvae. The tasar larvae in general can withstand temperatures as high as 34°C and as low as 10°C, for short duration. Prolonged exposure to extreme temperature is often fatal. However, the range of sensitivity to temperature varies between different species inhabiting different regions from tropical to temperate regions. Relative humidity also influences the growth of larvae. The larvae can independently tolerate relative humidity as low as 30-40% and as high as 90-100%. But higher temperatures of above 30°C when coupled with higher humidity of above 90%, enhances the rate of mortality as this facilitates a faster multiplication of viruses and thus the diseases caused by them. The larvae prefer bright sunlight, which is congenial to their healthy development. However, they become restless in scorching sun and take shelter in shady places. Conventionally the eggs are placed in leaf cups and mounted on the host plant positioning the tender leaves in the cups. The hatched larvae crawl on the leaves and start feeding. Sometimes the rain water remains in the leaf cups and spoils the eggs and new bourn larvae. It is, therefore, suggested to watch them regularly and after the worms crawl on top the twig, it is advised to break it and tie at different places on the tree near to tender leaves to ensure uniform distribution of larvae. The larvae hatched within a span of 48 hr. only are retained for rearing to ensure a uniform growth and healthy crop. The time suitable for mounting is early morning and evening, or any time in a shady place. Mounting during heavy shower, gusty wind or storm should be avoided. The younger larvae relish green juicy leaves but the tender leaves of red or pale green colour are injurious to their health. The tender leaves lose their turgidity quickly after they are separated from the main tree. Therefore, the leaves of moderate maturity are preferred. In case the rearing spots are situated far off from the forest, the cut branches are transported early in the morning wrapping and preserving them in wet gunny cloth in cool and shady place. Constant dawn to dusk watch is indispensable to protect the worms from pests and predators. The dead larvae are collected twice a day, in the morning and evening and disposed of by incineration or by putting them in lime powder or formalin to prevent spread of disease causing germs through contamination to healthy worms. They should not be allowed to remain hanging on the braches. The materials thus collected are buried in deep pits away from the rearing spot.

Regular microscopic examination of the dead and diseased larvae is essential to detect the disease and to take necessary steps to prevent further contamination. Special attention is required when the worms are infected with highly contagious diseases like, pebrine, virosis, Bacteriosis, Mycosis or muscardine, etc. The moulting larvae should not be disturbed. Disturbance or transfer of moulting larvae may cause rupture of the skin or loosening grip. Transfer of worms can be affected only 2-3 hr. after moulting and resuming normal activities. In outdoor rearing it is difficult to provide a particular quality of leaf suited for a particular stage of larvae. At the same time it is possible to remove bad leaves or shoots having poor quality leaves from dwarf or economic plantations. Spring sprouting leaves suit better for the larvae during the first crop. When foliage on a plant is consumed, the larvae are shifted to another plant. Rearers handpick larvae; place them on cut branches of the food plant and then hung on a fresh tree. Forcible pulling of larvae should be avoided to prevent injury to the larvae. Instead it is advised to separate them along with the twigs on which they rest. Spinning larvae should also be maintained on plants with sufficient foliage, as they require support for hammock formation as well as spinning. The foliage also gives protection from severe climatic conditions and camouflage against predators. Spinning larvae should not be disturbed especially at the early stage, which prompts them to desert the hammock or adversely affect spinning. Cocoons are harvested on the 6th day after initiation of spinning. Harvesting is conducted by cutting the twig near the cocoon ring. After harvest, cocoons are cleared of the adhering leaves and sorted for reeling purpose. Healthy cocoons from a healthy crop with known history of health only are used for seed production. Even there, sorting was done and poor quality cocoons are rejected which are stifled and used for reeling. All the rearing appliances are disinfected 24-48 hr. before the start of rearing, and after every use they are thoroughly washed and disinfected. Similarly, every worker in the field must clean their hands with soap and germicidal solutions before and after handling the larvae or cocoons. For disinfections 2-5% formalin solution or 5% bleaching water is used.

OAK TASAR CULTURE

India is bestowed with a rich forest based floral and faunal wealth of oak tasar with the stretches from Kashmir to Kanyakumari in south and Manipur in the far east. There are several species of oak tasar silk producing insect. *Antheraea roylei* is the most important native natural oak tasar silk producing insect available in the north-eastern and western region of India. The cocoon produced by this insect is double layered and unreelable and, therefore, they are not exploited commercially. But the introduction of fertile hybrid species, *Antheraea proylei* Jolly through the hybridization between the indigenous species *Antheraea roylei* Moore (male, n= 30) and the exogenous species *Antheraea pernyi* Guerin (female, n=49) of China has open a new aspect of sericulture in India (Jolly *et al.*, 1969; 1973; 1974). The discovery of this species (*A. proylei*) in the sub-Himalayan belt of India during late sixties has been one of the most significant achievements, which aided a new dimension in the history of Indian Sericulture. Oak tasar culture is practiced generally in the entire Sub-Himalayan belt right from Manipur, Nagaland, Arunachal Pradesh in the North-East to Jammu & Kashmir, Himachal Pradesh and Uttranchal in North-West touching the fringes of Assam and Meghalaya. Manipur and Nagaland of the North Eastern region. The climate of entire North-Eastern region is sub-tropical with excessive rain fall and shorter summer while the climate of North-Western region is more severe with extreme cold and low temperature with larger winter period. Due to different climatic conditions, the crop pattern in North-Eastern and North-Western region is quite different. In North-Eastern region, the spring crop is the main seed/ commercial crop which starts from the first week of March when the natural sprouting takes place in *Quercus serrata* plants. However, an additional subsidiary crop can also be taken up during summer or autumn by breaking the pupal diapause subjecting seed cocoons to photoperiodic treatment and by giving light pruning to plants well in advance.

The rearing schedule in north western region is different due to different climatic conditions which directly effects the sprouting behaviour of the silk host plants. The sprouting behaviour and maturity of various oak plant species is different. Sometimes, even the same species sprouts in different times at different altitudes *Q. Semicarpifolia* which is a slow maturing variety is abundant at higher altitude (7000-9000 ASL) and *Q. himalayana* is found at an altitude of 5000–7000 ASL. To raise a successful second crop rearing concept of low altitude, middle altitude and high altitude rearing has been adopted and it is found to be economically viable. Generally the first crop rearing starts from the first/second week of March in the spring season and hence termed as spring crop and seed crop in the autumn (Autumn crop). The first crop rearing is proponed and done on *Q. incana* at low altitudes in the month of March. The second crop is conducted at higher altitudes on *Q. Semicarpifolia* and *Q. himalayana* where sprouting is somewhat late (May–June) in comparison to the plants at lower altitudes (March). Middle altitude rearing is conducted during April/May. The cocoon harvested during middle altitude rearing can be used for reeling purposes.

Fig. 2.9 : Antheraea proylei larvae feeding on Quercus plant.

Life Cycle

The eggs are oval in shape and measure about 2.5 x 2 mm. in length and width and weight about 7 mg. In the beginning, the colour is brownish due to deposition of meconium and after washing it becomes light yellow. The eggs are incubated at 20 ±3°C for better hatching. In order to delay the hatching, the eggs are refrigerated at 5 ±2°C that ensures planned release of eggs and its supply in bulk. The newly hatched larvae are mounted on the oak bushes and they start feeding immediately. The suitable temperature and humidity for rearing are 20 to 25°C and 70 to 75 % relative humidity respectively. The newly hatched larva is very active during the morning hour and migrate immediately just after brushing on the twigs It starts feeding from the margin of the leaves and proceeds towards mid rib. The larvae eat up the midrib and even the petiole is consumed during the fourth and fifth instars. The period between hatching and the initiation of feeding varies from 2-3 hours The younger larvae prefers soft and tender leaves but the mature larvae prefers semi-hard leaves. The larva is about 6.5 mg. on hatching. The larva on hatching is black with red capsule. The consumption of leaves during first four instars accounts for only 20 % of the total consumption through out the larval period, the rest 80 per cent being consumed during the fifth instars alone. On an average the moulting period are 26-28, 28-30, 28-32 and 36-38 hours depending upon the atmospheric conditions. The first three larval instars require soft, juicy and green oak leaves for proper growth. Feeding of larvae on mature leaves during these instars inhibits the growth and development. However, the tender leaves droop, curl and dry up quickly even under shade after the branches are cut from the mother plant.

Therefore, early instar silkworm is reared indoor on branches of oak rested in pot with water. The larva after hatching are mounted on the leaves as per the controlled rearing technique and allowed to feed up to third moult with change of feed as and when required. The larvae after coming out of third moult are transferred to twigs having moderately mature leaves. The twigs are then hung on the strings. The subsequent transfers are affected by simply putting the fresh branches over the old ones. This system requires two feeding a day, one each in the morning and evening. Oak tasar larva passes through four moults and the larval period varies from 35 –50 days depending upon the atmospheric temperature and humidity. The larvae start spinning the cocoons about 17-20 days after coming out of fourth moult. The mature larva is green and attains a weight of about 15-20 g at maturity. Oak tasar larva completes its life cycle from cocoon to cocoon in a period of 60-65 days The larvae spin their cocoons on the hanging branches itself. The cocoon is oval in shape and light brown in colour. The peduncle is thin, weak and short. The cocoon weight, shell weight, shell ratio, filament length and denier are about 6.5 g. 0.9 g, 13.84 % 748 mtr and 5 respectively The average size of the cocoon is 4 x 2.5 cm. The pupae are brown, about 5.6 g in weight and 3 x 1.8 cm in size. The adult male is having a wing expense of 14 cm..The fore wing is light grey but the outer margin below the post median is having brownish tinge. The costa is dark brown. The ocellus and hyline area are oval. The wing expanse in female is 16 cm. The forewing is brownish but below the post median line it is light brown. The costa is deep brown in colour.

Silkworm rearing

New technique of rearing has been found to give better ERR and it has been practiced in initial years. Indoor rearing up to second instar and then transfer to out door has been found to improve ERR. Indoor rearing on twigs inserted in bottles/pitches/tins filled with water can be efficiently practiced on small scale. Rearing under nylon net protects the worms from pests and predators in the filed. Preponment of spring crop rearing so as to brush the worms on or before 10th of March and completes this rearing by April saves the worms from fly pests infestations and hazards of the hailstorm. Being reared outdoors, the caterpillars are exposed to the attacks of pests, heavy rainfall, hailstorm, etc., resulting in heavy losses. The following new rearing techniques give higher effective rate of rearing as the high rate of loss of worms in the young stages are minimized to a great extent. Young stage silkworm rearing can be conducted indoors on twigs inserted in bottles/tins/pitchers/trays, etc. After first and second moult the worms are transferred to oak bushes outdoor. It improves the crop out turn by 15-20% as against rearing outdoors directly after hatching. It has been observed that 3rd stage onwards-outdoor rearing gives better ERR than releasing outdoor immediately. During spring season indoor rearing up to Ist and IInd stage and after that out door rearing is highly successful and more than 95% E.R.R has been recorded. Rearing under nylon nets are also conducted to reduce the mortality in early instar rearing. The worms up to I moult are reared indoor and followed by outdoor under nylon net. ERR increased up to 50% and gives 40-50 cocoons/ laying as against 25 cocoons respectively during prepond spring rearing. However, in case of rearing without nets sometimes there is a complete failure due to hailstorm or any other severe natural hazard. Generally indoor rearing is conducted to avoid cold and hail storm attack on the rearing

High yielding varieties

Studies on the rearing performance of the interspecific hybrids (F1) between *A. proylei*, and *A. pernyi* have shown better performance of F1 hybrids over the better parents of *A. proylei* in all the economic characters. Although the F1 crosses of *A. roylei* are superior, the maintenance and multiplication of *A. roylei* at the captive condition is facing problem. Five breeds of Oak Tasar silkworm viz. PRP2, PRP3, PRP5, PRP12 and PRP4 have been evolved through hybridization and back crossing of *A roylei* and *A. proylei* followed by selection of the segregating population. These breeds have been mulitlocationally tested during both spring and summer seasons. During spring crop, these breeds have shown better performance over *A. proylei* in all the economic characters at the low altitude (650-780 m AMSL). These breeds are also found successful during summer/autumn seasons at higher altitude (1100 to 1860m AMSL). Of these 5 evolved breeds, PRP5 and PRP12 are found better in all the test places.

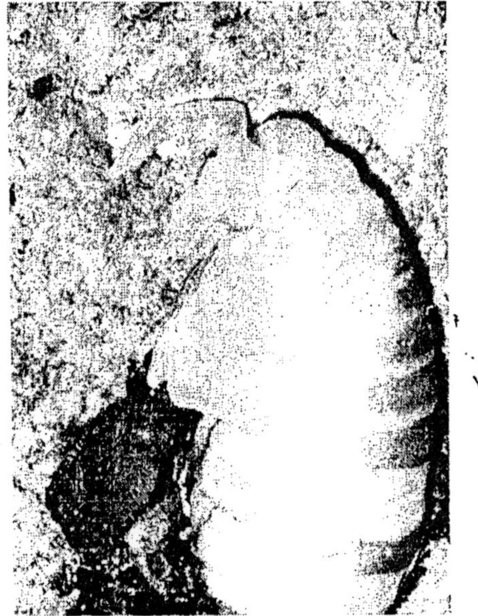

Fig. 2.10 : Pupa of Antheraea proylei during spinning

Seed cocoon preservation

Seed technology involves selection and preservation of cocoons, preparation of dfls, incubation of eggs and some times preservation of eggs at low temperature to postpone hatching. Cocoon production is not consistent during summer and autumn crops, therefore, cocoons harvested during spring crop is preserved under normal room temperature and relative humidity. Preservation of cocoons may be done at high altitude (6,000-6,500 AMSL) to delay emergence so that hatching may be synchronize with sprouting leaves. Preservation at low altitude is also conducted in well-ventilated room where room temperature is below 20°C from June to February. The cocoons are preserved for 8-9 months which leads to the physiological deterioration in vitality of the pupae. Prolonged preservation of seed cocoons leads to low fecundity, low hatching despite loss of a large number of seed cocoons due to sporadic emergence of moth (15 to 18%) and pupal mortality (8 to 10%). Among the emerged moths, a pretty number of crumpled moths, deformed and even eggless female moths are detected from the lot. It is presumed that loss of water in the pupae is the most remarkable cause for all these problems. Studies on the comparative grainage performance of the seed cocoon lots preserved from the spring to spring, autumn to spring and low temperature treatments indicated the adverse effects of prolonged preservation (spring to spring) on the economic characters like coupling percentage, fecundity, grainage duration and cocoon dfl ratio which are found improvement in the autumn to spring preserved lot and low temperature treated lots.

Production

During late seventies the Oak tasar cocoon production was very high. Manipur alone produced 2.5 crore of cocoons in 1979. Thereafter the production level went down steeply and now the production is dwindled to about few lakh (15-20 lakh). Much of the attempt has been made to increase the cocoon production by applying various technologies evolved by Research and Development wings of Central Silk Board. The analysis of the rearing data from the year 1970 till date revealed that the average cocoon production ranges from 18 to 54 cocoons per dfls during the spring season while the production ranges from 12 to 30 cocoons per dfls during summer/autumn season.

Oak tasar in China

Rearing of Oak tasar silkworm *Antheraea pernyi* has a long tradition and next to mulberry silkworm in China. About 90% of the total world tasar production is from China. Oak tasar silkworm *A. pernyi* is economic insect suitable for rearing in temperate conditions. The main tasar producting provinces are Lianing, Sandong, Henan and Guyzhou. Lianing province alone contributes about 80% of the total production of tasar silk. Two kinds of silkworm breeds are reared in China i.e.univoltine and bivoltine. Univoltine breeds are generally reared in South China whereas bivoltine silkworm breeds in North China. The bivoltine province have a colder climate than univoltine region. In the bivoltine zones, quality leaves are available in autumn season as compared to spring and autumn is humid and congenial. Though in China, univoltine and bivoltine regions have been clearly demarcated, the bivoltine breeds when shifted from bivoltine to univoltine zones, they tend to turn univoltine in their behaviour. This clearly shows the environment particularly the photoperiod plays an important role in the facultative diapause of *A. pernyi* (Beck 1968). The Northern China lies at the latitude of 45°N where the day length is 14-15 hours a day. When the pupae of *A. pernyi* is brought to this zone, the diapause is terminated automatically, thus an autumn crop is raised. The performance of the two Chinese breeds viz. B6 and BY1, which are sent from China under the Germplasm Exchange programme, is conducted in India. The performance of the three years has a declining trend in the economic characters, which may be attributed to the difference in environmental conditions that are not conducive for rearing these breeds.

Constraints

The main constraints, which hinder the development of Oak tasar culture, are low productivity, uncertainty of second crop, low disease free laying (dfl) to dfl multiplication ratio and uzi fly menace. Apart from this loss of seed cocoons due to sporadic emergence of moths and pupal mortality during seed cocoon preservation under room temperature due to high relative humidity are the major problems. Further, lack of infrastructure facilities like cold storage for seed cocoon preservation, lack of suitable technology for disease control and the termination of pupal diapauses decrease the productivity potential of Oak tasar silk in India. The present population of *A. proylei* has a stable karyotype of n = 49 similar to its *pernyi* parent. It may now be considered as a distinct breed of *A. pernyi*. Degeneration of the hybrid with respect to silk yield component characters is obviously due to loss of heterosis. Much of the attempt has been made to increase the productivity level of cocoons and raw silk of oak tasar but it has not reached the expected target of commercialization and industrialization. This is probably due to fast deterioration of the breeds caused by drastic changes in karyotype and genome composition in each generation. India does not have *A. pernyi* germplasm of wide genetic variability rendering it impossible to produce improved breeds by interspecific hybridization. Production of fertile hybrid with high heterosis from crossing between *A. roylei* and *A. pernyi* despite wide disparity in chromosome number clearly shows that the two species have close phylogenetic relationship and they can be considered as breeding stock with high genetic variability. Since we have limited germplasm of *A. pernyi* we can use only *A. roylei* germplasm as source of genetic variability. However, selection programme of interspecific hybridization between *A. roylei* and *A. pernyi* cannot follow the standard programme of breeding of oak tasar silkworm viz. production of selected material, family wise multiplication for two generations to produce super elite and elite groups and production of industrial material through hybridization of elite groups.

Theoretically the F1 hybrids of *A. roylei* and *A. pernyi* could have 19 trivalents and 11 bivalents. The F1 gametes would carry 30 roylei chromosomes in one extreme and 49 proylei chromosomes in another extreme and roylei-pernyi chromosome mixture ranging between 31 and 48 units. F2 progenies could have roylei karyotype (n=30), pernyi (karyotype (n=49) and roylei-pernyi chromosome constitution of 31 to 48 bivalent and trivalent mixture. Selection from such a population having variable complex karyotype cannot be carried out following standard practice of selection based on phenotype alone, for the selected material with unstable karyotype family selections should operate. Phenotypically selected materials should be examined for genomic composition through DNA analysis. Individuals having either of the parental genomes are retained and those having mixed genomes are rejected. Karyotypes of the selected individuals are confirmed by cytological testing of their progenies. Karyotype or genomic genetic composition of the family lines are monitored, zet lines with superior characters are retained irrespective of the karyotype and genomic composition and allowed to propagate. Monitoring continues and family lines having larger number of bivalents are selected. This process goes on until stable karyotype is attained. Another reason for family breeding is that in silkworm, only 5 to 10% of the parents pass on their good features to the progeny while others do not, therefore, breeding of silkworm in families is carried out. The hybrid of *A. proylei* and *A. pernyi* Jolly (Jolly *et al.*, 1973; Jolly 1976) reveals 30 elements at metaphase I in males in F1 generation, but the number of such elements varies from 32 to 49 in F2 and backcross individual (Jolly *et al.*, 1973). The hybrid has been reared by self-mating for the last several years and constant chromosome number i.e.2n=98 has been recorded in majority of male individuals (Narayan, 1980). It is however necessary to confirm this point Tazima (1974) put forward the view that the increased number of chromosomes of *A. pernyi* evolved from the chromosomes of *A. roylei* by splitting because F1 hybrid shows complete pairing of the chromosomes of both the species and has invariably 30 elements (bivalent and trivalents) at metaphase I. The remarkable fertility of *A. roylei - A. pernyi* hybrids despite variable chromosme number known to be existing so far (Jolly *et. al.* 1973), further shows closeness of *A. roylei* with *A. pernyi*. Since the population of hybrids is a mixed one having variations in chromosome number and is existing sympatically with *A. roylei* without attaining reproductive and or ecological isolation from the later, its status as a new species seems to be doubtful. An overview of the status of Oak tasar silk production out side India reveal that production in China has come down from 6,000 in 1980 to 4,000 tonnes in 1998. The other countries producing Oak tasar are North Korea (50 tonnes) and South Korea (negligible) The major cause for decline production of Oak tasar silk seems to be physio-genetic degeneration of Oak tasar silkworm species (*A. proylei*) which is a hybrid species and unstable voltinism. Failure of Autumn crop and depending upon the spring crop cocoons for both seed and for reeling has caused decline in production. The decline of hybrid vigour may be due to long storage of spring crop cocoons under artificial conditions. *A. proylei* being a hybrid (between *A. pernyi* x *A. roylei*) naturally looses the hybrid vigour when it is breed continuously within a limited population. There was no opportunity for improving the genetic vigour by introducing more genetic variability. It is common knowledge that any hybrid (plant or animal) looses its hybrid vigour sooner or later and *A. proylei* is no exception. One of the parents of the hybrids the *A. roylei* is occurring in nature but not given much attention that it deserves. If this species was used in a mission mode approach for improving the hybrid vigour, the productivity of hybrid could have been maintained at the desired level.

MUGACULTURE

The shimmering golden yellow silk is produced by muga silkworm *Antheraea assamensis* Westwood. Muga silkworm is multivoltine species and can be reared through out the year, yielding 5-6 crops annually. It. is a polyphagous insect feeding mainly on som (*Litsaea polyantha*) and soalu (*Persea bombycina*) trees, which thrive in the Brahmaputra valley as also Baraksurma (Assam) and foothills of Naga, Khasi and Garo hills of Meghalaya. The moths emerge from the cocoons at dusk and may continue till dawn. The male and female moths exhibit distinct sexual dimorphism. They can be easily identified with their curvature in the wing, bi-pectinate antenna and size of abdomen. The tip of the forewings of male moth has sharp curve, which is easily distinguishable from that of the females. The antennae of the male moths are broader than those of females. The abdomen of the male moths is narrow and small, whereas in females the abdomen is broad, larger and swollen. The colouration of the wings and the body of the male moths are copper brown to dark brown while in the female it is yellowish light brown. Some wild muga moths exhibit color variations. They are blackish brown in colour. The male moth flies actively and copulates with the stationary female. The female moths seldom fly. Coupling lasts overnight. After decoupling, the female moth lays eggs preferably on dry twigs called "Khorika". The moth is a non-feeding stage and dies within 7 to 12 days after its emergence. Maximum fecundity of 250 has been reported during May, June, July, September, November, January and March. From the seasonal variations in the oviposition index, it appears that a mean temperature of 26-27°C and 73-77% RH (September-October) is optimum for maximum oviposition in muga silkworm where as the higher temperature and low humidity

(November-December) adversely affects the same. The incubation period ranged from 8 days in early summer (average temperature around 29°C (24-36°C) and 77% RH (Aherua crop) to 15 days during winter (average temperature 19°C (9-27°C) and 72% RH.

Life cycle

The egg is oval, dorso-ventrally flattened, bilateral, symmetrical and creamy and brownish grey in colour but after washing it appears light green or creamy. On an average, size of the egg is 2.3 to 2.8 mm and 2.4 mm in length and width and 0.0075 mg in weight. The micropylar end of the egg is flat and caudal end is narrow. Hatching usually commences ın the morning hours and is completed within three days. The maximum hatching (about 75 %) occurs on the first day while the remaining eggs hatch on the second and third days respectively. Maximum hatching 82±5% occurs during the autumn (Kotia crop) with a mean temperature of 26°C and 77% RH. Almost similar trend of hatching is found during spring (Chotua) crop. The temperature directly affects hatching of egg. When the temperature increases from 17° to 25° C, the hatching increases but when it increases more than 28-35°C, it decreases. Thus the temperature plays vital role in the hatching of egg mass of *A. assama*. The rearing of muga silkworm (*Antheraea assama*) is similar to tropical and temperate tasar silkworm (*Antheraa mylitta* and *Antheraea proylei*). After hatching larvae starts feeding by nipping through egg shell and then migrates towards leaf for feeding. The larvae undergo four moults and pass through five instars. The different larval instars are identified by colours of tubercles, which changes at every instar. The newly hatched polypodus larva is black in colour with distinct yellow lines at the inter-segmental region. The body tubercles are yellow in colour and are provided with setae. The head is small but distinct and black in colour. The newly hatched larva is 0.6 to 1.3 cm in length and approximately 0.005 to 0.007gm in weight. The larva prefers to feed on the soft and tender leaf of som or soalu plant. By the end of 1st stadia, the larvae become stronger and organs became visible. The head bears a pair of mandibles. The mandibles have a prominent muscular base. Ist instar larval period varies from 3-4 days in summer and 5-7 days in winter. Larvae feed continuously on the soft and succulent leaf and by the end of first moult transform itself into second instars larva. The second instars larva is light yellow in colour and measures about 1.5 to 1.9 cm in length and approximately 0.085 to 0.095gm in weight. The colour of the head changes from black to dark brown. The tubercles are blue in colour. The second instars lasted for 3 - 5 days in summer and 7-10 days in winter. When the larvae entered 3rd instars, the colour turned to green measuring 1.9 – 2.5 cm in length and approximately 0.45–0.65 gm in weight. It lasted for 5-7 days in summer and 10-13 days in winter. The tubercle is violet in colour in the third instars larva. A faint yellow lateral line is visible in the lateral region of the body. The claspers become triangular in shape and yellowish green in colour. Fourth instars larva is dark green in colour and measures about 2.5-3.5 cm in length and approximately 2-3.5 gm in weight. Epicranial suture on the head of caterpillar is clearly visible. There is a small black eye spots on the lateral side of the head. There is a distinct yellow lateral line present on both side of the body. At the posterior side a brown colour line is also present above lateral line. The clasper is triangular in shape with a copper black margin. The tubercles are red in colour and provided with setae. It lasted for 7-10 days in summer and 12-15 days in winter. The newly transformed 5th instar larva is approximately 4-5.5 cm in length and weighs approximately 4.15-5.25 gm. The dorsal surface of body is light green but the ventral surface is deep green in colour.

The head is triangular in shape and colour is cupric. Eye spot are prominent on the either side of epicranial suture. The body tubercles are brick red in colour and provided with setae. The newly transformed 5th instar larva is approximately 4 - 5.5 cm in length and weighs approximately 4.10 – 5.15 gm. The dorsal surface of the body is light green but the ventral surface is deep green in colour. The head is triangular in shape and copper in colour. On either side of epicranial suture, eye spots are prominent. The body tubercles are brick red in colour and provided with setae. Each segment of the body bears a pair of dorsal tubercles, a pair of upper lateral tubercles and a pair of lower lateral tubercles. The lateral line is prominent, which extends from the posterior part of the body to the anterior abdominal segment. The lower portion of the lateral line is yellow and upper portion is chocolate in colour. In some hibernating forms only yellow colour lateral line is visible. The clasper is chocolate in colour. Three pairs of thoracic and four pairs of abdominal legs are well developed. The terminal portion of the thoracic legs is provided with claws. The abdominal legs are flat which help the larva to hold firmly to the branches of food plants. Spiracles are black in colour and are well developed in fifth instars larva. This stage lasts for 10-12 days in summer and 16-19 days in winter. The fully mature larva attains 10-15 gm weight. Generally, the female larvae are larger and heavier than the male larvae. The male and female larvae could be easily distinguished from the sexual markings, which develop distinctly in the eighth and ninth abdominal segments on the ventral side of fourth and fifth instars larva. In the female, the sexual markings appear as a pair of milky white spots (Ishiwata's glands) in each of the eighth and ninth segments of the male, a small milky white body (Haroldi gland) appears at the center on the ventral side between eighth and ninth segments. In the wild population colour polymorphism has also been reported during the larval period.

Fig. 2.11 : Morphotypes of muga silkworm

Three different colour morphs in the larval stage viz. yellow, blue and orange colour morphs were observed among the normal green muga silkworm population. The ratio of yellow to green larvae varied from 1 : 50 to 1 : 1000. The orange and blue colour larvae were, however rare. The ratio of orange and blue green larvae is 1 : 6000 and 1 : 3000 respectively. The blue colour larvae are bigger in size (wt. 18.5 gm) than the normal green larvae. The orange colour larvae have a coating of white powdery scales on the body surface. Unlike the tubercles in green larvae, the tubercles of yellow are white in colour. The wild larvae have lower survival rate in the plains. It is known that the allelic expression of colour in insect is largely due to alteration of extrinsic and or intrinsic factors influencing pigment metabolism. Similar processes might operate in the expression of colour polymorphism in muga silkworm also. On maturity the larva spins the cocoons with silk filaments around its body after selecting a suitable site for pupation. Generally, it prefers the space between two to three dry or semi dry leaves. The females spin larger cocoons than the males. Inside the cocoon the larva undergoes a final moult and passes through a short transitory pre-pupal stage. In the pre-pupal stage the dissolution of the larval organs takes place and this is followed by the formation of adult organs during the pupal stage. The pre-pupa is soft and light green in colour but gradually turns brown to dark brown and the pupae skin becomes harder. Sex markings in the pupa are prominent and it is comparatively easier to determine the sex in the pupal stadium than in the larval stage.

The female pupa has a fine longitudinal line on the eighth abdominal segment whereas such marking is absent in the males. The cocoon weight, shell ratio, pupal weight and silk filament length vary in different seasons. Pupal period is around 18 days during summer (28-35°C; 76% RH). These variations in the pupal period in different seasons are attributed to the increase/ decrease of metabolic rate with rise/fall in temperature. The spinning of cocoon takes 3 days in summer and 7 days in winter. The pupal stage lasts for 14 days in summer and 40 days in winter. A small population undergoes diapauses at the pupal stage in the hilly regions. During the early summer, moth emergence has been recorded 91-99% (temp.24-33°C and 70% RH). Moth emergence is governed by various a biotic factors. Most of the adult is normal but 10-15%-crippled moths also emerge in the grainage conducted in June, July, August, September, October and January. The cocoon : dfl ratio in the autumn crop is 1 : 67 as against 1 : 18 to 1 : 26 in the summer crop and 1 : 34 and 1 : 45 in the winter and spring crops. The overall low production in the pre-seed and seed crops was due to high fluctuation in the temperature 19-36°C, relative humidity (72-80%), heavy rainfall, hail storm and attack of parasites and predators. Maximum coupling (90%) is obtained during autumn (November). During the period average temperature and humidity varies form 23°C and 73% RH respectively. During summer season when the temperature and humidity increases 25-36°C; 77% RH and winter period when it goes down (14-22°C; 65% RH), there is heavy reduction in the coupling of moths. It reveals that coupling is adversely affected at low and high temperature and relative humidity.

A biotic factor plays a very vital role on the proper growth and development. The muga silkworm requires high humidity (75-80%) and moderate temperature preferably within 24°C to 30°C. The larval period ranges from 46-55 days during the winter crop with a mean temperature of 19°C (19-27°C) and 72%RH. During the summer crop (Aherua) the larval period ranges from 20-27 days with mean temperature of 29°C (24-36°C) and 72 % RH. Temperature plays vital role in the increasing/decreasing of the muga cocoon production. Egg, larva and pupa are highly influenced by the slight fluctuation in the temperature during the various seasons whereas relative humidity within the range of 72-81% is very useful.

Mature worms crawl down on the tree trunk - indication of their semi-domesticated nature

Fig. 2.12 : Mature muga silkworm larvae coming down for spinning

Commercial rearing

The commercial egg production technique in muga silkworm is quite different than other wild silk moths. The fertilized female moths are tied to kharikas with about 10 cm of cotton thread, there by allowing them movements that facilitate laying. Kharika is a bundle of dried grass, which varies from 30 cm to 1 m in length depending upon the number of moths to be accommodated. The Kharika carrying fertilized female moths are hung vertically in rows with the help of string in grainage room. Eggs are laid in darkness and the entire oviposition is completed till mid night. The eggs are laid in masses of 8 to 10 held together by miconium. The oviposition period varies from 3-4 days. The maximum eggs are laid on the second day. The average fecundity is 150 to 250 from May to October but it declines sharply to 80 to 100 during the winter. Muga silkworm is a mulativoltine species and can be reared through out the year, yielding 5-6 crops annually. Out of these, there are two commercial crops (Jethua and Kotia), two seed crops (Chotua and Bhodia) and two pre-seed crops (Jarua and aherua) in a year. These crops are designated by local names based on rearing months. The larval span varies by season mainly because of difference in temperature (Table 2.9).

Fig. 2.13 : Muscardine (fungal disease) infected muga silkworm larvae

Table. 2.9 : The detail crop cycle in muga culture

Sl. No.	Crop	Nature crop	Months covered	Larval span
1	Kotia	Commercial	October/November	26-32
2	Jarua	Pre-seed	December/January	50-55
3	Chotua	Seed	February/March	30-35
4	Jethua	Commercial	April/may	27-33
5	Aherua	Pre-seed	June/July	22-28
6	Bhodia	Seed	August/Sepetmber	25-31

The climatic conditions during the commercial crops remain with in optimum limit with a temperature range of 20-31°C and 65-95% RH and therefore yield more cocoons per dfl (1:50 or 60). Bhodia and late Jarua are two important seed crops while Aherua and Chotua are the pre-seed crops. The Bhodia seed crops falls during hostile climatic conditions i.e. very humid and hot climate conditions with 26-34°C and 76-95 % RH accompanied by heavy rainfall and hail storm. The Chotua crop falls in the post winter season with a temperature ranging from 18-30°C and 40-95 % RH. The commercial rearers of Assam require around one crore dfls annually for rearing during the Jethua and Kotia crops. However, due to shortage of basic seed for multiplication at various levels, there is always a wide gap between the demand and supply of commercial seeds. The rate of multiplication during the seed and pre-seed crops is poor which results in a low dfl ratio ranging from 1 : 5 to 1 : 15 thus affecting the production. During the Bhodia crop in spite of very humid and hot climate, the yield of cocoons is usually 30-40 per dfl. Bhodia crop suffers from heavy loss of worms due to adverse climatic conditions as well as attack of disease and pests and results in fluctuation in df : cocoon ratio. The seed cocoons are of good quality but the crop generally suffer from heavy virosis resulting in a dfl : cocoons ratio of 1 : 12. There is very high incidence of viral diseases during the period. The late Jarua crop climatic condition is also not favorable due to low temperature and relative humidity (42-60%). The cocoon quality is small and uzi infestation is very high.

The weather condition during Aherua crop is very hot and humid with high rainfall. Due to high fluctuation in temperature and humidity during the crop heavy larval mortality occurs due to virosis. In addition there is heavy loss of worms due to hailstorm. Thus the production from this crop is very low which ultimately affect the seed production.

Potential of muga sericulture

In muga, there are two commercial crops in a year viz. Jethua (April-May) and Kotia (October-November). There is a gap in production levels between what is achieved and what is attainable. For example, leaf production of 24-tons/hectare/year is attainable in muga host plant som and soalu but the current level of production is assessed at 16 tons/hectare/year leaving a gap of 33.3% in leaf yield. This is further reflecting in respect of silkworm rearing capacity and cocoon and raw silk production per hectare, which are currently at 70.8%, 45.9 and 36.7% respectively of the attainable productivity levels. The efficiency with respect to egg hatching, effective rate of rearing and cocoon production per laying are also low, the achievements being 77.7%, 83.3% and 64.2% respectively of the attainable productivity or performance.

ERI CULTURE

The colloquial name, 'eri silk insect' has been derived from its most commonly used food plant, castor (*Ricinus communis*), which was locally known as 'eri', 'endi' or 'eranda'. Eri culture is one of the oldest professions adopted by the people of North-Eastern region of India for production of eri silk as well as use of eri pupa as food material which is highly nutritive. Eri culture is purely traditional and a leisure time occupation limited to meet sericulture family's clothing and food need, partly. Almost all the states in Northeastern region of India has traditional concept for rearing of eri silkworm. Mostly the silkworms are confined to the Brahamaputra valley of Assam in the tribal inhabitant districts viz. Karbiaanglong. North Cachar Hills, Kokrajhar, Golpara, Kamrup,, North Lakhimpur, Dhemaji, Morigaon, Nagoan, Golaghat, Jorhat,, Sibsagar, etc. The culture is also practiced in the neighbouring states mainly in the districts of East and West Garo Hills and Rivoi districts of Meghalaya, Wokha, Mokokchung, Kohima, Tuensang and Zunheboto of Nagaland, Aizwal and Lunglai of Mizoram, Imphal, Thoubal and Churachandrapurof Maniopur, Lohit, Khonsa, Popumpere and Tirap of Arunachal Pradesh. Eri culture is practiced in Bihar, West Bengal, Orissa,, Anadhra Pradesh. Recently it has also been introduced in Kerala, Karnataka, Pondicherry and Tamil Nadu. Unlike mulberry, the eri silkworm is also domesticated and reared indoors. Eri culture has concentrated mostly on *Samia ricini* (Boisdual) (*Philosamia ricini* Hutt) that feeds on the leaves of castor plant (*Ricinus communis* linn, Euphorbiaceae). *Philosamia ricini* and its wild counterpart, *Philosamia cynthia.* is probably the original species from which many sub species have evolved. However, there has been considerable confusion in its binomial nomenclature and its systematic position.

Peigler (1991) made considerable efforts in clarifying and applying the correct name to the insect without making total generic revision, which is presented below for the benefit of the readers. Many authors used the generic name, '*Philosamia*' for this group. *Philosamia* Grote, 1874, is a junior objective synonym of *Samia* Hûbner, 1819, and therefore; *Philosamia* is not an available name according to the International Code of Zoological Nomenclature (1985). The type species for both generic names is *Phalaena cynthia* Drury (Fletcher and Nye, 1982). The name Drury must be placed in parentheses after *cynthia* because he originally assigned the species to the generic name, *Phalaena*. The Chinese *Samia watsoni* (Oberthûr) has been assigned to the genus *Desgodinsia* Oberthûr, 1914. However, this name is a junior homonym of *Desgodinsia* Senna, 1894, in Coleoptera. Since *watsoni* is *congeneric* with *cynthia*, it does not need a replacement of generic name (Fletcher and Nye, 1982). Therefore, the correct generic name for all species in this group is *Samia*. North American species now known under the genus, *Hyalophora*, were earlier placed under *Samia*. Some species of *Samia* have also been classified under the genus, *Attacus* in older literature (Peigler, 1989).

Species-group names

The insect has long been considered as a subspecies of *Samia cynthia* (Drury), a native of China (Peigler, 1992). The origin (probably India or Japan) of the domesticated silkworm, *Samia ricini* (Boisduval) is obscure (Peigler, 1991). Esaki *et al.* (1973) showed similarity between a Japanese *S. ricini* and *S. preiri* in the pink colour of the postmeridian band and the shape of the hind wing crescent. Peigler (1991) observed that *S. ricini* from Assam resembles *S. canningii*. He also suspected that having cultivated for many centuries in China, India and Japan, the genes from *S. cybthia, S. walkeri, S. pryeri* and *S. canningii* must have been bred into *S.ricini*. *Samia cynthia* (Drury): Ferguson (1972) concluded that there were no significant taxonomical differences between the specimens from Sichuan Province of China and those from the United States. Zhu and Wang (1983) reported the occurrence of this species to be Hebei, Shandong, Jiangxi, Zhejial Jiangsu, and "South China" provinces. Peigler (1991) believed that *S. walkeri* was combined with it. Fauvel (1895) mentioned that *S. cynthia* lives in all provinces of north and central China. The specimen from Hong Kong that Seitz (1926) called true *cynthia* was identical to the taxon in Paris and the United States. The fresh moth is dark olive green in colour and fades to a light brown colour on preservation. The crescents are broader than those of all other species except *S. watsonii*. The primary host plant is *Ailanthus altissima* (= *A. glandulosa*) popularly known as the tree-of-heaven in China. The host plant was first introduced to Europe and North America and this was followed by the introduction of the insect. *Samia walkeri* (C. & R. Felder): The type locality is Ning-po Mountains, Chekiang province (Pinyin: Ningbo, Zhejiang Province). It is distinguished from true *P. cynthia* by a more golden yellow ground colour (instead of olive green), more slender crescents, and small differences in wing shape, pattern, and colour of other markings. Fauvel (1895) considered *S. walkeri* and *S. cynthia* to be different species and cited Zhejiang as the source of *S. walkeri*. Lemaire (1978) considered *walkeri* to be a synonym of *cynthia*. Ferguson (1972), following the remarks of Rebel (1925), considered *walkeri* to be a subspecies of *cynthia*, and that the former was associated with camphor (*Cinnamomum camphora*) and the latter with *Ailanthus*. Peigler (1991) believed that the two were separate species. Camphor is a host in Hong Kong (Hill and Cheung, 1978), whereas camphor and other plants are hosts in Taiwan (Wang, 1988). Many European and some American authors used the name *walkeri* to refer to the populations introduced into Europe and North America. Jordan (1911) stated that *walkeri* is the form that was introduced into Europe in 1845. Peigler (1991) mentioned that specimens from the American and European populations do not agree with the type specimens of *walkeri*. Both *walkeri* and *cynthia* have a chromosome number of 13 (2n=26). The chromosome numbers for *S. canningii*, *pryeri*, and *ricini* are 14 (2n=28) Chromosome numbers are not known for *watsoni* or for any other of the tropical species.

Fig 2.14 : Matured Eri silkwrom larvae

Samia tetrica Rebel (=*borneensis* Rebel) has been cited as the dark olive green species and *cynthia* as the golden yellow one (Barlow, 1982 and Lampe, 1984). The latter is actually, *walkeri*, which was being compared to *tetrica*. Ironically, dark olive green coloration separates the true *cynthia* from *walkeri*. but would not separate *cynthia* from *tetrica* (Peigler, 1991). *Samia canningii* (Hutton): This has long been considered a subspecies of *cynthia*. The genitalia differ from those of *cynthia* (Lemarie, 1978 & Lemarie and Peigler, 1982). This is a larger species with a cinnamon brown ground colour and a dark reddish post median band (light pinkish purple in *cynthia*). It is sub-Himalayan, common in Assam and ranging into Thailand (Pinratana and Lampe, 1990). Peigler (1991) stated that Chowdhury sent the material of this species to him under the name of *cynthia*. He also mentioned that other Indian authors (Arora and Gupta, 1979 and Jolly, *et al*, 1979) have called this *Cynthia* and that the confusion was a result of the fact that *canningii* had been considered in literature to be a subspecies of *cynthia* and Indian workers have not had Chinese material of true *cynthia* for comparison. Bloodlines of this wild form are consistently bred into the domesticated *ricini* (Chowdhury, 1982). *Samia watsoni* (Oberthûr): This species occurs from Sichuan to eastern China and Taiwan, apparently always at higher altitudes than either *cynthia* or *walkeri*. Mell (1935) first reported it from eastern China, and Lemaire and Peigler (1982) summarized earlier literature on this species. This species is the largest in the genus, and it has the darkest colouration (Peigler, 1989). The host plants and immature stages of watsoni remain unknown.

Samia ricini (Boisduval) is distinctive, "species" although it is a domesticated form and not really a species; the correct identity has apparently always been clear to all authors (Peigler, 1991). It is a domesticated form and not available in the wild and the origin is obscure. The larvae are reared in open trays similar to those of *Bombyx mori*, and the large puffy cocoons indicate a long history of selection by humans for silk production (United Nations, 1980). The worms are commonly reared on castor (Ricinus communis) leaves resulting in naming the insect after the vernacular name of the plant, 'eri' or 'endi'. The grey brown colour and whitish abdomen make it easily recognizable. Peigler (1991) opined that it was probably derived from either the Japanese *pryeri* because of the gray coloration and the chromosome number or the Himalayan *canningii*. Having been cultivated for many centuries in China, India, and Japan, he suspects that genes from *cynthia*, *walkeri*, *pryeri*, and *canningii* have continually been bred into the *r-icini* stock at various times and localities through centuries. Esaki *et al.* (1973) showed a Japanese *ricini* and a *pryeri*, which agree closely with each other in the pink colour of the post median band and the shape of the hind wing crescent. Peigler (1991) observed that the specimens of *pryeri* from Sakai, Osaka Prefecture, and *ricini* from Fukuoka, showed similarities, which were not shared by *ricini* from Cuba; and Assam and the *ricini* he had from Assam look more like *canningii* (Chodhury, 1982). Thus, the systematic position of the domesticated eri silk moth stands as superfamily: Bombycoidea Latreille, 1802; Family :Saturniidae Boisduval (1837) 1834; Genus:*Samia* Hûbner, 1819 ; Species :*ricinii* (Boisduval).

Taxonomic History

There has been lot of confusion regarding taxonomic status of the genus *Samia* Hubner (= *Philosamia* Grote) especially regarding its monoculture. Packard (1914) treated *Samia* and *Philosamia* separately, where as Seitz (1926) considered *Samia* as a valid genus and treated *Philosamia* as a separate genera representing Africa, America, and Oriental fauna, respectively. Michener (1952) set at rest the controversy and regarded *Philosamia* as a synonym and reported. This is an oriental genus represented in the Eastern United States by a single introduced species" the type of species, *S. cynthia* Drury. But, it had not been naturalized there for over 100 years (Ferguson, 1972). Of late, Peigler (1992) observed that the genus *Philosamia* is a junior objective synonym of *Samia* and therefore *Philosamia* is not an available name according to International code of Zoological Nomenclature (1985). This nomenclature has been accepted by the recent works of Nassig *et al.* (1995) and Narimann and Peigler (2001). The genus *Samia* is currently being revised and it contains more than 15 species from tropical and temperate eastern Asia (Narimann and Peigler, 2001). Some of the species are superficially similar to one another and some are entirely different. Due to different ecological habitat there are certain marginal difference in the morphological characters of same species and therefore grouped subspecies of *Samia cynthia*.

This confusion has persisted into recent literature, in which many species were erroneously included. Likewise, Arora and Gupta (1979) reported single species of *Samia cynthia* with several sub-species including ricini in India, which is a domesticated race. However, Peigler (1992) reported that *Samia cynthia* does not occur in Assam and other Sub Himalayan areas where it is replaced by *S. canningi*. While, *S. ricini* is not a species but it is a domesticated forms, that has been or still cultivated on a large scale in many places including Cuba, Uruguay, Egypt, France, Italy, Japan, China and India. Further, it does not occur in the wild, and the origin is obscure, possibly it was deviced from the Japanese *S. pryeri* because of the gray colouration and chromosome number, but the Himalaya *S. canningi* is also a possible origin. *S. ricini* of North East of India look more like *S. canningi* (Chowdhury, 1964). Arora and Gupta (1979) reported distribution of *S. ricini* in its wild form in the Himalaya region and North Eastern states. Hamson (1892) reported morphology of the adult *P. ricini*. Domini (1941) described certain characters of larval instars. Reddy and Narasimasamy (2000) reported the external morphology of eri silkworm eggs during embryonic development. Several lines/sublines of *S. ricini* are available in North Eastern region of India (Table 2.10).

Table 2.10 : Important lines of *Samia ricini* in India

1.	White plain	Nongphoh, Lakhimpur, Demor
2.	White spotted	Diphu, Demori
3.	White spotted with black band	Demon
4.	White spotted with zebra and semi zebra markings	Musalpur and Bodiwar
5.	White spotted with alternate black	Musalpur and Dordnwar
6.	Greenish blue plain(deep and light colour)	Demon
7.	Greenish blue spotted with zebra and semi zebra markings	Musalpur Berduar
8.	Spotted with big spots and double spots on dorsal line	

Ecological Distribution

Genus Samia has dozens of species in oriental countries and only the species *ricini* is reared for production of silk. Rebel (1925) reported ecological distribution of species as follows.

	Species	Distribution
I. Group	**Cynthia**	
	cynthia advena walkeri elduvaria pryeri	S. China, South-East Asia N. America Korea Korea Kyushu, Taiwan
II Group	**Ricini**	
	ricini obscura luniloides	East India Himalayan foot hills East India
III Group	**Insularis** vanderberghi luzonica tetrica mindanaensis	**South-East Asia** South East Asia South East Asia Philippines Philipinensis
IV Group	**Canninghi** canninghi Fulva	South-East Asia South-East Asia

Peigler (1980) reported that *S. ricini* is evolved due to artificial selection after century's efforts as a consequence of continuous stress on yield of silk. The species may have been derived from the Japanese species *S. pryeri*, as both the species have equal number of chromosomes (2n=14). Genes from other species might have been introduced in a gradual manner to *S. ricini* with the passing of time after several generations inter breeding. China has able to evolve a fixed race by hybridizing *S. cynthia* with *S. ricini* after two decennials efforts which have great communal importance (FAO-UN, 1980). Seitz (1928) reported the hybrid of the crosses between various species of Samia.

CROSSES	OFFSPRING
Cynthia x canninghi	Xanthoxylon Wts
Cynthia x ricini	Walacei tutt
Canninghi x advena	andrei wts
Advena x canninghi	lasttours wts

CROSSES	OFFSPRING
ricini x pryeri	rothchildi wts
ricini x advena	vesta wts.
vesta x ricini	balli wts
vesta x pryeri	lefroi wts
advera x pryeri	pryadvena
pryeri x advena	oberthuri wts

Peigler (1984) observed crossing of inter-generic nature between different genus.

Samia Cynthia x Callosamia angulifera Sterile moth obtained.

Samia Cynthia x Styloplewacecropta Full grown larvae but died.

Stocks

The twelve varieties of eri silkworm, *Philosamia ricini* Boisduval have been reported (Sarkar, 1980). White Plain (WP), Green Plain (GP), Yellow Plain (YP) larval colours; White Dotted (WD), Green Dotted (GD), Yellow Dotted (YD) dotted larval markings; White Semi Zebra (WSZ), Green Semi Zebra (GSZ), Yellow Semi Zebra (YSZ), and White Zebra (WZ), Green Zebra (GZ), Yellow Zebra (YZ) larval markings were isolated from wild type *P. ricini* through selection and breeding over 10 generations and reared under standard laboratory condition.

White Plain (WP)

It is a multivoltine pure race of eri silkworm (fig.2.1) and it is reared 5 to 6 crop in a year. It is popular variety used for commercial production of eri silk in India. It undergoes four moults and five instar with a larval duration of 561 hrs. The larvae are resistant to diseases like flacherie and grasserie. The cocoons are white in colour and the quantitative characters are better than any other varieties.

During mass rearing of wild type White Plain (WP), the larval colours and larval markings were isolated and through selfing pure lines were evolved.

Green Plain (GP)

It is pure multivoltine eri silkworm variety without undergoing diapause. This race undergoes four moults and five instars with larval duration of 570 h. The male and female larvae are Green Plain (GP) in colour and spins white colour cocoon and the quantitative characters of this variety.

Yellow Plain (YP)

It is pure multivoltine eri silkworm variety without undergoing diapause. This race undergoes four moults and five instar with 568 h of larval durations. The male and female larvae are Yellow Plain (YP). The cocoons spun by this race are brick red.

White Dotted (WD)

It is a pure multivoltine variety undergoes four mults and five instars with larval duration of 550.31 h. The phenotype of male and female larvae exhibit White Dotted (WD) markings. The cocoons of these varieties are white in colour.

Green Dotted (GD)

This is a pure multivoltine variety; this variety undergoes four moults and the instars with larval duration of 568.25h. The phenotype of male and female larvae posses Green Dotted (GD) markings. It spins white cocoons.

Yellow Dotted (YD)

It is also a pure multivoltine variety without undergoing diapause. This variety undergoes four moults and five instar with larval duration of 577.24 h. The male and female larvae posses Yellow Dotted (YD) marklings. It spins brick red colour cocoons.

White Semi Zebra (WSZ)

It is a pure race of multivoltine eri silkworm and it is reared 5 to 6 times in a year. This race undergoes four moults and five instar with 551.10h. The phenotype of the male and female larvae posses White Semi Zebra (WSZ) markings. It spins white colour cocoons.

Green Semi Zebra (GSZ)

It is a pure multivoltine variety without undergoing diapause undergoes four moults and five instars with 546.24h of larval duration. The phenotype of the male and female larvae exhibits Green Semi Zebra (GSZ) markings. It spins white cocoons.

Yellow semi zebra (YSZ)

It is a pure multivoltine eri silkworm variety without undergoing diapause. This race undergoes four moults and five instars with larval duration of 578 h. The male and female larvae possess Yellow Semi Zebra (YSZ) markings. It spins brick red colour cocoons.

White Zebra (WZ)

It is also a pure multivoltine variety without under going diapause. It undergoes four moults and five instar with 535 h larval duration. The phenotype of malel and female larvae exhibit White Zebra (WZ) marking. It spins white cocoons

Green Zebra (GZ)

It is a pure multivoltine variety without under going diapauses. It undergoes four moults and five instar with 551.30 of larval duration. The phenotype of male and female larvae posses Green Zebra (GZ) markings. It spins white cocoons.

Yellow Zebra (YZ)

It is also pure multivoltine eri silkworm variety without under going diapauses. This variety undergoes four moults and five instar with larval duration of 556h. The male and female larvae posses

Yellow Zebra (YZ) markings. The cocoons are brick red in colour. Different agro climatic zones produced several regional silkworm races with different characteristics. Since the birth of silkworm genetics extensive investigations have been carried out with reference to the improvement of economic characters of different varieties of silkworms. These results together with those of silkworm breeding experiments have contributed much to the establishment of contemporary silkworm breeds of high economic value (Tazima, 1944). Scientific breeding methods evolved during the last few decades have enabled the breeders to synthesize desirable genotypes of known genetic constitution with objective to increase the productivity. Further application of genetic principles in understanding the hereditary nature of the quantitative traits coupled with appropriate selection procedures have contributed a great deal in increasing the productivity of silkworms (Yu cheng Wu *et al.*, 1994).

Life cycle

The eri siikworm *Philosamia ricini* is multivoltine in nature and 5-6 crops can be raised in a year. However the behaviour of the wild silkworm, *Philosamia cynthia* has been found to be univoltine, Bivoltine, as well as Trivoltine. The life cycle of eri silkworm has four stages namely egg, larva, pupae and moth. It ranges from 45 days in summer to 90 days in winter (Table 2.11). The duration of the life cycle is as follows

Table 2.11. Details of the duration of the developmental stages of *Philosamia ricini*

Stages	Summer	Winter
Egg stage	9	18
Larval stage	17	45
Spinning stage	3	5
Pupal stage	13	19
Adult stage	3	3
Total	45	90

The eggs are ovoid, candid white, measure 1.5 x 1.0 mm. and weigh 6 mg. As the embryo develops inside the egg, the colour of the shell changes from whitish to yellowish, yellowish to ashy and ashy to blackish just before hatching. Then, a tiny worm comes out, leaving the shell once again, all white and empty. Eggs, generally, hatch in the morning, i.e., between 7 and 9 a.m. To get uniform hatching, they should be incubated at temperature 22°C to 26°C. The incubation period varies from 8–20 days. This differences is due to the degree of environmental temperature. With the help of a refrigerator or cold storage, the hatching can be retarded, just as it can be accelerated and regulated in an incubator. The eggs laid by the female moths on the stick or kahikas or the cloth or in a small split bamboo basket, as the case may be, are scarped off, tied in a piece of cloth and hung up under the roof until they hatch. The larva on hatching is greenish yellow. It measures about 5 x 1 mm. and weigh 1.5 mg. The body colour changes gradually to pure yellow by the end of the third day. From third instars onwards the body colour segregates into yellow, cream,, green , blue or white.

The fully mature larva, which measures about 7.0 x 1.5 cm. and weigh 8 g. is translucent and covered with a white powdery substance. Both spotted and unspotted larvae are found. The spots are of various types: single, double, zebra and semi-zebra. The prothoracic hood of the first instar larva has a black dorsal band, which splits up into a pair of crescent shaped markings in the second and third instars. These markings disappears at the fourth instars. The area around the antennae acquires a black dumb - bell-shaped mark in the fourth instars, which in the fifth instars splits into two oval dots. The planta, anal flap and claspers are light yellow through out the larval span The planta has a horizontal blue band at the top. Early in the first instar the markings on the anal flaps ands claspers are black and appear as a continuous band. In the second and third instars the black marks are triangular on the anal flap and rectangular on the claspers.These marks disappear in the subsequent instars.The lateral line is creamy white and extends the length of the body. The tubercles are very conspicuous and tubular in shape. They are bluish at the base and cream colour at the tip. The setae, which are pointed and blackish, number 180 at the third instars.

The Pupa is obtect adectious and does not depart from the basic saturniide pattern. It measures about 2.8 x 1.5 cm. and weigh about 2.6 g. The cocoons can be easily distinguished from those of Antheraeas, they are elongated, soft, wooly, peduncle less, open mouthed and unreelable. They measure about 4.0 x 2.5 cm. and weigh about 3 g and exhibit colour polymorphism, being brick red and creamy white. The shell weight and shell ration are about 0.4 g and 13 %. The adult male moth is about 2.5 cm. and the female 3.0 cm The average wing expanse of male and female moths is 13 cm and 15 cm. respectively. The fore wing of both the sexes are more or less similar in structure and colour pattern. The ground colour is brownish, blackish or chocolate. The characteristic antemedian line (AM) is bright chocolate coloured with white border on either side and almost runs through the center. The post meridian line (PM) is black with a single dull grey border on either side. The ocellus in both the sexes is crescent shaped. There is a conspicuous black spot, pterostigma, with a whitish tinge, present at the top of the wing apex. The hind wing measures about 890 mm^2 in males and 1037 mm^2 in females. The black PM, unlike that of forewing, is split in two. The AM and ground colour are that of similar of fore wing. The shape of ocellus is the characteristic of the insect. The hyaline area (HA), almost invisible, is located in the most anterior region of the ocellus. The area of ocellus in the fore and hind wing is about 36 mm^2 and 30 mm^2 in the male and 40mm^2 and 35 mm^2 in the female. The HA of the fore and hind wings is 10 mm^2 and 8 mm^2 in the male and 12 mm^2 and 10 mm^2 in the female.

The colour pattern is identical, however in both the sexes. The scales are generally triangular with one to four spines at the top, although long bristle like scales with pointed ends are also found. As for wing venation, the wavy median cross vein (Mc) joins both A1 and A2. The anal veins An2 and An3 fuse in the forewing.

Samia cynthia (Drury 1773) is primarily univoltine in the United States. Moths fly from late June to early July. These insects are not overly attracted to lights, but cocoons are readily found in the fall, after leaf drop, hanging by strong peduncles from ailanthus compound leaf stalks. These moths mate readily in captivity, even in small (one cubic foot) cages usually on the evening of eclosion. After dusk the female extends a scent gland from her abdomen. The wind blown pheromone can attract males from several miles away. Mating occurs when the male grasps the scent gland between his claspers. The pair usually stays coupled until the following evening. It is more difficult to sex adult cynthias than it is to sex most other saturniidae. Wing shape and colouration is almost identical for both genders and antennae structures are very similar. The abdomen of the female is plump and rounded while the narrow abdomen of the male is more conical. Ova are small (as many as 400 from a single female) and white and are usually laid in short rows. Incubation is eight to twelve days, and the newly emerged larvae are gregarious for the first two instars. They appear striped about the girth much like *Promethea* larvae due to tiny black protuberances. Larvae are still gregarious in the second instars and have relatively small black heads. In the third and fourth instars, larvae are creamy white with yellow heads and legs. In the final instars, larvae begin to look much more like atlas larvae, pale blue-green and covered with a fine white powder. There are fleshy protuberances along the back and sides.

Grainage

Grainage operation in eri is quite different form tasar and muga due to its domesticated nature. The seed cocoons are stored by spreading them thinly on circular split bamboos–trays on wooden racks. The cocoons are kept at room temperature and normal humidity is maintained for emergence. Depending on climate, the period from cocoon formation to moth emergence ranges from 13 to 18 days. Generally moths emerge early in the morning and continues up to 7 AM. It takes place under natural light at room temperature and humidity. Moth takes about 5-8 minutes for emergence from the cocoon. Soon after emergence they emits a creamy excretion, which dirties the cocoon shell. Within 45 minutes to an hour they become fully active and prepare for mating. The freshly emerged males are tied to kharika with cotton thread. Kharika is hung on the strand of wire stretched across the room 60-90 cm above the floor. Either the male moths are positioned by hand or they themselves approach the virgin female moths. Eri Moths have very good coupling aptitude. Maximum pairing takes place at 4 to 7 PM and the moth may remain coupled for 24 hours, although 3-4 hours are adequate. The male moths can be utilized for a second coupling. The response to hand coupling is also encouraging. The decoupled fertilized females are left on the kharika. Egg laying starts one to two hour after decoupling and continues for 3 to 4 days, but only the eggs laid on first two days are considered for rearing. On an average, each moth lays 450-500 eggs, most of them on first day. The eggs are deposited on a single layer at two or three places on the kahrika. Specially design eggs carriers are used for distribution of the laying. Layings are examined microscopically and diseases free layings are disinfected in 5% formalin for five minutes.

The eggs are incubated either individually or collectively in the grainage house at normal room temperature and relative humidity, preferably at 22-24°C and. hatching starts at 6.00 to 9 Am in the morning within 9 to 10 days in summer and 14 to 15 days in winter. The newly hatched larvae are about 4 x 0.7 mm. and weigh about 1.3 mg. Eri silkworms are reared both in plains and hilly regions. The temperature ranges from 15°C in winter to 35°C in summer and from 50 % to 100 % relative humidity. The optimum range for temperature and relative humidity is 24-26° C and 75 – 85% . The larval span varies from 20 days in summer to 50 days in winter. The mature larvae are about 8.5 cm. long and 1.7 cm. wide and weigh about 6.2 g. Unlike other silkworm eri silkworms do not eat the empty egg shell on hatching nor the cast off the skin after moulting. They do not even consume the mid rib and petiole of leaves. They have a very poor gripping power. The mature larvae produce a very a rustling sound when rolled between the fingers and have a tendency to move upward and away from the foliage. The mature larvae search a suitable cocooning place on the Chandrika and remain there 4 to 5 hours after discharging the larval excreta. First moving the head to trace a horizontal figure eight forms the base of the cocoon. This is followed by formation of sides and finally the upper, part but during these operations the head movement is irregular. After about 13-15 hours of spinning the larvae are invisible. The cocooning is completed in three days during the summer and in 5 days in during the winter. The cocoons are peduncle less and open at one end.

Rearing House and Disinfections

Rearing house should be well-ventilated along with devices to make it air-light to facilitate effective disinfections. Appliances and rearing house should invariably disinfect by spraying 2% formaline solution before the commencement of every rearing. The doors and windows should be kept closed for a day immediately after spraying. Later, the doors and windows should be kept open at least for a day to allow the formaline vapors to go out. If the last rearing has suffered from pebrine, the disinfections should be carried out with 4% formaline solution at the rate 7-8 ltrs. /100 mtrs. and the room should be closed for two days.

Rearing appliance

Generally appliances required for mulberry silkworm rearing may be used for eri silkworm rearing except the thread net. The most common appliances are small and medium sized bamboo trays, spinning bamboo mat, a maximum and minimum thermometer, a dry wet bulb hygrometer, knife, a small wooden platform, basket, thick gunny cloth, bucket, glass cylinder, glass beaker, funnel, scissors, measuring cylinder, old newspapers, rearing register, microscope with oil emulsion, glass slides, cover glass, pestles and mortars, country balance, chemical balance, magnifying glass, sprayer, stone, lantern, aluminium pot, fine cloth, fine cloth, khorika, wire net covered wooden box or gage, small basket, table, thread ball, glass jar, forceps, scalpel etc. Apat from this, disinfectant such as formalin, copper sulphate, bleaching powder, lime is also required for successful rearing. For spinning cocoons, the cheapest material are dry plant leaves, mango twigs with green leaves and bamboo twigs. In some hill areas, a kind of bamboo is used for mountage. A whole bamboo, about 6' long, is split in the middle. Ripe worms are placed in the split bamboo and tied together by means of a rope creeper. Cocoons mounted this way are uniform and clean.

Climatic Conditions

A temperature of 24-28° C and 70-85% relative humidity are ideal for eri culture. Sudden and high fluctuations in temperature are always deleterious. The first and second instars larvae need slightly higher temperature and relative humidity. Free airflow keeps the worms healthy. Thatched buildings with clay walls can provide a cooler atmosphere in tropical areas; moist gunny cloth hung at the windows allowing free passage of air further brings down the temperature. Burning of coal or firewood can be done to increase the temperature in colder climates. But in such cases, free airflow should be ensured to decrease the carbon dioxide concentration.

Silkworm Rearing

Eri silkworm is reared indoors. The rearing house should be well-ventilated and free form the dust. Rearing of eri silkworm is taken up when castor leaves are available in plenty. There are two methods of rearing to eri silkworms. First tying 10 to 12 leaves together into bundles and second are spreading the leaves over the worms on the tray. In some places combined methods of rearing are conducted. The worms hatch out, generally, in the morning. The cloth containing the eggs is opened and placed on a tray. A few tender top leaves plucked from the cator plant are spread over the worms. Worms hatching for first two days are taken for rearing and transferred to rearing trays along with the leaves on which they crawl. The eggs are kept in the rearing tray on a piece of paper for 24 hours before the hatching. They are covered with a few tender top leaves when they start to hatch. The larvae crawl gradually on to the underside of the leaves and remain very close together in groups. During the early stages the larvae are feed on the chopped leaves to maintain the hygienic condition. The larvae are fed on trays or on hanging leaf bundles. If feed on tray, whole or chopped leaves are spread over the larvae. Considerable time is saved, but the rearing bed becomes dirty with excreta. In any case, this method is necessary until the second moult. If the hanging – bundle method is used, eight to ten leaves are tied together and hung on a stick resting across the two parallel bars of the stand. The vertical position of leaf lamina not only facilitates almost complete consumption of the foliage, but also permits the litters to drop directly onto the tray underneath without soiling the leaves. The feeding method thus combines foliage economy and cleanliness. Upto the third instar the larvae are given four feedings a day, and the late instars fed five times a day at regular intervals. The first feeding must be served at 5:00- 6:00 and the last at 21:00-22:00 hours. Moulting larvae, however, are served only when about 85% of them have cast off their old skin.

The rearing bed should always be cleaned before noon, else the excreta start decomposing with the rise in temperature. Not only the litters, but also the midribs, petioles, dry and unconsumed leaves as well as other waste must be removed. While cleaning, fresh leaves are put on the rearing tray to attract the larvae. Dead and infected larvae are removed as soon as they are noticed. When the batches are preparing to moult, cleaning should be conducted before they settle.It is desirable to rear early – and late – instar larvae in separate rooms because of their substantially different requirements. The rearing is generally conducted under normal room conditions in all seasons, but in case of necessity the temperature and relative humidity of the rearing room can be crought to the desirable level by artificial means. As soon as the larvae mature they are placed in the chandrika for spinning. The quality of the cocoons greatly depends on the food plant. The various eri food plants, in order of efficacy, are castor-oil payam, kesseru and tapioca. As to the seasonal effect, the cocoons produced in late spring and in late autumn are the best.

Experimental rearing

Castor and kesseru were found to be the best food plants for eri silkworms followed by tapioca and barkesseru. Castor varieties having non-powdery leaves are better than those with powdery leaves. About 650-750 kg. of castor leaves are required to rear 100 layings. It should be noted that quantity as well as quality of leaves are important to the growth and hygiene of silkworms. Since eggs are the future worms, they should be obtained from reliable sources like Govt. grainages or extension centers to ensure better hygiene and better rearing performance. The eggs generally hatch in 8-10 days at optimum temperature. But in winter it may take as many as 15 days or more. Hatching can artificially be delayed for 5 days by preserving the eggs at 5-10°C. The newly hatched worms should be provided with tender leaves on to which they crawl and then be transferred to the rearing trays. Feeding the young worms with the leaves chopped into bits of about 1 cm 2 is ideal. First four instars last about three days each while the last instar lasts for 5-6 days. The moulting duration between each instar is about 20-24 hours during which the worms do not feed. During moulting the rearing beds must be kept clean and dry allowing free passage of air. The entire larval duration lasts for 22 – 25 days at optimum temperature. However, higher temperature reduces the larval period while lower temperature increases it.

The first and second instar larvae are fed with tender leaves while the late age worms are given mature leaves. The experiments on bunch and tray feeding methods showed that bunch feeding yields a better crop due to hygienic conditions existing there while conserving the man power utilized in cleaning the beds. Overcrowding should be avoided since it leads to competition among the worms for food and space with results in under nourishments followed by restarted growth. A tray of 7.5 x 2 provides sufficient space to rear 10 layings (3, 000 worms) until second instar while 600 larvae can be reared upto 4th instar and only 300 in the final instar. Mature worms empty their guts, become translucent and move in search of a place for spinning cocoons. At this stage, they are to be transferred to the chandrike. Since these worms prefer shady corners, the mountage should be prepared accordingly. Dried twigs with large curved leaves can also be successfully used as mountages. Spinning takes place for about 4 days and cocoons are harvested on 5 th day. Since these cocoons are not reelable, they can be kept until moth emergence and spun at latter stage. For this reason, the eri silk is also known as Holy Silk.

Commercial exploitation

Eri silkworm is reared indoors with two different methods of rearing like bunch feeding in general (Choudhary, 1982, Sarkar, 1988) and tray feeding. Combined method of rearing is also used in some places. The worm can be reared up to third stage in the tray which save more time but rearing bed becomes dirty with excreta. The later stage worm can be reared in the bunch feeding method which combines foliage economy and cleanliness. The population density study on the eri sikworm indicate that effect due to crowding results in more larval mortality and has no influence on larval period. Studies on the bunch feeding and tray feeding of eri silkworm revealed that bunch feeding was found to be advantageous over the tray feeding (Annual Report 1983-84). Seven eco races of eri silkworm are available in North-eastern region of India. The most common and abundantly available in North-eastern region of India are Borduar, Khanapara local, Sille, Titabar, Nongpoh, Dhenubhanga, Mendipathar local. Eri worms are generally reared on castor, kesseru, tapioca, gulancha and payam during July-August, September and October. Average larval weight was maximum in castor (6.49g) followed kesseru (5.12g), payam (5.08 g) and tapioca and gulancha (4.7 g). Cocoon and shell weights were higher in castor (cocoon weight 4.08g; shell weight 0.56g) Castor also showed highest ERR % (95.79%) followed by kesseru (74%), tapioca (71%), payam (64%) and gulancha (56%).Rearing performance of six homozygous strains viz. Yellow spotted, G.B. Plain and G. B. spotted isolated from titabar local eco race and Yellow plain and Yellow Zebra, G.B. Plain and G.B. Zebra isolated from Borduar local eco race were reared at Regioanl Eri Research Station Mendipathar (Meghalaya). Average ERR%

ranged form 92.41% in yellow Zebra to 94.41% in Yellow plain of Borduar Local ecorace. The yellow plain strain of Borduar local eco race also showed maximum cocoon weight (4.32g). Other cocoons characters were better in strains isolated form Borduar local eco race (Table 2.12).

Table 2.12 : The detail of rearing performance of *Philosamia ricini*

Particulars	Borduar	Khnapara	Titabar	Dhenu	Nongpoh	Mendipa	Silli
Fecundity	451	418	455	423	395	421	413
Larval weight	6.73	6.31	6.00	5.57	5.29	5.64	5.42
ERR%	90.05	88.86	89.42	87.66	86.85	87.20	86.57
Cocoon weight	3.87	3.79	3.60	3.40	3.41	3.39	3.45
Shell weight	0.52	0.51	0.46	0.43	0.42	0,42	0.44
SR%	13.67	13.48	13.00	12.80	12.60	12.52	12.56

Seven eri stock viz Borduar, Khananapara local, Sille, Titabar, Nongpoh, Dhenubhanga, Mendipathar local were reared during different season. ERR ranged from 86.57 in Sille local to 90.05 in Borduar. Titabar stock showed highest fecundity (455) followed by Borduar stock (451). Average shell ratio was highest in Borduar stock being 13.67 % followed by 13.48% in Khanapara stock and 13 % in Titabar stock. Eri worms are reared during April, May-June, June-August, August-September, Septermber-October and November-January respectively. Average fecundity ranged from 514 per dfls in Borduar local to 405 per dfl in Dhenubhanga local and ERR% from 92.4% in Dhenubhanga local to 95.72% in Titabar local. Cocoons character were better in Borduar local (Cocoon weight : 4.12g; shell weight 0.63g).

CHAPTER-3

Pests of Mulberry and their Control

Several insects are known to cause economic loss to mulberry (*Morus* spp.), the silkworm host plant. Mealy bug, papaya mealybug, white fly, leaf roller, thrips and hoppers are the major pests and gives serious threat to mulberry silk production throughout south India (Ullal and Narasimhanna, 1978). Loss of mulberry crop yield due to these pests is estimated at between 10 to 30%. Serious damage usually occurs during the early stages of plant growth due to intensive sucking by the insect. Some of the insects are vector in transmitting viral diseases also. An insect population always fluctuates according to the dynamic condition of its environment. Both physical (a biotic) and biotic factors are believed to be the factors responsible for the change in a population. *Andrewartha and Birch* (1954) reported four components of the environment that could influence insect population's viz. weather condition, food, other insects or organisms causing diseases and a place in which to live. Climatic factors such as temperature and rainfall and relative humidity have known to greatly influence the insect population change (Muhamad and Chung, 1993; Way and Heong, 1994; Zhu, 1999' Heong *et al.*, 2007, Siswanto *et al.*, 2008; Singh *et al.*, 2000). Knowledge of the seasonal abundance and trends in the population build up of a pest has become important for its effective control schedules (Siswanto *et al.*, 2008). Food supply and improved irrigation in the mulberry field are conducive to the growth of pest populations. Mulberry plantation in temperate region have greater pest problems than plantation in tropical areas. Some of the major pests visiting mulberry plantation frequently and build up its population and causing extensive damage are reported in detail.

Pink Mealybug; *Maconellicoccus hirsutus* (Green) (Hemiptera: Pseudococcidae)

Among all the sap sucking insects of mulberry the pink mealybug, *Maconellicoccus hirsutus*, (Green) (Hemiptera: Pseudococcidae) is the major pest in tropical and temperate sericulture region of India. This pest has two common names (pink mealybug and hibiscus mealybug), but there was an effort to standardize the common name by calling the pest "pink mealybug," even though it attacks many plant species, including mulberry. Adult mealybugs are small (about 2.5 to 3 mm long) and pink in body colour but covered with hairs formed from a waxy secretion. The waxy filaments are short and females are usually obscured by white mealy wax. Adult males are smaller than females, reddish brown and have one pair of wings. Males have two long waxy "tails." Females die shortly after depositing eggs. Freshly laid eggs are orange, becoming pink before they hatch. Eggs are found in egg sacs. First instar nymphs (crawlers) of the pink mealybug disperse by walking and blowing away by wind. Nymphs also can walk considerable distances to find suitable host plants. It takes about 20 to 25 days to complete the life cycle. The pink mealybug has a high reproductive rate (> 600 eggs) and produces up to 13 generations per year. Pink mealybug eggs survive the winter by remaining in bark crevices, leaf scars, under bark, in the soil, tree holes and inside crumpled leaf clusters.

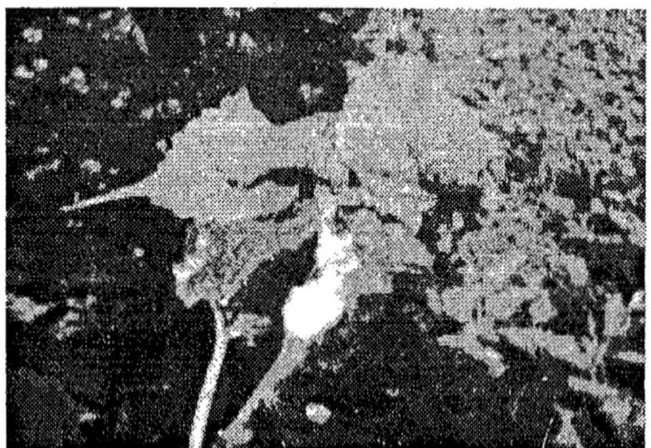

Fig. 3.1 : Mealybug on mulberry plants

It is a highly polyphagous that pest feeds on the soft and succulent leaves of mulberry and injects a toxic saliva which causes curling and contortion of leaves. The entire plant may be stunted and the shoot tips develop a bushy appearance and are commonly known as "bunchy top" or Tukra disease. It usually infests young twigs, causing deformed terminal growth. The apical growth of the plant is arrested which results in the stunted growth. The curled leaves can resemble viral damage, but this pest is not known to be the vector of any virus. The mealybug excretes honeydew which encourages the black sooty mold to develop. Nutritive value of the leaves, leaf yield and plant height are drastically reduced. Leaves become pale yellow after severe infestation. Feeding silkworm with Tukra affected leaf causes significant decline in economic characters of silkworm rearing like larval weight, cocoon weight, shell weight, silk ratio and effective rate of rearing (Wang, 1992; Ahmad *et al.*, 2000).

Ecologically Based Pest Management (EBPM)

Earlier, much of the attempts have been made to control mealybug population and enhance the quality leaf production in mulberry by using various chemical methods. But the worldwide problems with environmental pollution and with increasing resistance levels in all major pests including mealybug to pesticides strongly advocate the need of changes in the current pest and disease management paradigms. Therefore, it is considered that cultural control practices, combined with the advanced biological control strategies are the most logical approach for pest management (Singh and Saratchandra, 2002). Many practices, such as crop rotation, tillage operation, intercropping, and incorporation of organic matter into soils have strong influence on the conservation and fostering of local populations of these biological control agents. These biological controlling agents play important roles in minimizing pest outbreaks and crop disease epidemics (Baskaran and Narayanswamy, 1995). Therefore, it essential to develop strategies based on empirical knowledge under integrated pest management program, to utilize them more effectively to control the menace of mealybug in mulberry. Recent studies have suggested a significantly increasing role of biological control in Integrated Pest Management (IPM) programs. The biological control models based on knowledge of ecological process rather than chemical pesticides is known as Ecologically Based Pest Management (EBPM). The ecological principles of natural pest mortality factors, pest-predator relationships, genetic resistance of the host plant, and impact of cultural practices on pest control are the strongest component of Integrated Pest Management programme (Cate and Hinkle, 1993). Although, the theoretical basis of EBPM is similar to that of IPM (Schwalbe, 1993), EBPM relies more on the improved knowledge base of the complex ecological processes that occur in the niche of the plant. Indeed, safety, profitability, and durability are three fundamental goals of EBPM (Flint,1992; Hoffman and Frodsham, 1993). EBPM is a total systems approach designed to have minimal adverse effects on none target species and the environment because biological controls are limited by their specificity with respect to the target pest organism or by their distribution or persistence in the environment. For wider acceptance of EPBM, it is essential to make it more cost effective with least economic risks (Caltagirone and Huffaker,1980). The strategies must be long lasting to have a positive impact on crop protection. It should be durable and much of the data should be on the pest biology and economic feasibility. New pest-resistant varieties are introduced while resistance of pests to chemical pesticides is minimized (Flint and Van den Bosch 1981). EBPM also uses knowledge of interactions among pests and other naturally occurring beneficial organisms to modify cropping systems in ways that reduce rather than eliminate damage caused by pests; thus, EBPM is designed to enhance the inherent ecological strengths of the system. Mealy bug pest management studies have been conducted using manipulations of crop production inputs, trap crops, behavioural chemicals and other ecological methods to manage pest by fostering the beneficial insect species. Key natural enemies of major pests are identified along with its efficacy, adaptiveness and potentials for introduction, conservation and augmentation. In order to increase the longevity of EPBM, it is also important to devise strategies to provide farmers with sufficient inputs on biological-control and conservation of valuable biological-control resources.

Cultural control

Cultural practices are referred to as management techniques which may be manipulated by sericulture farmers to achieve their crop production goals (Singh *et al.*, 2005). It is a process by which impediments are generated to pest colonization through creation of adverse biotic conditions to reduce the survival of pest population. The cropping system is modified in such a way that helps to minimize pest infestation. Inter cultivation of various crops with mulberry helps to reduce the pest population. Further, it also helps enhance the populations of natural enemies of the pest by manipulating the environment. In India, mulberry has been grown as intercrops with several agricultural and cash crops. Inter-cultivation of mulberry with potato, mustard, vegetable and pulses is practiced in some areas particularly in the tropical sericultural belts (Kaur *et al.*, 2002). Mulberry can also grow between rows of tea/coffee as shade trees. Intercropping of mulberry with safflower, methi. coriander, french bean, chrysanthemum, marigold, gladiolus, raddish, chrysanthemum and Bengal gram has also been a been mulberry common practice in several areas (Dhingra, 2002). Winter vegetables are also planted between rows of mulberry. The results from these practices indicate that mulberry leaf yield was not affected by these intercrops. Regarding the pest incidence and control, mealybug showed a distinct host plant preference to ornamental crops when exposed them to different crops simultaneously. Chrysanthemum seems to be a high-ranking host for mealybug.

Crop rotation is an important strategy of pest control. Crop rotation interrupts normal life cycle of insect pests by placing the insects in a non-host habitat. Rotation is generally most successful against arthropod pest species with long generation cycles and with limited dispersal capabilities. Mealy bug adults lay many eggs when fed on mulberry and cause heavy damage to this plant. However, in the grass crop, including corn, it doest not survive well due to some nutritional deficiency. Avoidance of those host plant on which mealybug thrive well around the mulberry in the absence of mulberry leaf is one of the efficient way to disrupt the life cycle of mealybugs, thereby controlling them economically. Mulberry gardens are major habitat for mealybug population, which are commonly interspersed with ground nuts, grape, jowar fields. When the groundnut is uprooted the bugs leave the field in large numbers to infest mulberry fields. By strip cutting the jowar field, the number of bugs migrating to mulberry can be minimized. It has also been observed that beneficial parasitoids and predators were extremely abundant in these strips and, as they moved to and from the adjacent mulberry, an added benefit resulted

Tillage operations are land preparations which include soil turning and residue-burying, seedbed preparation, and other cultural operations. Several forms of tillage can reduce pest populations indirectly by destroying wild host vegetation (weeds) and volunteer crop plants in and around crop-production habitats. Over wintering populations of mealy bug can be greatly reduced with spring plowing operations. The mealybug population can be reduced through trimming and pruning of twigs, leaves, branches, fruiting areas and even removal of plants, especially seriously affected ones. Whole trees, as the single most serious source of continued widespread infestations, should be carefully cut to avoid scattering of ovisacs and crawlers. In general, any cutting and pruning should not be done when strong winds or breezes are present, as this will also scatter the various life stages of the mealy bug. Phytosanitary is a very important parameter in IPM. Flooding of mulberry plantation is a valuable tool for controlling several insect pests and plant pathogens of mulberry. In areas where this is problem, overhead sprinkler irrigation is recommended to prevent the cracking of soil. Overhead sprinkler irrigation can enhance dissemination and infectivity of some entomopathogenic organisms, especially fungal pathogens. Sanitation of nurseries, farms, gardens, and other establishments where mealy bugs are present can also reduce mealy bug populations considerably.

Chemical control

Earlier, pesticides were used continuously to minimize the pest population. Commonly used fungicides, dichlorovos and chlorpyriphos are sprayed in the tukra infected plots to reduce the pest population. Spraying of 0.2 percent DDVP prepare in 0.5% detergent solution with fish oil resin soap (25 g/lit) twice at an interval of 7 days is highly effective. After maintaining safe period of 17 days the leaves may be utilized for silkworm rearing. But pesticides cannot easily penetrate the heavy wax layers on the mealy bug's body due to their biological characteristics like secretion of wax, waxy ovisac and habitation in bark crevices. Therefore, applying pesticides is an ineffective control technique against this mealy bug species. Further, chemical control is both uneconomic and impractical for mulberry because of the pest's broad host range, widespread distribution and presence in areas with high human inhabitation (Singh and Saratchandra, 2002). Further, many of the synthetic chemical pesticides are broad-spectrum, killing not only pests but also beneficial organisms that serve as natural pest control systems. Without benefit of the natural controls that keep pest populations in check, farmers become increasingly dependent on chemical pesticides to which pests may eventually develop resistance. Moreover, residual effect of most of the chemical pesticides has serious effect on the survival, growth and development of silkworm. Additionally, pesticide applications disrupt the key natural enemies of other crop pests, such as thrips, scale insects, and whiteflies leading to the application of additional pesticides to control these pests. The pesticide also can contaminate food, water and farm workers (Singh *et al.*, 2004). Thus, there is an urgent need for an alternative approach to pest management that can complement and partially replace current chemical based pest-management practices.

Biological control

Biological control is the use of natural enemies (parasites and predators) against pest organisms to reduce their population densities (DeBach and Rosen,1991; Lenteran, 1986; Caltagirone and Huffaker. 1980). Many natural enemies have been screened against mealy bug which includes parasitoids, predators and pathogens. Parasitoids include *Aenasius cariocus* Compere, *Aenasius colombiensis* (Compere), *Anagyrus ananatis* (Gahan), Anagyrus loecki, *Euryhopauus propinquus* (Kerrich), *Hambletonia pseudococcina* (Compere) and *Ptomastidae abnormis* (Girault). Few host-specific encyrtid parasitoids, *Anagyrus* kamali (Moursi), *Anagyrus dactylopii* (How) and *Gyranusoidea indica,* (Alam and Agarwal) have been used on trial basis to control mealybug in sericulture (Mani, 1988). It was reported that *A. kamali* and *Gyranusoidea* indica are effective biological control agents and they can be used against the mealy bug in new areas of infestation. Kamal (1951) reported that complete control of *M. hirsutus* was achieved in Egypt by introducing *Anagyrus kamali* Moursi from Java. Such introduction should be tried further in India to get permanent and effective control of the *M. hirsutus*.

Among predators *Cryptolaemus montrouzièri* Mulsant, *Lobodiplosis pseudococci* Felt, *Nephus bilucernarius* Mulsant, *Scymnus* (*Pullus*) *unicatus* Sicard and *Scymnus pictus* Gorham are most important. The coccinellid predator *C. montrouzieri* is an Australian ladybird beetle, popularly known as mealybug destroyer. It is a very efficient predator for suppressing mealy bug population. It feeds on about 1000 eggs or 300-500 mealy bug nymphs. Its life cycle is completed in 30 days. The optimum temperature for maximum development of *C. montrouzierie* was reported to be 30°C (Anonymous, 1996). The females lay eggs next to the mealy bug eggs in the white wool egg pouches and from this the larval stage of the *Cryptolaemus* emerges. It is very similar in appearance to the mealy bug but the *Cryptolaemus* larvae are mobile. All stages of the Cryptolaemus eat adults and its eggs. The host searching ability of *Cryptolaemus* is very high in comparison to other predators. This beetle is well-established in India and also available commercially (Venkatesan *et al.*, 2001). Apart from this a number of other coccinellids, viz. *Cycloneda sanguinea limbifer* Casey, *Coelophora inaequalis* (F.), *Diomus* sp., *Scymnus* sp., and *Zilus eleutherae* were also reported (Santha, 1995). *Scymnus coccivora* Ayyar is another important potential coccinellid predator against mealybug. The species can survive at low population levels of mealy bug and are not adversely affected by low temperatures. A single predatory larva consumes about 60-70 mealybug nymphs during a developmental period of about 20 days (Mani, 1989). *Scymnus gratiousus* (Coccinellidae) was also reported to be effective species for the control of mealybug. It can also survive at low population levels of mealybug and is not adversely affected by low temperatures. Further, some mites have also been reported to minimize mealybug population. *Hypoaspis* sp. is a tiny mite that prefers to feed on mealybug. It destroys both adults and eggs. *Hypoaspis* is generally used on small plants (1-2 ft height) whereas *Cryptolaemus* flies to the top of taller plants and does not preferred to feed in smaller plants.

Mani, (1989) reported biocontrol potential of *Anagyrus dactylopii*, *Scymnus coccivora* and *Cryptolaemus montrouzieri* against mealybug. Besides this 16 another parasitoids and 31 predators were also reported against mealybug and suggested to use them under Integrated Pest Management programme. Among the indigenous natural enemies, the encyrtid parasitoid *Anagyrus dactylopii* and the coccinellid *Scymnus coccivora* are of considerable importance. Gowda *et al.* (1996) reported Spalgis *epius* Westwood (Lepidoptera: Lycaenidae) - a potential predator of mulberry mealybug, *M. hirsutus*. Chakraborty, *et al.* (1996) studied field efficacy of exotic predator, *C. montrouzieri* in controlling the mealybug, The results indicate that release of the predator at five adults per plant suppresses the mealybug population effectively within 15 days and reduced it to zero in 42 days. But, the predator could not establish itself in the mulberry ecosystem under West Bengal climatic conditions. Therefore, it was suggested that inoculative releases of *C. montrouzieri* may be done in mulberry fields for the predator's establishment and control of *M. hirsutus*. Mani (1989) reported *A. japonica* and *A. mirzai* as potential parasitoid of the mealybug. Gautam, (1998) reported successful release of exotic Coccinellids (*Cryptolaemus montrouzieri*, *Scymnus coccivora* and *Nephus regularis*) against mealybug. Release of coccinellid predator (*Cryptolaemus montrouzieri* @ 750/ha or *Scymnus coccivora* @ 1000/ha.) is quite effective in mulberry garden. It was found to parasitize up to 70 percent of the third instar and adult female of the mealybug on mulberry. Williams, (2003) reported suitable methods for insect rearing of the mealybug and its parasitoids. This technique is applied for mass rearing of the most highly efficient parasitoids or predators for mass release in infested areas. It is generally used when natural enemies are absent, occur too late, or are in numbers too small to provide effective pest control when needed (Singh *et al.*, 2004). In order to avail the usefulness of bio-agent for controlling mealybug, studies on various attributes of the parasitoids have been made as described by Hall *et al.*, (1980). The main attribute of the effective natural enemies is searching capacity, specificity, power of increase and adaptability (Greathead, and Greathead,1992; Singh and Saratchandra, 2002). All these attributes are closely related to each other and influence the population density of parasitoids in natural habitat.

Papaya Mealybug (PM); *Paracoccus marginatus* Williams and Granara de Willink, (Hemiptera: Pseudococcidae)

The papaya mealybug *Paracoccus marginatus* Williams and Granara de Willink, (Hemiptera: Pseudococcidae) is believed to be native to Mexico and/or Central America. It has never gained status as a serious pest there, probably due to the presence of an endemic natural enemy complex. The first specimens were collected in Mexico in 1955. The papaya mealybug was described in 1992 from the Neotropical Region in Belize, Costa Rica, Guatemala, and Mexico (Williams and Granara de Willink 1992). When the papaya mealybug invaded the Caribbean region, it became a pest there; since 1994 it has been recorded in the following 14 Caribbean countries: St. Martin, Guadeloupe, St. Barthelemy, Antigua, Bahamas, British Virgin Islands, Cuba, Dominican Republic, Haiti, Puerto Rico, Montserrat, Nevis, St. Kitts, and the U.S. Virgin Islands. More recently, specimens have turned up in the Pacific regions of Guam and the Republic of Palau. Specimens also have been intercepted in Texas and California, and it is expected that papaya mealybug could rapidly establish throughout Florida and through the Gulf states to California. It is possible that certain greenhouse crops could be at risk in areas as far north as Delaware, New Jersey and Maryland. It has already been identified on papaya plants in the Garfield Conservatory in Chicago, Illinois in late August of 2001. A biological control program was implemented in December of 2001 with very successful results. In India, the papaya mealybug (PM), *Paracoccus marginatus)* is a serious pest of mulberry in south India. The invasion due to this pest was first noticed in 2006 in Tamil Nadu and subsequently spread rapidly to several other part in India. The papaya mealybug is a plyphagous pest and has been recorded on more than 55 host plants in more than 25 genera. Presently this invasive pest has spread to other crops in the region like, Okra, Cotton, Teak, Mulberry and Sunflower. Economically important host plants of the papaya mealybug include papaya, hibiscus, avocado, citrus, cotton, tomato, eggplant, peppers, beans and peas, sweet potato, mango, cherry, and pomegranate.

Fig. 3.2 : Developmental stages of mealybug on mulberry plant

Papaya mealybug infestations are typically observed as clusters of cotton-like masses on the above-ground portion of plants. The adult female is yellow and is covered with a white waxy coating. Adult females are approximately 2.2 mm long and 1.4 mm wide. A series of short waxy caudal filaments less than 1/4 the length of the body exist around the margin. Eggs are greenish yellow and are laid in an egg sac that is three to four times the body length and entirely covered with white wax. The ovisac is developed ventrally on the adult female. Adult males tend to be colored pink, especially during the pre-pupal and pupal stages, but appear yellow in the first and second instar. Adult males are approximately 1.0 mm long, with an elongate oval body that is widest at the thorax (0.3 mm). Adult males have ten-segmented antennae, a distinct aedeagus, lateral pore clusters, a heavily sclerotized thorax and head, and well-developed wings.

The papaya mealybug can easily be distinguished from *Maconellicoccus marginatus* (Green), because papaya mealybug females have eight antennal segments, in contrast to nine in the latter species. Papaya mealybugs have piercing-sucking mouthparts and feed by inserting their mouthparts into plant tissue and sucking out sap. Mealybugs are most active in warm, dry weather. Females have no wings, and move by crawling short distances or by being blown in air currents. Females usually lay 100 to 600 eggs in an ovisac, although some species of mealybugs give birth to live young. Egg-laying usually occurs over the period of one to two weeks. Egg hatch occurs in about 10 days, and nymphs, or crawlers, begin to actively search for feeding sites. Female crawlers have four instars, with a generation taking approximately one month to complete, depending on the temperature. Males have five instars, the fourth of which is produced in a cocoon and referred to as the pupa. The fifth instar of the male is the only winged form of the species capable of flight. Adult females attract the males with sex pheromones. The papaya mealybug feeds on the sap of plants by inserting its stylets into the epidermis of the leaf, as well as into the fruit and stem. In doing so, it injects a toxic substance into the leaves. The result is chlorosis, plant stunting, leaf deformation, early leaf and fruit drop, a heavy buildup of honeydew, and death.

Control

Papaya mealybugs potentially poses a serious threat to mulberry crops as well as other crops, if not controlled. Earlier a number of chemical control methods were adopted to control mealybugs, although none are currently registered specifically for control of papaya mealybug. Active ingredients in registered pesticide formulations include carbaryl, chlorpyrifos, diazinon, dimethoate, malathion, and white mineral oils. Typically, twice the normal dose were applied when treating for mealybugs because mealybugs are protected by thick waxy, cottony sacs, and often are concealed inside damaged leaves and buds. Thus, chemical controls were only partially effective and require multiple applications. Furthermore, problems with insecticide resistance and non-target effects on natural enemies make chemical control a less desirable control option to combat the papaya mealybug. Since silkworm is very sensitive to insecticides therefore it is desirable to avoid the use of any pesticides in mulberry and other hosts so as to encourage conservation of available natural enemy population. There have been quick studies to identify certain native natural enemies since it is not generally sustainable to pursue insecticide-based management alone to contain such out break species of sap sucking insect. Natural enemies of papya mealybug have been noted to have potential impact on reduction of mealybug populations.

Therefore, Biological control was identified as a key component in a management strategy for the papaya mealybug, Several potential predators and parasitoids were identified. Natural enemies of the papaya mealybug include the commercially available mealybug destroyer lady beetles (*Cryptolaemus montrouzieri*), lacewings, and hover flies, all which are generalist predators that have a potential impact on mealybug populations. In addition to predators, several parasitoids were recorded as potential biological control agents. The most important among them are: *Acerophagus papayae* (Noyes and Schauff), *Anagyrus loecki* (Noyes and Menezes), *Anagyrus californicus* Compere, and *Pseudaphycus* sp. Recently another parasitoids identified as *Pseudleptomastix mexicana* (Noyes and Schauff) has been reported to be highly effective at field level. All four species of parasitoids prefers to attack second and third instars of *P. marginatus*. However, *Acerophagus* sp. emerged as the dominant parasitoid species. Few other parasitoids are also being collected and released in the affected areas along with measures to conserve and enhance their activity. Attempts have also been made to import potential parasitoids from United States of America for evaluation and large scale release in mulberry and other crops of the affected areas in Inda. Proven integrated methods adopted in effectively containing sugarcane woolly aphid and Cotton mealy bug may be replicated in mulberry also to minimize pest population.

Whitefly; *Dialeuropora decempuncta* (Hemiptera : Aleyrodidae)

Whitefly (*Dialeuropora decempuncta* (Quaintance & Baker, 1913); *Aleurodicus dispersus* Russell; *Aleuroclava* sp., and *Bemisia tabaci* Genn are the most common species reported on mulberry plants (Bandyopadhyay *et al.*, 1999; Douressamy, *et al.*, 1997). Geetha *et al.* (1998) recorded spiralling whitefly Aleurodicus dispersus Russell in Tamil Nadu. *Aleurolobus bardensis* Maskell, and *Neomaskellia bergii* Sign. have also been reported to damage mulberry plants (Mound and Halsey, 1978). *Dialeuropora decempuncta* is a major pest in West Bengal (Bandyopadhyay *et al.*, 1999), whereas spiralling whitefly *A. dispersus* has been reported to attack mulberry plants in south India (Ramani *et al.*, 2002). A leaf yield loss up to 24% was recorded in western Bengal areas. It indicates the level of damage caused by the pest to mulberry plants. *Alerodicus dispersus* is a native of the Caribbean region and Central America (Russell, 1965), where it is known from a wide range of host plants, but not regarded as a pest. It is more commonly known worldwide as "spiralling whitefly" because it lays eggs in a typical spiral pattern (Prathapan, 1996). It was introduced and assumed pest status in the Canary Islands in 1962, in Hawaii in 1978, and in American Samoa and Guam in 1981, and then in most of the Pacific islands (Paulson and Kumashiro, 1985). The whitefly later spread westwards into several regions including Africa (Neuenschwander, 1994), Asia (Wen *et al.*, 1994) and Australia (Lambkin, 1998). In south Asia, it is presently found in Bangladesh, Sri Lanka (Wijesekera and Kudagamage, 1990), and the Maldives (Martin, 1990). It was first reported in India in 1993 from Kerala (Palaniswami *et al.*, 1995) and later from other parts of peninsular India (David and Regu, 1995) and the Lakshadweep islands (Ramani, 2000).

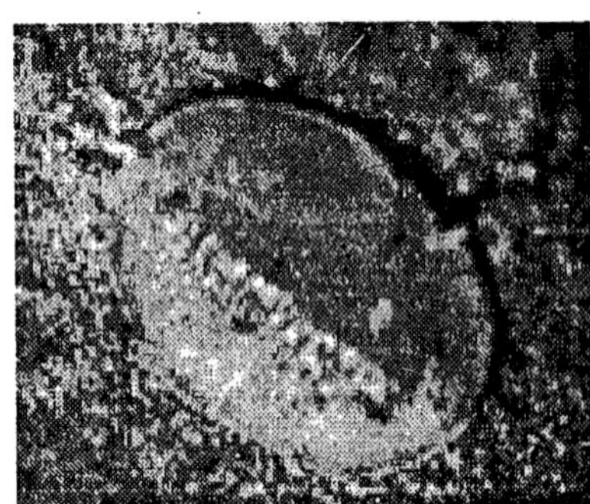

Fig. 3.3 : Adult and nymph of Whitefly (*D. decempuncta*)

Extent of Damage

The amount of damage caused by *D. decempuncta* and *A. dispersus* to mulberry plants is extensive because it causes huge economic loss to mulberry leaves which affects silkworm rearing. The damage is done both by nymphs and adults. Nymphs and adults suck the sap from mulberry leaves and cause yellowish speckling to appear on the leaves. Direct feeding damage is caused by piercing and sucking sap from the foliage of the plants. This feeding causes weakening and early wilting of the plant and reduces the plant growth rate and yield (Bandyopadhyay, 2001). It may also cause leaf chlorosis, leaf withering, premature dropping of leaves and plant death. Indirect damage is due to the accumulation of honeydew and white, waxy flocculent material produced by the whiteflies. Like other soft-bodied insects such as aphids, leafhoppers, mealybugs and scale, whiteflies produce honeydew. This sweet and watery excrement is fed on by bees, wasps, ants and other insects which, in turn, may tend upon and offer protection to the whiteflies. The honeydew also serves as a substrate on which sooty mould grows. The mould reduces the photosynthetic activity of the plant and serves as a medium for growth of bluish mycelia of sooty mould fungi, *Chaetothyrium* and *Curvularia affinis* Boedijin, causing mulberry leaf disease. Sooty mould blackens the leaf, decreases photosynthesis activity, decreases vigor and often causes disfigurement of the host and lessens the market value of the plant or makes it unmarketable (Berlinger, 1986). The flocculent material produced by the nymphs is scattered by the wind and becomes an unsightly nuisance (Waterhouse and Norris, 1989). Bandyopadhyay *et al.* (2002) determined the economic threshold level of whitefly, *D. decempuncta* in mulberry. The third type of damage is a result of the ability of this insect to act as a plant disease vector.

A small population of whiteflies is sufficient to cause considerable damage (Cohen and Berlinger, 1986). Plant viruses transmitted by whiteflies cause over 40 diseases of vegetable and fiber crops worldwide. Among the 1100 recognized species of whiteflies in the world, only three are recognized as vectors of plant viruses (Cohen and Berlinger, 1986). The mulberry whitefly has an extremely wide host range. It prefers to feed on S1 and S1635 varieties of mulberry plants. Douressamy *et al.* (1997) reported that Kanava-2, MR2, S36 and S54 are the most preferred varieties in southern Indian conditions. The warm humid conditions and abundant food are ideal for whitefly build-up. The whitefly population started appearing in higher numbers from August onwards and reached its peak during September–October. Correlation studies have shown that high humidity (75%-85%), moderate temperature (24-28°C) and moderate rainfall (180-380 mm) are influencing the maximum incidence of the pest in eastern and north-eastern regions of India (Bandyopadhyay *et al.*, 2000). The minimum whitefly number was recorded during March–June and December–February. The whitefly was present throughout the year in southern India, with high populations in summer (March-June) and low ones in winter (October-January). The population positively correlated with temperature and negatively correlated with humidity. Populations of indigenous predators remained low and did not have any impact on the whitefly populations (Mani and Krishnamoorthy, 2000). Narayanaswamy *et al.* (1999) found a high incidence of the pest on mulberry plants during April-June in and around Bangalore. Palaniswami *et al.* (1995) reported outbreaks in the post-rainy dry season between November and April and peaks in February in Kerala. Severe infestation was observed during March in the Lakshadweeps, which lessened during June with the onset of rains. Whitefly populations were higher during May–October in northern India whereas in southern India the incidence was higher in cooler months (November-February) on several trees (Ramani, 2000). This survey indicates that the adults are found on the topmost two leaves, whereas eggs and nymphs are found on 2–4th and 7–14th position of leaves respectively. The population of egg masses per leaf on an average ranged from 25–56. Eggs are mostly laid on the bottom and middle portion of the leaves and on matured leaves. The number of adults per leaf ranged from 40-60. They are mostly found congregated on the lower surface of the leaves. Leaf recovery is restricted to approximately 15%-25% of affected foliage.

Biology

The adults of *Aleurodicus dispersus* are similar in appearance to those of many other species of whiteflies but the behaviour is different. They hatch in 3–5 days in April–September, 5–7 days in October–November and 35 days in December–January. The nymphs feed on the cell sap and grow in three stages to form the pupae within 9–14 days in April–September and in 17– 82 days in October-March. In 2–8 days pupae change into whiteflies. The life cycle is completed in 15-125 days and 11 generations are completed in a year (Waterhouse and Norris, 1989; Douressamy *et al.* 1997). The adults of *Aleurodicus dispersus* are much larger than the most common *Bemisia tabaci* and are white with powdery waxy scales all over the body and wings. The wings of newly emerged adults are clear, but develop a covering of white powder over the next few hours. The eyes are dark reddish-brown and the forewings have two characteristic dark spots. Bandyopadhyay et al. (2001) reported the biology of *Dialeuropora decempuncta* and *Aleuroclava* sp. on mulberry plants in western Bengal climatic conditions. Adults congregate on the lower surface of the leaves where they lay yellow elliptical eggs. Adults are active during morning hours. The mean length of the adult measures 0.7 mm and wings expand 2.1 mm. Like other whitefly species it emerges during the cool hours. This pest is most active during September-October. The female lays approximately 120-150 eggs in circular or semi-circular masses on the under-surface of the leaves during an oviposition period of 2-3 days. The freshly laid white eggs are elliptical in shape and they turn black later on. The eggs hatch in 2–3 days. On hatching, the tiny first instar nymphs settle on suitable places. The second and subsequent instar nymphs usually remain feeding in the same place. Nymphs with reddish eyespots are covered with white waxy material. The pupal stage lasts 7-10 days and the life cycle is completed in 15–18 days in September and 18-25 days from October onwards.

Ecological control

Ecological control involves the removal, destruction, modification, or isolation of material that might favor the survival of an insect pest by affording food or making a site suitable for breeding. Ecological control refers to application of cultural control measures, judicious use of chemicals, biological control agents and resistant varieties (Singh *et al.*, 2000). This concept describes procedures for evaluation and consolidation of the available techniques into a unified program for managing pest populations. This leads to avoidance of economic damage and adverse side-effects on the environment (Southwood, 1978). In the past, several cultural practices were implemented to control the whitefly population in sericulture. Beneficial organisms and resistant mulberry varieties were screened through trial and error and natural substances were discovered. This laid the foundation for pest management of whitefly based on the biology of the ecosystem. Successful strategies were predominantly those that served to maintain the ecological balance of the region and the natural balance of whiteflies and their enemies (Goolsby *et al.*, 1996). It has been observed that such practices, combined with the advanced biological technologies now available, are the most logical approach to developing a profitable, safe, and durable (long-lasting and self-maintaining) approach to pest management. This system is identified as ecologically based pest management (EBPM). The concept of EBPM builds on the cultural and biological approaches to pest management and includes many practices, such as crop rotation, intercropping, tilling, trap crop, sanitation, host resistance/plant resistance. Apart from this, physical barriers have been used effectively for decades to reduce the impact of pests. Regular farm operations and cultivation of soil are conducted to destroy the insects or to prevent whitefly from causing injury. Generally, mechanical and physical control methods are popular with farm workers, and are adopted during the leaf-harvesting period. Mulberry leaves with eggs of whitefly are collected and destroyed. Mechanical control methods can be rapid and effective, but are mostly suited for small acute pest problems. Mechanical controls have relatively little impact on natural enemies and other non-target organisms, and are therefore well-suited for use with biological controls (Srinivasan and Mohanasundaram, 1997). Under physical control methods, insect pests are physically kept away from reaching their hosts. Crop rotation interrupts the normal life cycle of insect pests by placing the insects in a non-host habitat. Rotation is generally most successful against arthropod pest species with long generation cycles and with limited dispersal capabilities (Wang, 1992). Whitefly adults lay many eggs when fed on mulberry leaves and cause heavy damage to this crop. However, some leguminous crops, including cow pea, horse gram and pigeon pea are in some way nutritionally deficient to support feeding, and do not suffer damage from whitefly. So, a mulberry/leguminous rotation is effective and economical.

Crop rotation interrupts the life cycle of the whitefly, and both species can be effectively controlled by rotation throughout the temperate region of India. Further, crop monocultures are often damaged more severely by pests. There are cases where such diversity can aggravate pest problems. It is in these situations where trap crops can be important. Besides this, intercropping is another important method to reduce pest problems. Several field trials have been conducted using different intercrops such as tomato, garlic, onion, coriander and carrot. The intercrops were grown in alternate rows with mulberry, and their influence on the pest population was estimated. It was observed that the intercropped plots had significantly lower numbers of whitefly nymphs and pupae, and a higher yield of good quality leaf, as compared to the control. Similarly, garlic in inter-rows of mulberry also has been reported to decrease whitefly numbers (Anonymous, 1976). Two parallel tests, one on different crops, including tomato, bean and cucumber and one on different tomato cultivars, were conducted. Bemisia tabaci showed a distinct behavioural preference for cucumber when exposed to the different crops simultaneously. Cucumber thus seems to be a high-ranking host for *B. tabaci* (Botha *et al.*, 2004). This suggests that *B. tabaci* has no problem in choosing a host plant, that is, showing a preference when one of the plants in the test is a high-ranking host plant. However, when only low-ranking host plants giving similar, but not identical, stimuli were present, female whiteflies tended to have difficulty in making a selection, resulting in increased movement and reduced fecundity. If the above results can be confirmed under different conditions, the value of intercropping with different cultivars could contribute toward reducing pest population build-up in an integrated pest management program. Tillage operation is the most common practice in mulberry cultivation. It is conducted for soil-turning and residue-burying practices, seedbed preparation, and cultivation (Reddy and Kotikal, 1998). Some forms of tillage can reduce pest populations indirectly by destroying wild vegetation (weeds) and volunteer crop plants in and around crop-production habitats. Over wintering populations of whitefly may be greatly reduced by either fall or spring plowing operations. These characteristics also influence the quality of food, which determines the abundance of whitefly. Thus with the proper stirring and management of the soil, the whitefly population can be controlled. Phytosanitary is very important parameter in integrated pest management (IPM). Destroying dead and decaying materials from nearby rearing houses is important in reducing viruses and protozoans (Etebari *et al.*, 2004). Since crop residues can harbor whiteflies and virus inoculum, they should be rapidly and completely destroyed after the final harvest. The subsequent planting of susceptible crops should be avoided until migration has ended. This practice can reduce whitefly infestation as well as carryover of viral inoculums. Although these practices may not completely eliminate whitefly problems, they can help reduce pest populations and damage to manageable levels. These practices should be modified only to preserve known populations of natural enemies of whiteflies. Physical barriers, use of synthetic sex pheromones, repellants and confusants help in containing whitefly population (Heinz and Parrella, 1991). In general, host plant resistance to insect pests is likely to play an ever increasing role in pest management. Planting the mulberry varieties that are less susceptible to whitefly may also affect the level of infestation.

Biological control

Studies on the biological control of whitefly has been widely used in glasshouses, especially since the development of insecticide-resistant whiteflies, and is chiefly based on parasitoids (*Encarsia* sp. and *Eretmocerus* sp.), predators (*Anegleis* sp., *Coelophora* sp., *Cryptolaemus* sp.) and entomopathogenic fungi (*Paecilomyces* sp.). Attempts have been made to search for appropriate biological control agents in native natural habitats. Polaszek *et al.* (1992) reported 19 previously described parasitoids belonging to the *Encarsia* and *Eretmocerus* genera, as well as many more yet to be described that attack the mulberry whitefly. A key to the known *Encarsia* species is provided by Polaszek *et al.* (1992). Among important parasitoids, two aphelinid parasitoids *Encarsia haitiensis* Dozier and *E. meritoria* have the most potential (Srinivasan, 2000). *Encarsia sp. nr. haitiensis* and *E. sp. nr. meritoria* were first recorded from Kerala in 1998 (Beevi *et al.*, 1999). Ramani (2000) reported *Encarsia guadeloupae* as another important pupal parasitoid of whitefly. Besides this, *Eretmoceras delhiensis* Mani, *Encarsia neomaskelliae* Prasad and *E. isaaci* Mani have also been reported to attack up to 45% of whitefly pupa (Mani *et al.*, 2000). The parasitized puparia turn dark in color and can easily be distinguished from non-parasitized ones. For the control of *Dialeurodes* spp., *Encarsia lahorensis* (How) has been reported as the key parasitoid other than *Aphelinus fuscipennis* How (Wen *et al.*, 1995; Rosen and DeBach, 1981). Chien et al. (2000) reported introduction, propagation and liberation of two parasitoids for the control of spiraling whitefly (Homoptera: Aleyrodidae) in Taiwan. The life table characteristics of the parasitoids have been studied and it was found that it has a high potential for the biological control of whitefly in seri-ecosystems (Srinivasan *et al.*, 1999; Ramani, 2000). Among predators of whiteflies, true bugs (Hemiptera: Anthocoridae) and predators (Miridae), beetles (Coleoptera: Coccinellidae), lacewings (Neuroptera: Chrysopidae, Hemerobiidae, Coniopterygidae), flies (Diptera: Dolichopodidae, Syrphidae, Anthomyoodae), ants (Hymenoptera: Formicidae), spiders (Araneida) and mites (Acarina: Phytoseiidae, Stigmaeidae) have been reported (Ramani *et al.*, 2002). Some of these are opportunistic predators, others are general feeders and some are specific predators of whiteflies. More than 40 indigenous predators, mostly generalists and few specialist species have been recorded in India. *Coelophora unicolor* has also been recorded as a potential predator of whitefly. Hoelmer *et al.* (1994) reported the interaction of the whitefly predator *Delphastus pusillus* (Coleoptera: Coccinellidae) with parasitized sweet potato whitefly. *Anegleis cardoni* (Weise), *Anegleis perrotteti* (Mulsant), *Axinoscymnus puttarudriahi* (Kapur & Munshi), *Cheilomenes sexmaculata* (F), *Jauravia* sp., *Nephus regularis* (Sicard), *Pseudoscymnus* sp., *Pseudaspidimerus flaviceps* (Walker), *Pseudaspidimerus trinotatus* (Thunberg), and *Scymnus saciformis* (Motschulsky) are the most common predator (Hagen, 1962). Jallali and Singh (1989) studied the biotic potential of three coccinellid predators on various diaspine hosts. Mani and Krishnamoorthy (1997) observed that the naturalized Australian ladybird beetle, *Cryptolaemus montrouzieri* (Mulsant) preyed on the whitefly almost throughout the year, but had little effect in reducing the pest population. Mani and Krishnamoorthy (1999) studied the predatory potential and developmental period of *C. montrouzieri* on the whitefly in the laboratory. *Chilocorus nigrita* (F.) is another important predator feeding on the whitefly and it attacks almost all the developmental stages of whitefly. Hagen *et al.* (1976) reported biology and ecology of some predaceous coccinellids beetles. Several birds, ants and spiders have also been recorded feeding on *A. dispersus* in India.

Among pathogens, *Paecilomyces farinosus* (Holm.) has been recorded on *A. dispersus* from India (Mani et al., 2000). Although many fungi have been found in association with *Bemisia,* only *Verticillium lecanii*, *Paecilomyces fumosoroseus*, *Peacilomyces farinosus*, *Aschersonia aleyrodis*, and *Beauveria bassiana* have been demonstrated to be pathogenic. Studies on fungal pathogens indicate some success in controlling whitefly. However, these products are not yet commercially available, and reliable programs based solely on biological agents have not been developed (Botha *et al.*, 2004). To ensure biological control, selection of most suitable natural parasitoids is desirable. Sometimes the potential parasitoid in the laboratory may prove to be inadequate under field conditions. Such was the case with the commercially available natural enemy, *Encarsia formosa*, which has a proven track record of successful whitefly control in greenhouses but was less than satisfactory when used for control under field conditions (Parrella *et al.*, 1991). A search for new, effective natural enemies must be assessed with its host-searching capacity, specificity, power of increase and adaptability (Waage, 1986; van Lenteren, 1983). A high reproductive rate is important so that populations of the natural enemy can rapidly increase when hosts are available. The natural enemy must be effective at searching for its host and it should be searching for only one or a few host species. All these attributes are of course closely related to each other and influence the population density of parasitoids in the natural habitat.

Searching capacity, manifested by the ability to find the host even when it is scarce, is commonly regarded as the most important attributes of effective natural enemies. A parasitoid beneficial as a bioagent must successfully utilize the low-density population of the pest. *Encarsia haitiensis* and *E. meritoria* have been found capable of attacking more than 50% of the whitefly pupa within an hour. This indicates its high searching capacity, that is, the ability to find hosts, when the host density is low. They are attuned to the physiology, behavior, habitat preferences, and patterns of dispersion and phenology of whitefly (Kirk *et al.*, 2001). These parasitoids have a degree of biological adaptation to the pupa of whitefly and also have a greater degree of direct and rapid responsiveness to density changes in the whitefly population (Goolsby et al., 1998). Most of the whitefly parasitoids have received host regulation in a stable environment that has reached the hypothetical steady state and would only require a power of increase sufficient for replacement of the parent population in each generation. The actual power of increase of a natural enemy in the field may be affected by its fecundity and rate of development, as well as by other factors such as its searching capacity and adaptability to the conditions of the particular habitat (Clausen, 1978; Goolsby and Ciomperlik, 1999). Fitness and adaptability are other important attributes. To be effective in a new habitat, an introduced natural enemy should preferably be pre-adapted to it. However, the possibility of possible gradual post-colonization adaptation cannot be discounted. A well-adapted natural enemy should not require any essential requisites that are not present in the new habitat. It should be able to tolerate the prevailing climatic conditions, and should be effectively synchronized with the biology and phenology of its host in the habitat. Ideally, a natural enemy should be adapted to all the habitats and niches occupied by the target pest. It should frequent all the host plants and tolerate all the same climatic regimes as its host does, and should be equally effective in all of them. Just as the natural enemy should be well adapted to the various natural aspect of the ecosystem, it should also be adapted to cope with man-made hazards, such as pesticidal treatments (Rosen and Huffaker, 1983). Although the outlook for biological control through natural enemies is promising, several situations limit the effectiveness of predators and parasites of the whitefly. These instances were discussed by van Lenteren (1983) and outlined by Gerling (1986).

In a well-design integrated pest management programme, the pest population is maintained at a lower level by the action of three basic approaches to applied biological control: conservation, importation and augmentation. Specific techniques within these approaches are constantly being developed and adapted to meet the changing needs of pest management. Improvements in rearing and release techniques and genetic improvement of natural enemies have resulted in more effective augmentation programs. Application of new ecological theory is transforming the methods for conservation of natural enemies. Continued refinement and adaptation of biological control approaches and applications are necessary if the full potential of this biologically based pest management strategy is to be fulfilled. The conservation of natural enemies is probably the most important and readily available biological control practice available to farmers. Natural enemies occur in all production systems, from the backyard garden to the commercial field. They are adapted to the local environment and to the target pest, and their conservation is generally simple and cost-effective. Conservation involves identifying the factor(s) which may limit the effectiveness of a particular natural enemy and modifying them to increase the effectiveness of the beneficial species. In general, conservation of natural enemies involves either reducing factors which interfere with natural enemies or providing resources that natural enemies need in their environment. Many factors can interfere with the effectiveness of a natural enemy. Pesticide applications may directly kill natural enemies or have indirect effects through reduction in the numbers or availability of hosts. Various cultural practices such as tillage or burning of crop debris can kill natural enemies or make the crop habitat unsuitable. In orchards, repeated tillage may create dust deposits on leaves, killing small predators and parasites and causing increases in certain insect and mite pests. Ensuring that the ecological requirements of the natural enemy are met in the cropping environment is the other major means of conserving natural enemies. To be effective, natural enemies may need access to: alternate hosts; adult food resources; overwintering habitats; constant food supply; and appropriate microclimates (Rabb *et al.,* 1976). Conservation involves permeated action to protect and preserve existing parasites, predators and pathogens: basically not taking actions that would be detrimental to natural enemies (Nordlund, 1984; Rabb *et al.*, 1976). IPM programs that result in a reduction in pesticide use generally contribute to conservation. Conservation and augmentation in biocontrol involves two phases: first, the maintenance of existing parasitoids by avoiding harmful practices; and secondly, the augmentation of parasitoids, either directly releasing them in the field or by indirectly making the field environment more favorable for them. Although a number of species of eulophid parasitoids have been reported, it is rather easy to identify more effective parasitoids based on their parasitization capabilities. *Encarsia haitiensis* Dozie and *E. meritoria*, both multivoltine parasitoids, have been identified as such potential parasitoids. These parasitoids which appear after the arrival of whitefly pupa need to be conserved from the use of pesticides which is very intensive on mulberry plants. Importation (classical biological control) has been bv far the most important and most promising approach to date, and has accounted for the great majority of outstanding successes in applied biological control. It is also the least expensive method of natural enemy utilization.

However, in certain instances an already established natural enemy, whether indigenous or introduced, may show considerable promise but fall short of fulfilling its potential due to inadequacies of its own attributes or the environment. There are several potential natural parasitoids of whitefly but their exploitation has not been achieved due to lack of sufficient information on the behavioural response of the parasitoids. Even well known natural enemies of proven high efficiency still await transfer into many areas where whitefly is a serious problem. There are several hymenopteran natural parasitoids which can be utilized to control the whitefly population. These unknown species are considered as potential weapons and may be applied as biological control tools (Kerrich, 1960). The first step in the process of classical biological control is to determine the origin of the introduced pest and then collect appropriate natural enemies (from that location or similar locations) associated with the pest or closely related species. The natural enemy is then passed through a rigorous quarantine process, to ensure that no unwanted organisms (such as hyperparasitoids) are introduced, then reared, ideally in large numbers, and released. Follow-up studies are conducted to determine if the natural enemy has successfully established at the site of release, and to assess the long-term benefits of its presence. There are many examples of successful classical biological control programs. Introduction of *Trichogramma ostriniae*, from China to control the European corn borer is one of the most successful examples of classical biological control. Classical biological control is long-lasting and inexpensive. Other than the initial costs of collection, importation, and rearing, little expense is incurred. When a natural enemy is successfully established it rarely requires additional input and it continues to kill the pest with no direct help from humans and at no cost. Augmentation is the direct manipulation of natural enemies to increase their effectiveness.

This can be accomplished by mass production and periodic colonization; or genetic enhancement of natural enemies. The most commonly used of these approaches is mass production, in which natural enemies are produced in insectaries, then released either inoculatively or inundatively. Augmentation involves actions to increase the populations of beneficial parasitoids, predators or pathogens (Ridgway and Vinson, 1977; Alphen and Vet, 1986). There are two basic approaches to augmentation: environmental manipulation and periodic releases (Stinner & Bradley, 1989). Periodic releases can be inoculative or inundative. Inoculative releases are releases of a relatively small number of biological control agents, often on a seasonal basis. The control in inoculative release programs is expected to come primarily from the progeny of these agents being released. Inundative releases are releases of relatively large numbers of biological control agents, and the control is expected to come from the released agents, not necessarily from their progeny (DeBach and Hagen, 1964). Inundative releases programmes usually involve a number of releases during the season, while inoculative release programs may involve only one release. It is possible by sustained release of laboratory-reared parasitoids. Our ability to use periodic releases of the parasitoid to control whitefly, is to rear, transport and effectively release large numbers of high quality biological control agents. Periodic release requires continuous release programme, and thus has commercial potential and fits IPM programmes well.

The growing number of commercial suppliers of biocontrol agents is evidence of this potential (Thomson, 1992). Adoption of biological controls has had positive economic impact. An example of the inoculative release method is the use of the parasitoid wasp, Encarsia formosa Gahan, to suppress populations of the greenhouse whitefly, *Trialeurodes vaporariorum* (Westwood), (Parrella, 1991). The greenhouse whitefly is a ubiquitous pest of moriculture and floriculture crops that is notoriously difficult to manage, even with pesticides. Releases of relatively low densities (typically 0.25–2.00 per plant, depending on the crop) of *E. formosa* immediately after the first whiteflies are detected on crops can effectively prevent populations from developing to damaging levels. However, releases should be made within the context of an integrated crop management program that takes into account the low tolerance of the parasitoids to pesticides. Habitat or environmental manipulation is another form of augmentation. This tactic involves altering the cropping system to augment or enhance the effectiveness of a natural enemy. Many adult parasitoids and predators benefit from sources of nectar and the protection provided by refuges such as hedgerows, cover crops, and weedy borders. Mixed plantings and the provision of flowering borders can increase the diversity of habitats and provide shelter and alternative food sources. They are easily incorporated into home gardens and even small-scale commercial plantings, but are more difficult to accommodate in large-scale crop production. There may also be some conflict with pest control for the large producer because of the difficulty of targeting the pest species and the use of refuges by the pest insects as well as natural enemies.

White fly ; *Bemisia tabaci* Gennadius (Hemiptera : Aleyrodidae)

It is polyphagous species mostly available in mulberry plantation during rainy season in the eastern region of India. Winged nymphs are 1.0 to 1.5 mm long and their yellowish bodies are slightly dusted with a white waxy powder. They have two pairs of pure white wings and have a prominent long hind wing. The female lay eggs singly on the under side of the leaves, averaging 119 per female. The eggs are stalked, sub-elliptical and light yellow at first, turn brown later on. They hatch in 3–5 days in April–September, 5–7 days in October–November and 33 days in December–January. The nymphs feed on cell sap and grow into the three stages to form the pupae within 9–14 days in April–September, 17–81 days in October–March. In 2–8 days, the pupae change into adult white flies. The life cycle is completed in 14–21 days during April September and 40-50 days during November February. The nymphs on emergence look elliptical and soon fix their mouthparts in the plant tissues. They feed on cell sap causing damage in three ways: (a) the vitality of the plant is lowered through the loss of cell sap, and (b) normal photosynthesis is interfered due to the growth of sooty mould on the honey dew excreted by the insect. From a distance, the attacked plant gives a sickly, black appearance. Consequently, the growth of the plant is adversely affected and when the attack appears late in the season, the yield is lowered considerably. It is a vector of number of virus diseases including the leaf curl disease of mulberry. The losses in yield at 250 white fly population level of *Bemisia tabaci* Gennadius was recorded 2.50 g/ plants as compared to 30.20g/plant in uninfected plots. The white fly infested plants remained stunted, produced lesser number of monopodial and sympodial branches, retain fewer number of leaves.

Control

Earlier chemical control measures were developed to control the white fly population. Insecticides viz. Dimethoate 30EC (0.05%), Malathion 50 EC, (0.07%) Fenitrothion 50 EC (0.05%) were used to control during February–March (Spring flush), May–June before rainy season and July–August after rainy season. Whenever there is incidence of white fly in mulberry gardens, these insecticides are sprayed and after maintaining safe period the leaves are used for silkworm rearing. Apart from this some natural predators and parasitoids have also been screened which kill the white fly nymphs. The coccinelids, *Cryptognatha flavescens* Motsch, *Brumus suturalis* Fab. and *Sticholotis sp.*, a drosophilid fly *Acletoxenus indica* and a lygaeid bug, *Geocoris tricolor* are predacious on the nymphs of this pest. The nymphs are parasitised by *Prospaltella lahorensis* How and *Aphlinus fuscipennis*

Leaf hopper

15 different genera of leaf and plant hoppers are reported, which causes extensive damage to mulberry plants. The major leaf hopper associated with mulberry plants are *Amrasca biguttula biguttula, Balclutha* sp., *Cicadulina bipunctata, Cicadulina* sp. *Cofana spectra, C unimaculata, Empoasca* sp., *Hecalus porrectus, Hishimonus Phycitis, Kolla,* sp, *Kolla ceylonica, Nephotettix nigropictus, N. vircscens, Nirvana* sp., *Nirvana pallida, Peregrinus maidis, Orosius albicinctus, Racilia dorsalis* and *Sophonida bakeri.* The most common and widespread leafhopper is *Empoasca flavescens*

Jassids; *Empoasca flavescens* Fab. (Hemiptera : Cicadellidae)

Leaf hoppers (*Empoasca flavescens*) commonly known as Jassids have been reported as serious pest of mulberry in India during October to May (FAO Manual 1987). The little greenish hoppers feed on the underside of the leaf, sucking the sap from veins and causing the characteristic symptoms known as hopper burn. The hopper burn is caused due to toxic virus for which the insect is a vector. The first symptom of this appearance appears as triangular brown spot at the tip of the leaf. Similar triangle appears at the end of each vein or the entire margin or the entire margin may roll upward and turn down at one time, as though scorched by the fire or drought. These brown margins increase in width until only a narrow strip of the leaf along with the mid rib remains green, the rest is shovel and dead, with the leaf veins much distorted. The leaves below the growing tips usually burn first. The plant becomes stunted and often in highly susceptible varieties it causes complete mortality of the plants, if left unprotected. Owing to the loss of plant vitality the leaf quantity and quality is reduced. Adults are pale green in colour. They measure 2.5 to 4 mm in body length. Eggs are yellowish white and laid on underside of the leaves, embedded into leave veins. The female lays 13 to 18 eggs. Incubation period varies from 9–11 days. There are six nymphal instars and the adult stage is obtained after 13 to 21 days. The adults and nymphs suck the sap from lower side of leaf margin, showing hopper burn symptom. In this a triangular spot of dark brown appears at the tip followed by such patches along the margins of the vein. It starts from the periphery and extend towards midrib of the leaf. In the final attack the leaf become cup shaped, withers and drop off from the plant. It is reported widely in Karnataka, Tamil Nadu and Andhra Pradesh.

Control

Collection of adults and nymphs by netting and sprinkler irrigation is effective in controlling the pest. Use of light traps is conducted to minimize the population during infestation. Under chemical control approach, spraying of 0.05% DDVP or dimethoate is quite effective (safe period 11 days). *Chrysopa cymbele* and a spider, *Distina albida* L. are predaceous on the insect.

Defoliators

Leaf roller; Diaphania pulverulentalis Hampson (Lepidoptera : Pyralididae)

The leaf roller is a sporadic pest. Its greenish caterpillars are very agile and they feed inside the fold made by fastening together the edges of the leaf. The pest is active during November to December in south India but it feeds mainly on the mulberry throughout the year. It is a monophagous pest and causes extensive damage to mulberry plants. The moths are golden or yellowish brown and measure 8-10 mm in length and 15-20 mm in wing expanse. The wings have 2-3 wavy lines characterized by dark bands. The moths are nocturnal rest on the undersurface of the leaves during the day. The eggs are flat and pink in colour and hatch in 5–7 days. After hatching the larvae moves to the apical shoot of the plant and starts feeding on the unopened leaves, resulting in drying of terminal portion of shoot. Larval period lasts 15–20 days and pupal period is completed 8-10 days during the active season. The life-cycle is completed in 25-35 days. The young larvae feed on tender leaves without folding them. The older larvae fasten the longitudinal margins of leaf together with a sticky substance and feed inside the fold by scraping the green matter. The scraped leaves become membranous, turn whit and finally wither. A single larva may damage a number of leaves as it migrates from one leaf to another. The infested leaves are bound together by silken web. Sometimes single leaf is folded and completely eaten by late stage larvae, resulting in considerable decline in the leaf yield. This act of feeding causes drying of the apical shoot leading to stunted growth of the plant, which causes reduction in leaf yield.

Control

Clipping and destroying of affected apical shoots and burning of the fallen leaves minimizes the pest population in the field. Weeds should be removed from the plots. Foliar application of 0.2% DDVP in 0.5 per cent with soap solution is quite effective (waiting period 17 days) .*Apanteles* sp. a larval parasitoid has been recorded in Karnataka and Andhra pradesh parasitizing 0.5 % of the field population of the pest. *Bracon hebetor* and *Brecon breveicornis* a larval parasitoid was also recorded to parasitise roller caterpillar. *Calosoma*, a predatory carabid is an important predator. The release of *Trichogramma chilonis* and pupal parasitoid *Tetrastichus howardii* on 3^{rd,} 10th and 30th days after pruning minimize the pest population in the field.

Bihar Hairy Caterpillar; Spilosoma obliqua walker = Diacrisia obliqua (Lepidoptera : Arctiidae).

The pest is polyphagous. Full-grown caterpillars are profusely covered with long grayish hairs and measures about 40 – 45 mm in length. The moth measures about 50 mm across the wingspread. The head, thorax and the under side of the body is dull yellow. The antennae and eyes are black. The adult is a medium sized moth. The forewings are white with brown streaks all over and yellowish streak along the anterior. A yellow band is seen on the head. In summer (May–June) and in winter (December–February), the pest remains in pupal diapauses and is active only in March–April and again from July–November. After moth emergence, they mate in the night and lay 412–1241 light green eggs in clusters on the underside of the leaves. Eggs are laid in 5–6 days. The incubation period is 8–12 days. The larvae after passing seven instars become full–grown in 1 to 2 months. As they grow the feed gregariously the whole leaf behind the petiole and midrib. On completion of feeding, they migrate to other plant. The life cycle is completed in 45–50 days. The affected leaves look like dead and dried and easily fall off. Occurrence of pest is reported through out the year but its peak period of attack is recorded during May–September The last instar larva spins a loose cocoon and pupates in it in plant debris or soil. The pupal period and the longevity of the moths are 1-2 weeks. The total life cycle occupies 6 –12 weeks. There is three or four generation in a year. The pest is sporadic in nature. The tiny caterpillars of first two instars feed gregariously, but as they grow older, they disperse widely in search of food. The larvae eat on the leaves and soft portion of the stems and leaves. In case of sever infestation, the whole crop is defoliated and give meshed appearance.

Fig. 3.4 : Bihar hairy caterpillar feeding on leaves

Control

Cultural and mechanical control methods are the best approaches to control the pest population in the beginning itself. Use of light trap to attract adults and collection and destruction of egg masses by mechanical method is the most common method, which is generally practiced by the farmers. Soil trenching to prevent migration of caterpillars and summer ploughing of the mulberry gardens provides relief to the farmers. The earlier instars of caterpillar are easier to control while it is difficult to kill the grown up larvae and a very high dose of pesticide is required. Generally the crop is dusted with 10 per cent BHC, or malathion 5 per cent. Apart from this, spraying of methyl parathion 0.04 % or endosulfan 0.05 per cent is quite effective to control the pest population in the field.

Cut worm ; Spodoptera litura = Prodenia litura Fabricius (Lepidoptera : Noctuidae)

It is commonly known as the tobacco caterpillar or fall Army worm or cluster caterpillar or cutworm. Pest is polyphagous and distributed through out India, Pakistan, Bangladesh, Sri Lanka The full grown larvae is stout, cylindrical pale greenish brown with dark marking which measures about 35 40 mm in length. The moth is stout, dark with wavy white marking on the forewings and white hind wings, margin having a brown colour. They are about 22 mm long and measures 40 mm across the spread wings.

The pest is active throughout the year. The female moth after mating at night lays about 30 eggs in clusters. These clusters are covered over by brown hairs. The incubation period is about 3–5 days. After hatching the larvae feed gregariously together and later on they feed individually. There are six instars completing in 2–4 weeks. The last instars larvae enter in the soil for pupation. The pupal period and the longevity of moths are 7–15 days and 7–10 days respectively. The total life cycle covers 32–60 days. There are 8 generations in a year. The caterpillars feed on leaves in the nursery and in the main field and cause serious damage

Fig. 3.5 : Adult *Spodoptera litura*

Control

Cultural and mechanical control measures are adopted to control the pest. Hand picking and mechanical destruction of caterpillars during early stage of attacks can reduce the infestation. Spraying 400 ml malathion 50EC or, 370 ml endosulfan 35EC in 250 litres of water per acre is effective to control the pest population. If the pest population is above economic threshold level, the spraying is repeated at 10 days intervals if necessary.

Thrips ; *Psudodendrothrips mori* (Thysanoptera : Thripidae)

Thrips are tiny, slender insects with fringed wings. Thrips are available all over mulberry growing area is India. Most adult thrips are slender, minute (0.5 to 15 millimetres), and have long fringes on the margins of both pairs of their long, narrow wings. Most thrips range in color from translucent white or yellowish to dark brown or blackish, depending on the species and life stage. Thrips are poor fliers but can readily spread long distances by floating with the wind or being transported on infested plants. The mouthparts are stylet like and are used for rasping and sucking plant sap. The tarsi are short, each ending in a vesicle. In the life history of this insect there is an incipient pupal instar. They are minute insects with 6 –9 segmented antennae. Immatures (called larvae or nymphs) are similarly shaped with a long, narrow abdomen but lack wings.

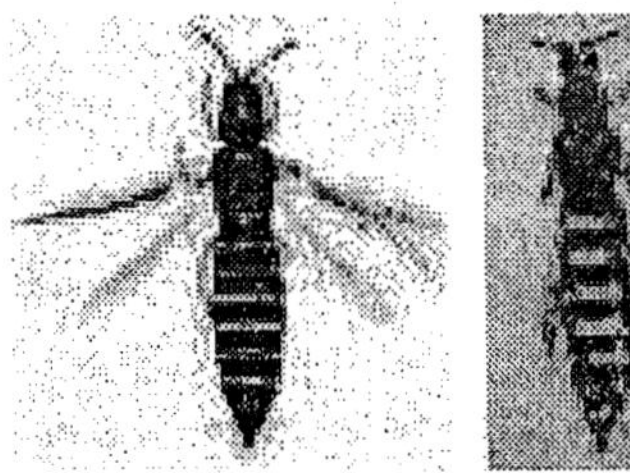

Fig. 3.6 : Thrips in mulberry garden

Thrips is polyphagous insect affect the mulberry leaves by damaging the epidermal tissues. The major thrips which caused damage to mulberry are *Thrips tabaci*, *Caliothrips indicus*, *Scirtothrips dorsalis*, etc. In species *Heliothrips haemorrhoidalis*, *Aptinothrips rufus*, etc., the males are extremely rare. *Taeniothrips glycines* (Okamoto), *Taeniothrips melanicornis* (Shumshir) and *Taeniothrips calaratris* (Shumshir) also damage the mulberry plants. Sexual reproduction is more common among thysanoptera, the female being larger than the males and there is overlapping of generations in *T. tabaci*. Adult male is brownish and female is dark brown in colour. Adult female lays about 30 50 bean shaped yellow coloured eggs on the ventral side of the leaf. The incubation period is 6–8 days. Nymphs are yellow coloured which moults four times in 15–16 days. Pupation occurs in soil. There are generally 5–10 generations per year. Nymphs are saffron coloured and abdomen exhibits pinkish yellow colour on the inter–segmental region. Occurrence of the pest is reported through out the year but its peak period of infestation is during summer seasons. Both nymphs and adults cause extensive damage by feeding on the upper side of the leaf. The leaves quality is also deteriorated due to shortage of moisture, reduction in crude protein and total sugars in the infected leaves. In many species, thrips feed within buds and furled leaves or in other enclosed parts of the plant. Their damage is often observed before the thrips are seen. Discoloured or distorted plant tissue or black specks of feces around stippled leaf surfaces are clues that thrips are or were present.

Control

For the control of thrips in mulberry garden, use of sprinkler irrigation is quite effective. In case of severe infestation, malathion 50 EC 250 ml in 250 liters of water per acre is quite effective. Waiting period of 7 days before harvest the leaves may be maintained.

Wingless grasshopper; *Neorthacris acuticeps sp. nilgriensis Uvarao* (Orthoptera : Acrididae)

This is a wingless grasshopper known to cause appreciable damage to mulberry in south India. Adults are green in colour. Female lays 6–8 ootheca in soil at depth of 2–3 cm. Nymphs are light brown in early stage and green in late stage. The life cycle is completed in 5–6 months. Nymphs and adults feed voraciously on leaves and leaf yield is reduced considerably. Pest occurrence is observed mostly in July and August.

Control

Deep ploughing in previously infested fields may expose the egg masses to be destroyed. Dusting with 10% BHC, chloradane or dieldrin have also been practiced. The maggots of a bombylid fly *Systaechus nivalis* feed on the eggs of the grasshopper in the soil. The grubs of *Zonabris* sp. also attack the egg masses in the soil. Spraying of 0.05% fenitrothion (Safe period 15 days) is also effective to control the pest population in the filed.

Long horn beetle : *Batocera rufomaculata* (Coleoptera : Cerambycidae)

Batocera rufomaculata (Coleoptera : Cerambycidae) is the most common stem borer of mulberry plants in temperate region of India. It is commonly known as long horn beetle and widely distributed in Darjeeling, Kumaon hills, Kullu valley, Jammu and Kashmir, Dehradun and Simla hills. *B. rufomaculata* causes extensive damage to the growing branches and main stem of mulberry plant. The grubs are pale yellow and measure 90 to 150 mm long. The beetles are grayish brown with white lateral band all along the length of body. There are two orange yellow spots on pronotum, numerous yellow spots on elytra and yellow or cream colour of scutellum are distinctive. The beetles emerge in June and July and live for about four months. During this period, they mate and lay eggs after finding their suitable sites. The eggs are oval in shape and brown in colour and laid singly in the bark of mulberry. A female may lay 50–60 eggs in her life span. The eggs hatch in 8–15 days. The grubs at first feed on the bark and than bore inside the wood; completing their development inside the stem. Prepupal stage lasts 50–70 days and makes puparia. Pupal stage lasted for 5–75 days. The adult longevity is 4–5 months. The young grubs feed on the inner side of the bark making zig zag tunnels. Later on, the grubs bore down to the surface of the sap wood and up to the center of heart wood. In case of severe infestation the plants may die and subsequently may be invaded by termites/fungus.

Fig. 3.7 : Long horn stem borer *Batocera rufomaculata*

Control

Hand picking and mechanical destruction of grubs and adults are the easy method to minimize the pest population. Plugging the live holes with cotton soaked in Kerosene, petroleum or monochrotophos 0.05.% is highly effective in case of severe infestation.

***Apriona germari* Hope (Coleoptera : Cerambycidae)**

The adult beetles are ashy-grey and measures about 35–40 mm long. With the onset of monsoon, the beetles appear in July–August. After mating, the females excavate an oval patch on a shoot and lays eggs inside the cavity. While laying number of eggs, she moves from tree to tree. The incubation period is 7–10 days and after hatching the grubs start feeding. The feeding activity becomes very slow in autumn and by October stops completely. It remains quiescent during winter and resumes feeding in March, reaching the tree trunk by autumn (September–October) again go in hibernation during winter. As the summer season comes, it starts activity and when full fed it pupate inside the tunnel and in the woody tissue. Pupal period is more than a month. The life cycle is completed in about two years mostly depending on climatic condition. Pupation takes place inside the tunnel. Severely damaged plants may die. The holes are made by the pest in the main stem/branches can be easily identified by the presence of frass along with the holes. Pest attack is noticed throughout the year. The grubs bore through the shoots and make circuitous galleries downwards, leading to main stem and trunk of the trees. Due to feeding by the grubs, the affected trees may not die for many years but their vitality and productivity is greatly impaired. The adult beetle feed on bark.

Control

Pruning and burning of attacked shoots and branches are the best cultural control methods. The heavily infested branch of the tree should be cut and burnt in the field itself. Cotton soaked in 0.05% monocrotophos solution may be inserted in the holes made by the borer. It will kill the grubs inside the galleries.

Chafer beetle ***Holotrichia consanguinea*** *(Blanchard)* (=***Lachnosterna consanguinea*** *Blanchard).* (Coleoptera : **Melolonthidae)**

This is one of the most common defoliator of mulberry leaves in northern part of India. The insect appeared as major pests just after onset of monsoon and causes maximum damage to leaves. The full-grown larvae are white, having brown head and prominent thoracic legs and measure about 35 mm long. The adult beetles are dull brown and measures about 18cm in length and 7 mm in width. The beetles are active at night and hide in the soil during day time. The female lays eggs in the soil. The incubation period is 7–10 days. Larval development is completed in 50–65 days. After monsoon, the full grown larvae migrate to a considerable depth in soil for pupation. The pupa is creamy white and semicircular. Pupal period last about 15 days. The beetles conceal themselves in the soil and come out for feeding at the night. The adults and grubs damage the mulberry plant. The grubs eat away the nodules, the fine rootlets and may also girdle the main root, ultimately kill the plant. The damage becomes evident only when the entire plant dries up. During night beetles feed on foliage and may completely defoliate whole plant.

Control

Collection and destruction of adults using light trap is the most common methods to collect the pest population in the field. Deep ploughing of soil is also conducted to expose grubs which is eaten by the birds. Soil application of 4 kg Phorate 10G or 13 kg Carbofuran 3G per acre is quite effective to control the pest population. Spraying of 200g carbaryl 50 WP or 50 ml of fenitrothion 50 EC in 50 litres of water per acre on host plant is effective to control the adult beetle. Apart form this some agents of biological control have also been reported. *Scolia aureipennis* parasitises the grubs of this pest. The fungus, *Metarrhizium anisopliae*, parasitizes the adults. The common toad (*Bufo melanostictus*) and the wall lizard, *Giecko gecko* feed on the beetles.

Leaf beetle ; *Mimastra cyanura* Hope (Coleoptera : Chrysomelidae)

Mimastra cyanura is the most common chrysomelid that damages the mulberry plants. The insect is nocturnal in habit and adult and grub both damages large number of mulberry plants in tropical region during summer season. The grubs damage the mulberry roots where as the adult beetles defoliate mulberry trees. The damaged leaves present a characteristic sieve-like appearance on account of the shot holes. The life history is completed in one year. The beetles resume activity as soon as the season warms up. Adult is small with prominent antennae, elytra dull yellow with bluish posterior end and the abdomen is black vertically. They attack mulberry plantation in swarm and completely defoliate the plants leaving only the skeleton. Tender shoots are also eaten occasionally. Adult beetles appear in swarm and feed voraciously on tender, medium and coarse mulberry leaves and move to another mulberry plant. The severity of attack is such that the beetles eat up of all the leaves including veins and mid ribs. With a slight disturbance, the beetle fly away from one plant to another. During hot sunny time these beetles were observed to settle on the ground at the base of the plants. It occurs in Dehradun, Udhampur, West and border district of Jammu and Kathua.

Control

Cultural and mechanical control is effective to minimize the pest population. Light traps are used for collection and destruction of adult beetle. Deep ploughing of soil is conducted to expose grubs. Soil application of fensulfothion 5% G @ 30 40 kg/ha controls grubs. Spraying 0.25% metacid is also effective in minimizing the beetle population.

Leaf caterpillar ; Margaronia pyloalis Walk. (Lepidoptera : Pyralididae)

The moths are small and larvae feed on leaves. The female lays about 200 eggs on the margin of the veins of mulberry leaves on its ventral side. The eggs are green in colour and spherical. The female is nocturnal laying egg within six hours of emergence. The incubation period of the egg is 5-6 days. The larvae are green in colour and they moult four to five times to reach final instar. Full-grown caterpillar are bluish green, cylindrical with small dark black spots on either side of the segment. The larvae make leaves unsuitable for silkworm feeding by accumulating excreta on it, which is subsequently rolled and woven by silky threads. Young larvae skeletonize the leaves where as mature larvae consume complete leaves. The larvae feed on the leaves of the mulberry and generally bind joining the leaves together causing great loss. The moth is dark brown in colour and measures about 2 cm. at the wing expanse.

Control

Cultural and mechanical control is adopted to check the population of this insect. If the population is above economic threshold level, 0.02% DDVP or 0.05 % malathion may be sprayed in the infected plots.

White ants ; (Isoptera : Termitidae)

The termites or white ants are social insects, living together in organized communities composed of matured males, females, sterile workers and soldiers. They abound in the tropics and warm temperate regions. Most of the termites feed on the cellulose of wood; they cause serious damage to wood, whether it is on the tree or in building and hence are economically important. A typical termite is having prognathous head and well-developed compound eye in winged forms. This compound eyes are absent or reduced in the wingless form. The mouth part is mandibulate and the legs are short and stout. The most common termites attacking mulberry plants are *Odontotermes obesus* Rhamb. O. wollonensis, *O. horni Wasman, Microlentrotermes* sp., *Coptotermus heim.* All these termites are reported to attack mulberry but *Odontotermes obesus* is one of the common termites in tropical and temperate mulberry growing area of India. Since the termites are social insects and their colony organization is based on a caste system.

A. *Primary reproductive.* The queen and king, the queen has a creamy white abdomen marked with transverse dark brown stripes. The king is much smaller than queen and is constant companion of the queen living with her in the royal chamber

B. *Supplementary reproductive.* They are short winged or wingless creatures of both sexes. In event of ultimately death of the king or queen in a colony, the complementary castes replace them.

C. *Workers.* They are sterile, usually pale in colour slightly skelenotized and look more like nymphs than adults.

D. *Soldiers.* Sterile form, head and thorax well chitinized, abdomen is delicate and dirty white in colour. They are of four types (i) mandibulate soldiers (ii) nasute soldiers (iii) nasutoid soldiers and (iv) phragmotic soldiers.

The king mate with the queen from time to time in royal chambers and thus aids her in laying fertilized eggs from which the colonizing form of workers develop. The queen is a phenomenal egg laying machine, laying one egg per second or 70,000 to 80,000 eggs in 24hrs. The eggs hatch after one week during summer. Within 6 weeks the larvae develop to form soldiers or workers as the case may be. The reproductive castes mature in 1-2 years. There is only one queen in the colony and normally she lives from 5-10 years. Sometimes extending up to 20 years. The queen is fed by the worker on the choicest food and is always confined to the royal chamber with male (king, father of the colony) and performs the sole function of reproduction. The king's life is much shorter than that of queen, and when he dies, he is replaced by a new one (supplementary reproductive). They are induced by the worker to undergo sexual development. The workers developed from fertilized eggs but remain stunted as they are reared on ordinary food. They are abundant in colony but are smaller in size than soldiers. Except for the reproduction and defense of the community, the workers perform all other duties. They take the care of eggs and young ones and remove them to safer place at the time of danger. They also tend and feed the queen, collect food and cultivate a fungus food (ambrosia) in under ground gardens. The soldiers developed from unfertilized eggs are under developed comparatively to fertilized eggs. The soldiers defend the colony mainly against other insects like ants and vertebrate predators. The mandibulate type of soldiers defend the colony by fighting the intruders and nasute type soldiers repel them by spraying an abnoxious smelling fluid through the rostrum.

Termite infestation starts after planting and continues till pruning. In pre-monsoon period, the germinating saplings and young shoots are affected, while in post -monsoon period the mature plants are damaged. In the nursery the termite enters through the cut ends or through the buds and feed on the soft inner tissues. The tunnel thus excavated is filled with soil. The roots of seedling and young plants are attacked and eaten up resulting in the mortality of mulberry plant. The destruction of buds due to termite attack is reported from 20-30 per cent Though termite is found in all types of soil but more frequent in the sandy and red loamy soil.

The important termite species and their food plants/materials are are presented in Table 3.1 :

Table 3.1 : Common termites in sericulture

Sl.No	Species	Hosts	Reference
1	*Odonotermes obesus* (Rambur)	Fruit trees and forest trees	Kushwaha et al. (1980), Rajagopal (1983). Reddy and puttaswamy (1985).
2	*Odonotermes wallonensis* (Wasmann)	Jackfruit, cassshew	Rajagopal (1983), sudhakar (1983)
3	*O. brunneus* (Hagen)	Deadwood	Rajagopal (1983)
4	*O.horni* (Wassmann)	Cassshew, tamarind	Sudhakar (1983)
5	*Microtermes obesi* (Holmgren)	Fruit trees, plantation crops	Kushawa (1983), Roonwal (1983a), Kashyap et al. (1984).
6	*T. heini* (Wasshmann)	Fruit trees	Chhotani (1980), Rajagopal (1983)

Control

It is very difficult to mange termites as they are social insects, large in numbers, live in subterranean conditions and feed on a variety of materials. Destruction of termite colonies by breaking the mound and removal and destruction of queens from the colony is prerequisite factor for killing termites in the field. Frequent irrigation of mulberry plant helps to reduce the termite attack. Cleaning of rotting twigs, burning of crop residue also helps to decrease the termite population considerably. To control the termites during nursery stage, application of BHC 5 percent dust @ 1 kg a.i./ha or lindane 20 EC or heplachlor 20 EC @ 1 kg. a.i./ha is highly effective. Further, application of either aldrin 30 EC or lindane 20 EC @ 1 kg a.i./ha may be given in the mature plantation. Mounds can be treated with Captan 75 SD, Captafol 80 WP, Carboxin 95WP @ 30g/15 litre/ mound. Soil incorporation of 5% aldrin @ 1 kg a.i./ha is highly effective. In case of established garden swabbing or drenching at the base with 1% chlordane may be conducted (safe period 20 to 25 days). In the past large numbers of insecticides were applied for the control of termites (Beal, 1980; Kard *et al.,* 1989). They are, however, toxic to soil, persisting for long periods and pose serious environmental hazards. Many chemicals in the list of termiticides have been banned. There is vast scope of using non-chemical methods. Logan *et al.* (1990) reviewed the non-chemical methods for termite control in agriculture and forestry. Chemicals other than insecticides like juvenile hormone anlogues (JHAs) are also currently employed (Howard and Haverty, 1979). Attempt has been made to search for biological Control methods to control the termite population. Birds are the most common predator of termites under field condition. Termites provide abundant and easy prey to all sorts of animals, both invertebrates and vertebrates. Further, symbiotic fungi *Termitomyces sp.* is used to control the termite. It acts as food poison in the termites colony.

The swarming winged forms of termites preyed upon by insectivorous birds, bats, reptiles and amphibians. Some times, centipedes, scorpions and ants are also reported to feed upon termites. The use of ants like *Solenopsis geminalia rufa* is also reported for the control of termites in nurseries (Roonwal, 1979). Khan *et al.*, (1992) studied the mycopathogenic biopesticides for biological control of mound building termites in an agro-forest ecosystem They applied spores of white Muscardine fungus, *Beaveria bassiana* and green Muscardine fungus, *Metrahizium anisopliae* along with a food attractant of jaggery with a view to infect the workers with some success. As there is mouth to mouth feeding (trophalaxis) by workers to the queen, it may be possible to infect the queen with the fungus by this method.

Neem cake (125 kg/ha) and neem seek kernal extract controlled O. obesus population. The search for control agents at natural origin against termites has been going on mainly due to the possible pollution problems. Physiological and behaviour modifying chemicals have also been used to control the termite population. Insect hormones, especially juvenile hormone analogues, have been found to disrupt the colony organization and cause subsequent mortality in social insects including termites (Loscher, 1976; Hardy *et al.*, 1979; Varma 1982, 1985). Termite attractants (food attractants, trail pheromones, etc.) in combination with termiticides are used for the control of termites (Ritter *et.al.* (1977). Varma *et al.* (1984) reported the effectiveness of extracts of eucalyptus roots to control *Odontotermes guptai.*

MINOR PESTS

Morning Hairy Caterpillar; *Eupterote mollifera* (Lepidoptera : Eupterotidae)

Eupterote mollifera (Lepidoptera : Eupterotidae) is commonly available in all the mulberry growing area of India. Male moths are light yellow and females are larger and dark brown in color. Eggs are laid in masses either around the tender twigs or on the petiole of the leaves. Incubation period is 9–13 days. It takes about 68 days to pass through 5 instars. Late stage caterpillars are dark in colour with thick growth of hairy tufts and each measures 48 50 mm in length. The hairs are irritating to touch. Pupae are dark brown in color and measure about 18–20 mm in length. Pupal period lasts for about 1–2 months. They pupate in the soil. Caterpillars feed on mulberry leaves resulting in reduction of leaf yield. Occurrence of this pest is reported during August to February.

Control

Cultural and mechanical control methods are adopted to check the pest population. Burning the larvae which congregate on tree trunk or spraying fish oil, rosin soap or parathion 0.05 per cent may be useful in controlling the pest. Further, pest can be controlled as suggested for Bihar hairy caterpillar.

Tussock Caterpillar; *Euproctis fraterna* Moore (Lepidoptera : lymantriidae).

It is commonly known as hairy caterpillar and causes extensive damage to caster, tasar, eri and mulberry food plants. The full-grown caterpillar has red head darkish brown body with white hairs on the head and a tuft of long hairs at anal end. They measure 35–40 mm in length. The moth is pale yellow. The activity of this pest is throughout the year, with slow development in winter. The moths emerging in February, lay large numbers of eggs on mulberry leaves. The eggs are flat, circular and yellow and are laid in masses covered with yellow hairs on the undersurface of the leaves. As many as 150–300 eggs are laid by a single female. The incubation period is 5–7 days. After hatching, the young larvae feed gregariously for the first few days. Later on, they disperse and feed individually. The larvae are full-grown in 14–21 days after passing through six stages of their development. The last instars larvae make loose silk cocoons in the plant debris lying on the ground and pupate inside. The pupal period is about a week in summer. The egg, larval and pupal stages may last 10 days, respectively, during the winter months. The pest completes several generations in a year. The larvae cause damage by feeding on the leaves. In case of sever infestation; there may be complete defoliation of plants. The damaged plants remain stunted and there is a considerable reduction in foliage production.

Control

Freshly hatched larvae can be collected and killed. Dusting 10 percent BHC or 5 percent aldrin @ 10kg/acre. is also effective in killing pest population in the field. Spraying 250ml methyl parathion 50EC or 500ml endosulfan 35EC in 250 litres of water per acre is highly effective in case of severe infestation.

***Bark caterpillar ; Indarbela quadrinotata* Walker (Lepidoptera : Metarbelidae)**

Indarbela quadrinotata Walker (Lepidoptera : Metarbelidae), commonly known as bark caterpillar is a polyphagous pest. The newly hatched larvae are dirty brown while the full grown caterpillars have pale brown bodies with dark brown heads and measures 50–60 mm in length. The adults are pale brown moths with rufous head and thorax. The forewings are pale rufous with numerous dark rufous bands. Their hind wings are fuscous. The moths emerge and become active with start of the summer season. The female lays eggs in cluster of 15–25 eggs each under loose bark of the tree and continues throughout the summer. During her life span, it deposits as many as 2000 eggs. The incubation period is 8–10 days and after nibbling at the bark for 2–3 days, bore into the stem or main branch and feed within. The larvae have the habit of making webs along the feeding galleries and above the holes. The larvae complete development in 9–11 months and when full grown, they make a hole into the wood and pupate. Pupal period lasts 3–4 weeks. The adult emerges in summer having a portion of empty pupal case protruding out of the base hole and they are short lived. Only one generation is completed in a year. The larvae have the habit of making webs. The thick ribbon like silken webs are seen running on the bark of the main stems, especially near the forks extending from the bore hole. Only one larva is seen inside a hole which is used only as a shelter when it does not feed during day time. As many as 16 holes are seen on a tree. They eat through the bark into the wood and in case of severe infestation, sap movement is interfered and the tree ceases to flush, the growth is arrested and fruiting capacity is considerably reduced.

Control

The infected plants are treated with insecticides during February–March. The insecticide soaked in cotton (insecticides for 100 liters of water are 40 g BHC 50WP or DDT 50WP or 40 g Carbaryl 50 WP or 10 ml Fenitrothian 50 EC or 2ml Dichlorovos 100EC or 5 ml Methyl parathion 50EC or 10ml Monocrtophos.) are inserted into the bore holes. These chemicals are applied after removing webbings. The infected stem may be plaster on the outside with mud,

Wasp moth ; *Amata passalis* Fab (Lepidoptera : Amatidae)

Wasp moth is the tropical insect, small to medium sized, day flying and usually inactive. They resemble with sawflies and other hymenoptera and brightly coloured. The adult males are elongated with narrow and slender abdomen where as the females are stout and bulky. The thorax and abdominal segments are brick red in colour. Wings are brownish black in colour and there are seven transparent spots on the forewing. A single female lays about 500 eggs on the ventral side of the leaves. Hatching takes place in 6–7 days. The larva is short bearing numerous setae. The larvae passes seven moults and larval period lasts for about 32 days. The freshly emerged caterpillar is whitish in colour with brownish hairs all over the body. Caterpillars are active and scrap the chlorophyll content of the leaves. Full-grown caterpillar measures 20–25 mm in length and feed directly on leaves. Pupation takes place in the folds of leaves within silken web. Pupal duration is 10–12 days. Life cycle is completed in 48–51 days. Pest is active from February to August.

Wasp moth ; *Ceryx godarti* Bdv. (Lepidoptera : Amatidae)

Ceryx godarti Bdv. (Lepidoptera : Amatidae) is a tropical insect with small to medium size and diurnal in habits. The male moths are smaller in size with a narrow abdomen and wing. The female is stouter, larval stages are similar to *A. Passalis*. The adult female lays 125–230 eggs in batches on the lower surface of leaves of mulberry plants. Caterpillars pass five moults and full-grown caterpillars measure about 20–25 mm in length. The colour of the larva is almost black and thickly covered with hairs. The larval period varies from 25–30 days. Pupation takes place under dry leaves in secreted silk cocoons with hairs. The pupae are copper brown in colour and measures about 11–14 mm in length. Pupal period lasts 13 days. The young larvae skeletonised the leaves by scrapping of chlorophyll. Full-grown larvae feed on whole leaves.

Control

Collection and destruction of egg mass and destruction of young larval scrapping is highly effective. Spraying 0.2% dimethoate or DDVP is also conducted in case of severe infestation.

Looper; *Hyposidra talaca* walker (Lepidoptera : Geometridae)

Adults have a slender body with relatively large wings. The moths are not strong fliers and often keep the wings horizontal at rest. The eggs are pale blue or green, oval, about 0.7 mm long, laid in crevices in the bark. Larva moult five times taking about 25–42 days. Full grown larva is 5 cm in length. The larvae are generally elongate and slender and possess prolegs on the sixth and tenth abdominal segments. Locomotion is achieved by drawing the posterior segments close to the thorax, the body thus forming a loop, and then extending forward. The looping action is repeated in this way. Pupation takes place in soil ; pupa shiny brown and 18–20 mm long; pupal period is 14–21 days. Wing colour in adult is variable but generally grey with many dark grey markings. Adult female lays about 2000 eggs and live for 14 days.

Control

The adults may be collected by hanging light traps in the field. Summer ploughing is conducted to expose pupae to predatory birds to minimize the pest population. Spraying 0.05% malathion is highly effective (waiting period 10 days).

Stem girdler ; *Sthenias grisator Fabricius* (Coleoptera : Cerambycidae)

Stem girdler, *Sthenias grisator* Fab. is one of the most important stem borer of mulberry plants in tropical and temperate region of India. The emergence of beetles generally takes place around July–August. They become active at night and mate. A female beetle cuts transverse slits under the bark of girdled branch and thrust in between the bark and sap wood in cluster of 2–4 eggs. The eggs are oval in shape, white in colour and about 4 mm long and 1 mm wide in middle. The incubation period varies for 6–8 days. On hatching, the grubs feed inside the stem and complete their development. The freshly hatched grubs are tiny maggots having dark brown head, globular thorax and mouth parts with a pair of prominent mandibles each with two teeth. The abdomen is cylindrical sparsely covered with hairs. They measure 2-3 mm in length and when full grown, they measure 10–12 mm long. The period covered by the grubs is usually 7-8 months but it considerably depends on dryness and the nutritive value of the food material. The adults are grey brown with white and brown irregular marking resembling the bark colour, elytra have an elliptical greyish median spot and an eye shaped patch. They are about 24 mm long. The total life cycle takes more than a year. The pest hibernates in winter and again appears in late summer. The beetles are very active during night and during day time, they hide on the lower side of the leaves under forking of branches. Before egg laying, girdling of vine is essential event so that it dries up to enable the grubs to bore and tunnel the dry wood. The girdling is done 15 cm to 3 m above the ground and preferred branches are 1.25 to 2.50 cm in thickness.

Control

Cutting and burning of attacked branches below girdling point and hand collection and destruction of beetles may help in mitigating this long corn beetle. Application of 2 kg of BHC 50 WP or 1 litre of dieldrin in 20 EC 500 litres of water per acre and spray mulberry especially at ground level is effective.

Red hairy caterpillar ; *Amsacta moorei* Butler (*Amsacta albistriga* Walk (Lepidoptera : Arctiidae).

This is the most common polyphagous pest widely distributed in mulberry growing areas of India. *Amsacta moorei* is common in North India where as *A. albistriga* is the predominant species in south India. The adult moth is moderate-sized having white forewings with brownish markings and streaks and white hind wings with black spots. There is a yellow band on the head as also a yellow streak along the anterior margin of the forewings in *A. albistriga* and the marking are red in *A. moorei*. The colour of full grown caterpillars varies from reddish amber to olive green and body is covered with numerous long hairs arising from the fleshy tubercles. They measure about 25 mm in length. The adults are medium sized moths. The forewings are white with brown streaks all over and yellowish streak along the anterior margin and the hind wings white with black spots. A yellow band is seen on the head. The insect undergoes pupal diapause in the soil and becomes active from June to August. The moth starts emerging after the first shower of the monsoon. The emergence continues for sometimes depending upon the subsequent rains. The pest is nocturnal in habit. After mating the female lays light yellow spherical eggs in cluster of 700–850 each on the under surface of the leaves of the mulberry plants. A single female may lay up to 15,00 eggs. The eggs are light yellow in colour and spherical in shape. The incubation period is 2–3 days. The larval period varies from 15–25 days after passing through six stages. The larvae enter the soil, shed their hairs and make earthen cocoons at a depth of about 10–15 cm. The pupation takes place in the soil and remain in diapause for many months till they emerge next year from the cocoon. After emergence the young caterpillars feed gregariously and as they grow older, they march in bands destroying filed after field of various crops. Thus the field is completely destroyed by moving army of caterpillars as if the entire area has been grazed by the cattle.

Control

Light traps are used to attract the moths. It should be put up at night following the first showers of the monsoon and continued throughout the period of emergence for about one month. The pupae may be collected at the time of summer ploughing and destroyed. The growing larvae found in the field on infested plants be collected and destroyed. The first instar larvae which are vulnerable at this stage should be dusted with BHC 10 per cent. The grown up larvae can be controlled by spraying in one acre 0.5 litre of endosulfan 35 EC, quinalphos 25 EC or one litre of toxaphene 80 EC or 200 ml of dichlorvos 100 EC in 100 – 200 litres of water or by dusting 15 kg of 5 per cent aldrin per acre.

Hairy caterpillar ; *Spilarctia casignata* Koll (Lepidoptera : Arctiidae)

Spilarctia casignata is a tropical insects with stout body and moderately broad brightly coloured wings with spots or bands on them. They are nocturnal in habit. The female lays more than 1000 eggs on the underside of the leaves. In warm and moist weather, the egg stage lasts about 7–8 days. The tiny caterpillars on emerging congregate and skeletonize the under surface of leaves by feeding on the epidermal tissues. Late instars feed on whole leaves. The larvae pass through six instars on 23–42 days and pupate in the soil. Pupal period is 12–16 days. There are five generation in a year.

Control

Collection and destruction of egg mass and young larvae, ploughing of the soil and spraying of 0.2% Dimethoate or DDVP is effective to control the pest population.

Hairy caterpillar ; Utetheisa pulchella Linn. (Lepidoptera : Arctiidae)

Utetheisa pulchella was recorded in mulberry garden at number of occasions. The full-grown larva is about 3.8 mm in length and has red dark and white markings on its body and a brownish head. The adult moth is pale, whitish with red black spots on the upper wings and black marginal blotches on the lower wings. The female lays about 80–100 round, smooth yellow eggs on leaves singly or in small groups on tender leaves and shoots. On emergence from eggs, the larvae feed on the leaves. They pupate in the soil. The egg, larval and pupal periods are 3 to 4, 18 to 21 and 6 to 8 days respectively. The total life cycle is 27 to 31 days. The larvae first feed on the leaves and defoliate the plants. The damage is much more serious in later stage when the caterpillars feed on matured leaves.

Black scale insect ; *Saissetia nigra* (Nietm) (Hemiptera : Coccidae).

The mulberry scale is characterized by a grey waxy hamlet like a scale. Mature scales are about 2 –3 mm long. The immature females have an H-shaped yellow mark on their body. The dark brown adult has a strongly convex helmet-shaped carapace. The males have never been recorded. Adult female lays 300 600-elongated eggs, beneath her helmet, which remain attached to the plant even after the eggs have hatched. Eggs hatch within 6 days. Newly hatched crawlers are flattened oval in shape. They take up the position on leaf or shoot and begin to feed. It passes through three instars before becoming adult. Nymph starts crawling and feeding within few hours after hatching. It secrets a fibrous waxy material which hardens the scales. Female moult three times and male twice and during the process, it loses the legs. Reproduction takes place by parthenogenesis. Both nymph and adult suck the sap from the leaves and tender shoots, then affected shoots start drying from the distant. The adult females are immobile. It takes six months to complete one generation. The leaf blade shows yellowish appearance of the leaf. Besides feeding, they secrete honeydew, which develops into sooty mould.

Control

The affected plants are pruned to remove the infected shoots. Apart from cultural and mechanical control measures, spraying of 300 ml of malathion 50 EC or 200 ml dimethoate 30 EC or 150 ml methyl parathion 50 EC or 80 ml phophamidon 85 WSC in 200 litres of water per acre are applied in the infected field to minimize the pest population.

Chafer grub; *Schizonycha ruficollis* Fab. (Coleoptera : Melolonthidae)

Adult is medium sized golden brown in colour. In the male the antennal club is often elongate and horny process are present on its head. The mentum and ligula are fused. The metasternum laterally attains the coxae and the abdomen is with six tergites. Prothorax well developed, scuttelum small, pygidium visible, abdomen swollen posteriorly. Eggs are laid in the soil, especially where there is high organic content. The grub is fleshy, white with brown head, curved body and the terminal segment large and smooth. The larval stage lasts for about 5 6 months. Adults are nocturnal and defoliate young mulberry plants. These beetles are commonly attracted to light. The larvae feed on the roots of the plants and trees and are often destructive. The beetles both in their adult and larval stages are destructive to mulberry plants.

Control

Collection and destruction of adults using light trap and deep Ploughing of soil to expose grubs are effective to control the grub population in the field. Soil application of fensulphothion 5% @ 30 40 kg/ha. is highly effective.

Leaf beetle ; Maladera insanabilis Brenske (Coleoptera : Melolonthidae)

The pest appears during June July in the Jammu and Kashmir region. Beetles are nocturnal and defoliate the plants. Beetles are very small with dark brown colour. Elytra with longitudinal stripes, held closely to the abdomen. Prothorax distinctly covers meso and metathorax. The eyes are prominent and dark in colour.

Control

Collection and destruction of adults using light traps and deep ploughing of soil to expose grubs is effective . Soil application of fensulphothion 5% G @ 30 40 kg /ha control grubs.

Mites

Mites have world wide distribution. Plant feeding mites belong to families Tetranychidae, Eriophyidae, Tenuipalpidae, Tarsonemidae and Tuckerellidae. Important genera of mulberry plant feeders are *Eotetranychus suginamensis* Yokoyama, *Eutetranychus orientalis* MC G and Petrobia. For detail information regarding taxonomic characters of each of these genera, the reader should refer to Jeppson, Keifr and Backer's book entitled *Mites Injurious to Economic Plants*. The red spider mite *Tetranychus cinnabarinus*

(Boisduval) is also available in mulberry garden. The pest is polyphagous and found to damage local and hybrid varieties of mulberry plants. The nymph is light brown and has two eye spots, four pairs of legs and is quite active. A fully developed nymph is microscopic and measures about 0.33 mm in length. The adult male measures about 0.52 mm in length and 0.30 mm in breath. The body of female is oval, pyriform and variable in colour. After passing winter as a gravid female, the mite becomes active from March to October. It spins webs on the underside of the leaves of various host plants as the seasons warm in March. A female lays 60–80 spherical eggs. The incubation period is 2–6 days. The adult longevity is 7–12 days. During the active period, the life cycle is completed in 9–15 days. The damage is caused both by nymphs and adults. A large number webs are formed on the leaves under which the pest suck the cell sap. The affected leaves dry up and finally wither away giving an unhealthy appearance. Webbing also interferes with plant growth.

Control

Collection and destruction of adults and removal of affected leaves are common methods to reduce the mite's population in mulberry garden. Spraying of malathion 0.05% or kelthene 0.1% or 0.1% Phosalone is highly effective (Safe period 15 days).

Occasional pest of mulberry plants

Cocksafer beetle ; *Melolontha melolontha*, Holotrichia sp (Coleoptera : Melolonthidae)

The cock-chafers (*Melolontha melolontha*) are variously coloured, black, brown, green, blue or metallic coloured. This beetle has been observed in large numbers in mulberry farms located at Katra (Jammu & Kashmir). They occur mostly during April–July. For distribution, host range, life cycle, damage and control measures see under Major pests of mulberry plants. In the male the antennal club is often elongate and horny processes are present on its head. The mentum and ligula are fused. The metasternum laterally attains the coxae and the abdomen is with six tergites. These beetles are commonly attracted to light. The larvae fed on the roots of mulberry platns and are often destructive. The grub is fleshy, white with brown head, curved body and the terminal segment large and smooth. The beetles both in their adult and larval stages are destructive to a wide variety of crops and fruits trees. In Mulberry garden the beetle and *Holotrichia* sp. are very common pest.

Leaf beetles; *Mimastra cyanura* Flope (Coleoptera : Chrysomelidae)

The beetles are oval, convex, brightly coloured and measure 3.5 to 12 mm long. The antennae are widely separated at base, head not produced, eyes not prominent, prothorax laterally margined. Prosternum broad and pygidium not exposed beyond elytra. The third tarsal segment is not bilobed and the fore coxae are transverse. They are commonly known as almond chrysomelid. Pest in nocturnal in habit, defoliate large numbers of plants in summer in Jammu and Kashmir region. The damage is most severe on the marginal plants of gardens. The damaged leaves present a characteristic sieve like appearance due to short holes. The grubs feed on humus, roots of mulberry plants and other plants. There are one or some more two generations in a year.

Control

Collection and destruction of adults using, light trap and deep ploughing of soil to expose grubs is useful. Soil application of fensulphothion 5% G @ 30 40 ha is effective for the control of grubs.

Leaf caterpillar; *Hymenia recurvalis* (Lepidoptera : Pyraustidae)

Pyraustids are generally large moths having vestigial proboscis.The pest is active from March to November. Adult is small, light silvery cream in colour. The female lays the eggs in the grooves of leaf veins. As many as 150 eggs are laid by a female. Incubation period is 3–4 days. Larva undergoes four moults and become full grown in 12 to 15 days. Full grown caterpillar is bluish green, cylindrical with small dark black spots on either side of the segment. The larvae make leaves unsuitable for silkworm feeding by accumulating the excreta on it which is subsequently rolled and woven by silky threads. Young larvae skeletonise the leaves where as mature larvae consume complete leaves.

Control

Collection of young larvae and destruction of adults is useful. Spraying 0.02 % DDVP or 0.05. malathion is effective to control the pest population in the field. .

CHAPTER-4

Pest of Non-Mulberry Host Plants

Wild silkworms are polyphagous in nature and reared out door (except eri silkworm) on various food plants. The food plants are abundantly available in the natural forest and depending upon the acceptability of the foliage by the silkworm and rearing performance, it has been broadly classified into two groups namely primary and secondary food plants. Tasar silkworm (*Antheraea mylitta*) is reared out door on *Terminalia arjuna*, *T. tomentosa* and *T. robusta*. All these primary and secondary food plants are attacked by large number of insect pests. Gall insects, stem borer, shoot borer, soil inhabiting insects, defoliators and leaf minors are the most common pests. Insect pests representing coleoptera, lepidoptera, hemiptera, thysanoptera and isoptera are regular pests and warrant control measures. All parts of the tree and all stages of tree growth are subjected to insect attack. Different insect species infest different parts, but all stages of growth are subjected to all types of insect attack. Some types of injury however, are more common or more serious in certain age groups. For examples, insect that bore mostly within the inner bark usually caused most trouble to mature parts, where as root feeding insects are most injurious to young saplings.

A total of 50 major pests have been reported feeding on *Terminalia arjuna T. tomentosa* in various parts of the tropical tasar region of India (Singh *et al,* 1992, Singh *et al.*, 2000). Biology and behaviour of some key pests have been reported (Singh and Thangavelu, 1994). The information about the various pests infesting *Terminalia* spp. has also been described by Lefroy (1909), Imms

(1963), Atwal (1976), Pedigo (1989), Nayar (1976) and Panwar (1995). Oak tasar silkworm *Antheraea proylei* is reared out door on various species of oak plants viz. *Quercus incana, Quercus ilex, Quercus glauca, Quercus semicarpifolia, Quercus himalayana*, in the western sub-Himalayan belt and *Quercus serrata, Quercus semiserrata, Quercus dealbata* in the eastern belt. Of these *Quercus incana, Quercus serrata, Quercus dealbata* and *Quercus himalayana* are the primary food plants of temperate tasar silkworm. Various pests attack these food plants, which reduces the quality and quantity of the oak tasar food plants. The major categories of insect pest of oak have been grouped as sap sucking, defoliating, meristem feeding, acorn feeding and gall forming insects. Some of the information are available on the pest complex of Oak plants (Lefroy, 1909, Stebbing, 1914, Beeson, 1941, Anderson 1942, Peigler 1993, Seitz , 1933 and Singh *et al., 2000*) but biology and behaviour of these oak feeding insects in context with sericulture needs special attention so that a suitable control measures may be evolved. Muga silkworm *Antheraea assama* is reared on Som (*Persea bombycina*) and Soalu (*Litsaea monopetala*). The food plants are abundantly available in Assam and within 4–5 years it is ready for the silkworms rearing. The quality and quantity of the leaves are deteriorated due to attack of large number of pests from nursery to mature plants. Eri silkworm *Philosamia ricini* Boisduval is the only non-mulberry silkworm, which is fully domesticated. It is polyphagous and multivoltine and can be raised six broods in a year. It prefers feeding on castor (*Ricinus communis).* Kesseru (*Heteropanax fragrans* Seem), Cassava (*Manihot utilissima*), Papita (*Carica papaya*), Pyam (*Evodia fraxinifolia* Hook) and Maharukh (*Ailanthus excelsa* Roxb). Other secondary food plants reported are *Jatropa curcas* L. *Sapium* spp. and *Plumeria rubra* L. All these food plants are attack by large number of pests through out the year (Atwal, 1976; Imms, 1963; Panwar, 1995; Nayar *et al.*, 1976; Pedigo, 1989; Mishra, 1985 and Sarkar, 1988).

The losses insect causes to the plants are directly the result of primary attack. The losses include the killing the plants or parts thereof, reduction of foliage growth, woody destruction and seed destruction. The killing of the plants or branches thereof is usually the result of feeding on the leaves, twigs or roots or by concentrated boring in the growing tips. Growth impact loss (tree mortality plus the reduction of growth) results from insect attack on wild silk host plants are very high depending on different climatic conditions. The annual tree mortality caused by insect was estimated to be 15–20 percent during nursery period (Singh *et al.*, 1987).

Gall insect ;Trioza fletcheri minor (Hemiptera : Psyllidae)

Leaf gall *Trioza fletcheri minor* (Hemiptera : Psyllidae) is the most important gall forming insect on the leaves of primary tasar food plants (*Terminalia arjuna* and *Terminalia tomentosa*). *Terminalia* sp. distributed along the foothills of the Himalaya and Indo-Gangetic plain suffers more by the attack of these insects. These galls occur with equal frequency in the subtropical and deciduous forests of lower montana regions and plains of Peninsular India (Singh *et al.*, 1998). Unlike majority of gall inducing psylloids that are generally host and site specific, *Trioza fletcheri minor* is known not only from the leaf galls of *T. tomentosa* and *T. arjuna,* but also from *T. catappa* Linn., *T. penniculata* Roth and *T. tomentosa* x *T. arjuna* hybrids. Life history performances like the timing, rate of reproduction and survival, offspring quality, etc., appear to be well–specialized in gall insects since a major part of their life cycle is closely associated with the formation of gall guild, in which the gall insect almost completes its development while deriving its nutrition from the gall tissue (Singh and Thangavelu, 1994; Dhar *et al.*, 1998). The life cycle of this gall insect consists of an egg stage followed by a five nymphal stages. The final instar nymph become quiescent and moults to produce the adult. The adults are translucent white at emergence and their colour develops fully after feeding for some time.

Fig. 4.1 : Galls formed by Psyllids on the leaves

Cecidogenesis (Gall formation)

A subtropical plant louse, *Trioza fletcheri minor*, induces leaf galls on at least five species of *Terminalia* in the Indian sub-continent. Gall is the product of an active growth reaction of a plant to the feeding activity of an insect. Growth and differentiation of gall is contingent upon the continued presence of the insect (Rohfritsch and Shorthouse, 1982). A well-synchronized interactive process involving a high degree of mutual adjustment and extraordinary specific behavioural pattern in both the participants is the hall mark of galling. Choice of the appropriate developmental stage of the host tissue, highly specialized colonization and a oviposition behaviour of the gravid female, and triggering of timely physiological modifications in the gall, such as directed growth responses of the plant organ eventually leading to the enclosure of the gall inducers and dehiscence synchronizing with the escape of the mature developmental stage of the insect at appropriate time are characters that uniquely distinguish gall insects from related plant feeding insects (Shorthouse and Lalonde,1984; Raman , 1987; 1991; Mani, 1992).

Galling is the consequence of interaction between the offensive stimulus of the insect and the defensive response of the plant (Rohfritsch, 1992). Considerable information is available on the basic aspect of cecidogenesis in *T. arjuna* and *T. tomentosa*. Gravid females of *T. fletcheri minor* deposit eggs on the stem axis and petioles of young vegetative branches. When laid freshly, each stands perpendicular to the horizontal axis of the substratum with its pedicle buried in the host tissue. The short cylindrical pedicel penetrates the host tissue only up to the second or third layer of cortical parenchyma. In about 48 hours after oviposition, as a result of embryo growth, the egg inclines itself in such a way that the operculum touches the plant surface, facilitating easy migration of nymphs after the dehiscence of the egg.

Nymph I instar crawl from shoot/petiole axis towards venal angles along the abaxial surfaces of the differentiating leaves and settle with their mouthparts lying on the stomata. By inserting their stylets through stomatal apertures, these stationary nymphs feed on the primordial mesophyll parenchyma cells which become metabolically active as evidenced by hyper tropical cells and nuclei, besides enriched cytoplasm, indicating the activation of cellular metaplosia. Feeding activity triggers growth in the host leaf, especially around the nymph bed, thus making the gall recognizable. The adaxial epidermis of the leaf part accommodating the nymph responds actively by producing cylindrical trichomes that are filled completely by polyphenolic materials. Mesophyll cells adjoining these trichomes show intense hypertrophy and hyperplasia and thus contribute cells to labial growth around the nymph. In 4–6 days, labial growth in the form of a cone encloses the nymph, meeting along the abaxial side of the host leaf, though there is no fusion at the tip. By this time the nymphal chamber gets organized with well-defined parenchyma cells bordering its inner perimeter. Mesophyll cells lying around the developing gall, particularly along the adoxial side of the host leaf, include dense polyphenolic inclusions (Raman *et al.,* 1997). With the moulting of the nymph into second instars, growth is intense in the mesophyll region of the gall and the gall volume rapidly increases. The nutritive parenchyma cells are rich in cytoplasm and contain prominent nuclei. Following the moult to the third instar, the gall attains the final shape including the prominent conical covering growth along the abaxial side and distinct bulge along the adaxial side of the host leaf. At this stage, the nymph shifts its feeding site from parenchyma cells to vascular phloem. Moulting to fourth nymphal stage marks a number of vital morphogenetic changes in the gall. The erstwhile nutritive parenchyma and cells in their neighborhood begin to accumulate starch and polyphenolic materials either as crystalline or viscous inclusion. Such cells are distributed rather irregularly in the gall mesophyll. Cells bordering the narrow space in the conical edge proliferate at random points thereby the widening pathway is narrowed. This can be seen as an adaptation in the gall system to prevent the possible entry of predator. Synchronizing with the moulting of fifth nymphal instar, the adaxial region of the gall expands laterally either by producing new growth centres that divides or expands.

Gall ready for dehiscence includes 10–12 rows have horizontally distributed sclerites elements, and the sclerification of parenchyma cells snaps off the plasmodesmatic connections between the cells. Consequently, the parenchyma cells lying close to the nymphal chamber shrink and degenerate. This results in the widening of the ostiole in the abaxial cone and it enables the escape of the final nymphal instar for moulting into adult out side *T. arjuna* (Raman *et al.*, 1997). Adults' male and females at emergence are pale brown with black terminal antennae segments, but a progressive color changes to deep brown follows with maturation. Under laboratory and field rearing from both host plants, the nymphal instar duration is 20–22 and adult longevity is 10–12 days. Nymphal V instar expands from the gall for moulting into the adult outside. In 3–4 days, mating takes place and the gravid females lay eggs on tender shoot of host plants.

Big gall; *Phylloplecta hirsuta* and *Megatrioza hirsuta* (Homoptera : Psyllidae)

Phylloplecta hirsuta and *Megatrioza hirsuta*, commonly known as leaf roll gall are other important gall forming insects on the leaves of *T. tomentosa* and *T. arjuna*. The gall formed by *Trioza fletcheri minor* is epiphyllous, pustuloid, subglubose or hemispherical, unilocular, fleshy, sessile, dehiscent, deciduous, yellowish or brownish pouch galls where as *Megatrioza hirsuta* is a leaf roll gall on *T. arjuna*. The leaf roll gall is epiphyllous in rolling of the two margins towards the mid rib. The leaf is irregularly swollen, twisted pale green and sometimes pinkish in colour. One or both margins of the leaves are rolled into the mid ribs on the upper surface and the rolled up margins gradually become thickened and hard like a gall. The nymphs remain sheltered in these rolled leaves and are often covered with a copious secretion of white wax. Large round drops of liquid excrement with their surface coated with powdered wax and fragments of wax threads are also presents in these rolled leaves. The mature nymphs crawled out of the gall when it dehisces or leaves the shelter of the roll and shed their last skin and this moulting is more commonly performed on the surface of an older leaf.

There are five moults and the earlier shedding of the skin is done inside the roll gall. Adults are translucent white at emergence and their colour develops fully after feeding for some time. The mechanism of gall formation is almost similar in both the species.

The only difference observed between both the species was the development of meta coxal spurs. In fact *Megatrioza hirsuta* has been considered synonym of *Phylloplecta hirsuta*. The population dynamics of these gall insects are influenced by a biotic factors especially temperature, humidity and rainfall. The population of psyllid varies from year to year but generally they follow more or less similar trend. After each pruning and pollarding of tasar food plants, a fresh vegetative flush occurs during April-May, resulting in abundance of soft and succulent leaves. A steep rise of the psyllid population results during the period from the eggs laid by over wintering population. Nymphal population increases from June onwards and reaches its peak during July–August. At the end of rainy season, psyllid population gradually declines and from November to February it decreases steadily. Such fluctuation in the population is due to the change in a biotic factor. The climatic condition of June, July and August proved conducive for the growth and development of the nymphal population on the leaves of tasar food plants. The most active period of the gall population coincides with the rainy season.

Control

The population of gall insect can be minimized by mechanical and cultural practices. The nymphs of *Trioza* sp. undergo diapauses in winter (December-January). During winter period when the rearing of tasar silkworm is not conducted the gall infested leaves are plucked and burnt in the field. It has been found to be the most convenient way to nipping the pest infestation in the bud. Further, some regular farming practices viz. weeding, intercultivation, pruning and pollarding ensure further reduction in the gall population. In some cases even after successful adoption of these cultural and mechanical practices, gall infestation was recorded up to 45 per cent. To combat these populations Dimethoate (Rogor 30 EC, or Monocrotophos 36 EC or Malathion 50 EC or Fenitrothion 50 EC) at 0.09 per cent is prayed on the infested plants (Singh *et al.*, 1989).

As the *Trioza fletcheri minor* breeds from April to November and has several overlapping generations, spraying at least three times consecutively at 15 days intervals starting from first week of March is necessary to break the continuity of the life cycle and to ensure quality foliage for tasar silkworm rearing. On account of the effectiveness of dimethoate on gall insect and its harmless effect on silkworm larvae, dimethoate 0.09 per cent has been recommended for control of gall insect on primary tasar food plants. Similar strategy is also adopted for the control of big galls on primary tasar food plants. For biological control of gall insect two important parasitoids have been screened from the plantation of *Terminalia arjuna* and *T. tomentosa*. The parasitoids are *Trechnites secundus* Girault (1959) (Encyrtidae: Hymenoptera) and *Aprostocetus niger* Ranaweera (1927) (Eulophidae : Hymenoptera). The parasitization potential of these parasitoids are very high and after proper augmentation and conservation it may be used to control the gall insect population in tasar culture.

Gall wasp; (Hymenoptera: Cynipidae)

Oaks are the favorite hosts of gall wasps (Hymenoptera: Cynipidae). Generally, individual trees or small groups of trees become slightly to moderately infested by these *Callirhytis* spp. Widespread infestations are not common. However, severe gall wasp infestations have been observed in North eastern region of India. Extremely high numbers of twig galls of *C. quercusclaviger* have occurred on thousands of oak plants in several locations. These infestations have affected young to mature trees in north eastern and western region. The overabundance of twig galls has resulted in notable levels of branch dieback, crown thinning, and tree mortality.

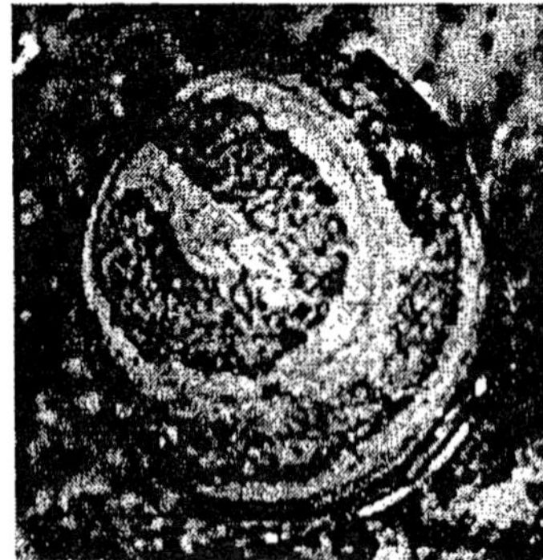

Fig. 4.2 : Cynipid galls on Quercus plant

Two kinds of galls are produced by cynipid wasp. The most notable are produced by the asexual generation that are spherical, corky, 0.5 to 1 inch in diameter and appear on twigs and branches of live oak in late summer and early fall. When first formed, they are pink to pinkish brown and later changes in yellowish-green. A number of cynipid wasps cause unique galls on oak trees; the gounty oak gall is formed by *Callirhytis punctata.* Mature (with horns) spine-bearing potato gall is caused by the gall wasp *Callirhytis quercusclaviger* (Ashmead). The horned oak gall is caused by *Callirhytis cornigera*. The hedgehog gall is formed by *Acraspis erinacei* on the leaves. The gall formed by *Callirhytis seminator is called as* wool sower gall, which is sticky and spongy on twigs with seed-like structures inside. *Andricus laniger* Ashmead forms leaf galls on several oak plant species. The oak apple gall formed by *Amphibolips confluenta* is spherical, spongy-filled galls on red oak. Galls on oak plants are also caused by other insects such as aphids, flies, phylloxera, psyllids, thrips and mites. Seven species of cynipid wasps are so far known to infest the oaks in India viz. *Callirhytis semicarpifolia, Neuroterus hassi, Neuroterus quercusbaccarum, Dryocosmus kuriphilus, Cynipis mukaigawae, Neosynergus subodhi, Saphonicrus* sp. *a*n unidentified species of the genus *Neuroterus* forms leaf galls in *Q. acutissima* which results in the curling of leaves. The infected leaves are not suitable for silkworm rearing and leaves cannot carry on their normal functions of food preparation, transpiration, etc. *Dryocosmus kuriphilus* forms gall on twigs and *Cynipis mukaigawae* make galls on the trunks of *Q. acutissima. Callirhytis semicarpifolia* occurs on acorns of *Q. semicarpifolia* as a deformed gland. Several gall of *Neuroterus hassi* often clustered on branches of *Q. spicata. Neosynergus subodhi* occurs on twigs of *Q. griffithii* as pseudo cone like, solid, spiny galls. Plant galls induced by the larvae of cynipid wasps are considered the most complex in the guild of gall inducing insects. The ability of each species of cynipid to induce structurally distinct galls on oak plant is unique.

Little information is available on the biology of *C. quercusclaviger*. The apparent life cycle of *C. cornigera* is complicated. The parthenogenic female wasps emerge from twig galls during spring. These females oviposit eggs on veins on the underside of leaves. Small vein galls appear from late May through June. Male and female wasps emerge from the vein galls in early July. Mated females oviposit eggs in young oak twigs. Twig galls appear the following spring. Two or more years are required for the immature gall wasps to complete development in the twig galls. The galls provide shelter, protection, and food for the immature wasps. Inside a gall, the larvae are surrounded by tissues (nutritive zone) rich in nutrients. As the larvae reach maturity small spines or horns become evident on the gall. An adult wasp emerges from each horn. Adults emerge from galls of the "asexual generation" during December. All adult females do not mate before laying eggs on swollen leaf buds. Eggs hatch in early spring as leaf buds begin to open. Larvae develop quickly in leaf tissue and stimulate the development of small, beige-colored galls resembling kernels of wheat. Adults of both sexes emerge from these galls or the "sexual generation" after a few weeks. After mating, females lay eggs in post oak twigs and branches. These eggs remain dormant for 3 to 5 months. Then they hatch and stimulate the formation of galls of the asexual generation. The gall forming insects live in the gall and feeds on plant tissues within it . As galls grow, the larvae within it, procure their nutritional requirements, and damage the host plants.

The oak galls have large number of commercial utilization and therefore most of the gall infected leaves are protected for its commercial utilization. The oak galls have long standing value as articles of international trade due to their relatively high tannin content (44 to 70% by weight). Certain galls have been used in preparing ink and dying wool. Oak galls have medicinal values also. It is used to prepare medicines for toothaches, gum disease, dysentery, ear and eye sores, abscesses and menstrual disorders; and as good sources of extractable tannins for leather industry.

Contr ol

The control of cynipid gall is difficult due to variation in the biology of gall forming insect. The insects of this group are the most difficult to control. The destructive stage of each species is within the plant tissue and in accessible for the most part to contact with insecticides. All of the species have one or more stages outside the plant tissue and causes a constraint in controlling these insects.

Application of Phosphamidon 0.05 % during the initiation of gall formation on the oak plant is highly effective and the gall population can be reduced.

Soalu gall ; *Pauropsylla besooni* (Hemiptera : Psyllidae)

Pauropsylla besooni (Hemiptera : Psyllidae) is the most common gall forming insect on the leaves of soalu plant (*Litsaea polyantha* Juss). The galls are oval, globular and spherical in shape. The most active period of this psyllid population coincides with the Bhodia crop of muga silkworm rearing and causes extensive loss to leaves of soalu plants. Eggs are laid on the leaves during September. The total number of eggs laid per female varies from 150-250. The eggs are elliptical, deep violet in colour possessing hook at the micropylar end. The colour of the egg change to black after 24 hours of oviposition. The eggs hatch on the leaf and start feeding by sucking the juices of the leaf and this becomes the initiation of the gall formation on soalu plants. There are five nymphal stages, which are identified by the increasing the number of their antennal segments. There are five segments in fifth nymphal instar. The final instar becomes quiescent and moults to produce the adults. These adults come out after dehiscence. The adult are translucent at the time of emergence and later becomes darker after feeding for some time. The adult has modified hind wings, which are larger and more muscular than fore legs. The mating starts within 24 hours and starts the next generation (Singh 1993). The population dynamics of the gall population is highly influenced by the various a biotic factors. Temperature, humidity and rain fall plays an important role in increasing the population of gall insect. The population of insect varies from season to season and year to year but trends of the population is almost similar. The galls are seen through out the year however the peak period of abundance has been reported to be July – October. During this period the temperature ranges from 25 to 30°C and relative humidity 80 to 85 per cent and it creates a very conducive atmosphere for the population build up of this gall-forming insect on soalu plants. The growth and development of the nymphal stages are also governed by various other extrinsic and intrinsic factors and a lot of physiological changes occur in the leaf tissues, which resulted in the formation of big gall on the soalu plants.

Cecidogenesis

Gall is the product of an active growth reaction of a plant to the feeding activity of an insect. Growth and differentiation of the gall is the contingent upon the continued presence of the insect. A well–synchronized interactive process involving a high degree of mutual adjustment and extraordinary specific behavioural patterns in both the participants is the hallmark of galling. Choice of the appropriate developmental stage of the host tissue, highly specialized colonization and ovipositional behaviour of the gravid female, and triggering of timely physiological modifications in the gall, such as directed growth responses of the plant organ eventually leading to the enclosure of the gall inducer(s) and dehiscence synchronizing with the escape of the mature developmental stage of the insect at the appropriate time are unique characters that distinguish gall insects from related plant feeding insects. Galling is the consequence of interaction between the offensive stimulus of the insect and the defensive response of the plant. Gall insect exhibit many traits well adapted to their specialized habitats. The host plants, in turn counter the insect interaction with diverse, localized morphological expressions modulated by physiological modifications. When the gall insect put the host plant under feeding stress, the plants respond by expressing several new morphogenetic phenomena like hypertrophy, hyperplasia and establishment of a physiologically active nutritive tissue. Levels of expression of these morphogenetic phenomena and the nutritive tissue vary with specific insect groups and this highlights how the response pattern of the plant tissue varies with the nature of insect action. The active feeding stages of insects synchronizes with the active growth phases of the host plant and the pattern of synchronization is well demonstrated, particularly at the time of dehiscence of the gall and dispersal of the mature stage of the insect. Cessation of feeding by the late larval stage prepares the host tissues to go through metabolic changes leading to dehiscence of the gall. Cessation of active feeding phase of the insect correlates with the reduced metabolic activity of the host on the one hand, and with climatic changes in the external environment on the other, leading to gall dehiscence and insect discharge.

Gall midges ; *Asphondylia sp.* (Diptera : Cecidomyiidae)

The univoltine gall midge species *Asphondylia* sp. induce very specific galls on the leaves of the som plant *(Machilus bombycina). The* gall forming cecidomyiid induces a simple spherical characteristic gall in lower surface and tapering in the terminal region of the leaves. It is distributed throughout Assam in muga growing areas. The leaves with galls become unsuitable for muga silkworm feeding. Galls are seen through out the year, however, it is more prevalent during June–November. In mid-July the mature gall present a central hole (the ostiole) through which the larva leaves the gall; around the hole the lignified trichomes form a clear halo. Adults are delicate insects with relatively long antennae and legs. The antennae are monoliform type adorned with conspicuous whorls of hairs. Ocelli are present. The adult fly inserts its eggs into the leaves. After hatching, the larva leaves the egg shell among the trichomes and the main veins of the leaves and crawl slowly towards lamina, remains glued to plant epidermis. Larva starts feeding by sucking the activated, turgid cells around the wound area and these cells becomes active food attractors and are stimulated to growth and photosynthesis. The gall midges attack the host epidermal cells with sharp mandibles and probably the saliva. The wounding pattern is very specific. The gall results from a continual interaction of the gall midge larva with the host cells. Gall specificity is produced by a specific interaction that each gall midge species has evolved with its host. The specific interaction is generated by a sophisticated and highly determined early attack and feeding behaviour. There are altogether three larval instars and their behaviour is dictated by the life strategy because the species is adopted to over come the constraints imposed by its hyper specialization on the host tree. The third instar larva pupates inside the malformed gall-like structure and emerges as adult. This pest is generally destructive to young plants and the attack is particularly severe from June to November. The leaves become unsuitable for feeding of silkworms and due to this the muga silkworm growers face a substantial reduction in the quantity of leaves, which affects their economy.

Control

Infected plant should be sprayed with rogor 0.05 per cent keeping in view the rearing schedule of the muga silkworms. The infested leaves in the nursery and young age plants should be plucked and burnt in the field itself. *Apantalis* sp. a braconid parasitoid is a very common parasite of this insect. It kills the gall insect in the chamber itself and reduces its population. But it cannot be applied as biological control agent because it is one of the dreadful enemies of muga silkworm larvae also.

Stem Borer

During the recent years, the plantation of primary silk host plants suffered a serious set back due to the damage caused by the coleopteran stem borers. Some cerambycid borers have seriously affected sal plantation where as arjun and asan plantation, the buprestid borers have damaged. The grubs damage the xylem and phloem portion of the silk host plants, and ultimately the plants dies. The damage is caused mostly by the grubs of stem borer (Singh *et al.*, 1987, 1989 and 1990) . The pest in its larval stage feeds on the bark and burrows into the stem causing dead hearts and ultimately loss of the plants. The round-headed stem borer, *Aeolesthes holosericea* is wide spread through out the tropical and temperate tasar region of India, causing considerable damage to *Terminalia* and *Quercus* sp. The flat headed stem borer, *Psiloptera fastuosa, Sphenoptera koenbierencis* are key pest of *Terminalia arjuna* and *T. tomentosa* in the tropical tasar region, causing as high as 40 per cent losses at some places. The major buprestid and cerambycid borers infesting silk host plants are *Sphenoptera Koenbierencis*, *Spehenoptera cupriventris* Kerremans, *Sphenoptera indica* Laporte and Gory, *Psiloptera orientalis* Laporte and Gory, *Psiloptera fastuosa* Fabricious, *Sternocera orientalis*, *Aeolesthes holocericea* Fabricious, *Chelidargentlum onium* and *Hiplocerambyx spiricornis*. *S. cupriventris*, *S. koenbierencis* and *S. indica* are quite common and abundant in Central India and attack new plantation of *T. arjuna* where as *P. fastuosa* and *S. indica* has been reported from tomentosa mostly. The sal plantation has been damaged extensively by the *H. spiricornis*.

Flat headed stem borer; ***Sphenoptera koenbierencis*** **(Coleoptera : Buprestidae)**

The beetle is widely distributed in the tropical and temperate region of India. The beetle that feeds on the foliage of the host plants are black-bronze and are 12-14 mm long, somewhat flattened or cylindrical and the elytra with a scutellar stria. They are bronzy coloured beetles. The beetles possess 11 segmented, short, serrate antennae. The pro, meso and meta thorax are well-developed. The metasternum has a well-marked suture. The thorax and abdomen are firmly united. The tarsi possess membranous expansions. The larvae are remarkable in having a greatly developed prothorax, slander hind body and vestigial legs, which may be sometimes absent. They bore into the stem of plant and make a flattened gallery in or beneath the bark of the tree. The grubs are dull brown, club shaped and attains 15-20 mm of body length.The beetle starts appearing by the middle of April. After mating they lay small spherical, white eggs singly scattered all over tree trunk and main branches. The eggs hatch in 20 days. The grubs on emergence feed on the bark and as they grow they bore inside. The larval stage are completed in two months during summer, but those which winter over take 6 months. The larva makes a small chamber in the woody tissue about 10 mm deep from the surface The pre-pupal and pupal period lasts for 1 to 2 and 8 to 12 days in summer. The winter is passed in the grub stage. Three generations of this pest are completed in a year. The grubs feed under the bark as well as bore deep into the wood. The young grubs feed below the bark making minute irregular galleries. A number of grubs feed near each others as a result of which the bark get loosened and splits. Gum globules ooze out of the entrance hole. Leaves of the attacked plant turn pale and their growth is arrested. The attacked branches dry up and quality of leaves deteriorates. With continued damage the tree dies ultimately.

Flat headed stem borer; ***Sphenoptera cupriventris*** **Kerr (Coleoptera : Buprestidae)**

The stem borer, *S. cupriventris* is becoming a serious pest through out the tropical and temperate tasar region of India. Besides *Terminalia*, the stem girdler feeds on *Shorea robusta* and Mulberry plants also. The beetle is dull bronze with coppery reflections. It is pentagonal, widest at mesothoracic region and tapering posteriorly. The antennae are eleven segmented in both the sexes. The head is bifurcated between eyes and slightly channeled on vertex. The elytra are hard and convex. In female, the last abdominal sternite is narrow with deep small concavity at the posterior end where as in male it is not prominent. The males are 12-15 mm in length and 3-5 mm in breadth while the female measures 14-18 mm. The beetles begin emerging in the late June and early part of the July, found sitting on the leaves or twigs mostly on *T. tomentosa* and less on *T. arjuna* plants in the field. The sexual activity of insect is more vigorous only during July-September. The mated female oviposits in the cracks or crevices of the barks of *T. ajuna* and *T. tomentosa*. Eggs are laid singly in the cracks and crevices of the bark. They select more often the injured area on the plant. The freshly laid egg is white in colour and pin head size. It measures 1-2 mm in length and 0.65-1 mm in breadth. The colour of the egg change to blue few days before hatching takes place. The young grubs come out from the egg in about 10-15 days. The newly hatched grub bore into the bark and feeds beneath it. The grubs are apodus and sluggish in nature. They are normally called flat headed borer with the thoracic region being enlarged and flattened. The grubs are light yellow or shinning white in colour, flat elongated and narrow with tapering abdomen. The head is small brown, somewhat depressed and expanded and the prothorax is swollen. It is dorsally sculptured. The abdomen consists of nine segments and there are no anal processes. There are nine pairs of spiracles, the first pair being situated between pro and mesothorax. The grubs never remain straight but lie with abdomen inclined to one side of the body giving it a hook shaped appearance. There are five larval instars. The first instars measures 4-7 mm in length and 1.5-2mm in breadth. The second instar measure 10-13 mm x 2-3 mm where as third, fourth and full grown grubs were 15 - 18 x 2-3 mm; 21-24 x 3-6 mm ; and 28-35 x 5-6 mm respectively. The grub stage is about 6-7 months. Pupation takes place in full grown pupal chambers made by the full grown grub in the sapwood, which is some what larger than the size of the future beetle. The grubs shrink antero-posteriorly, becomes wider through out and attains ovoid shape, which almost resembles the future beetle. The pupae are exerate. The colour of the pupa is white and body is soft wax like in appearance, naked in chamber, when freshly formed. But after few days, it turns black as the integument hardens. It measures from 12-16 mm in length and 4-5.5 mm in width.. The adult lives in the same tunnel and ultimately the wood dries up. The grubs feed in the bark, at first eating out irregular chambers and as it grows it goes deeper makes grooves both on the outer bark and sapwood. This gallery has no definite direction. When the grub is fully grown, it bores down into the sapwood at an angle for about half an inch and then eats it. It prepares small elliptical pupal chambers in the wood, somewhat larger than the size of the future beetle. The selection of the entrance of tunnel in the sapwood is narrow and elliptical. The presence of these entrance tunnels in the sapwood is very characteristic of the beetle and they are clearly visible when the decayed bark above has fallen off in patches. But in case of old hard bark trees the pupating chambers may be located in the bark itself. The larvae changes to pupae in the pupal chambers. The beetle on emergence crawls out of the tunnel, eats its way through the bark and escapes from the plants. The duration of pupal stage is about 25-45 days and the longevity of the adult life span is about 2-3 months There is only one generation in a year i.e. the larval stages generally last from December to June and the adults start emerging at the end of June.

Stem borer; Psiloptera orientalis Laporte and Gory and Psiloptera fastuosa Fabricious (Coloeptera : Buprestidae)

Buprestid borers such as *Psiloptera fastuosa* and *Psiloptera orientalis* are easily recognized by their metallic colour. Their bodies are hard and inflexible and usually appear as if may be of bronze. The integument is hard and strong, the head partly sunk in thorax, which is strongly fixed, to the abdomen, the elytra accurately adapted to the body, the antennae are readily concealed under the head. The beetles feed on the leaves and bark of the host plants. During the spring, the adult emerges and after 20–25 days the females deposits egg on the joint of the branches or under the bark. The larvae is of characteristic form, legless with thoracic segment what swollen into distinct bulb, the abdomen very long and slender. The swelling fits the bore made in the plants and gives the larva the necessary hold to move along the borer to work with mandibles against the hard tissues. Pupation takes place inside the bore, the pupa lying naked in a chamber made by closing the bore with debris. The larva prepares the hole for exit of the pupa, leaving only a thin covering of bark through which the beetle can readily emerge. The bronze borer makes compact zig- zag types of galleries in *T. arjuna* trees. At intervals these galleries departs from the inner bark region and dip into the sap wood.

Sal borer; *Hiplocerambyx spiricornis* (Coleoptera : Cerambycidae)

Sal borer *Hiplocerambyx spinicornis* (Coleoptera : Cerambycidae) is one of the major pest of sal plant (*Shorea robusta*) in tropical tasar region of India. The beetle's activity start by the middle of June. After mating, the adult female lays 300-500 eggs after 7-9 days of fertilization. The eggs are laid solitarily under the bark hole and piled skin in shady areas of the tree. Eggs are creamy white in colour and cylindrical in shape. It measures about 3.2 mm in length and 1.2 mm in width. The incubation period ranges from 3-7 days. The larvae develop in the inner side of the bark and bores the sap wood in the course of its development The average length of the mature larvae was 5.5 cm and width of 1cm. During February and March the mature larvae makes pupal chamber. Pupal stage is the inactive stage and the dormancy period ranges from November-December to May-June. The adults emerged after the onset of the monsoon. The antenna is long about 44 mm, which is bigger than body size. The male measures about 32 mm. The body is divided into three parts i.e. head, thorax and abdomen. The prothorax is small and thinner in comparison to abdomen. The legs are strong. The average length of the female varies from 30-35 mm and antenna are smaller in size in comparison to male.

Long horn Beetle; *Aeolesthes holosericea* Fabricious (Coleoptera : Cerambycidae)

Adult cerambycids are commonly known as long horn beetle because they have long antennae. Adult *Aeolesthes holosericea* is a typical cerambycid readily recognizable from their general form on the sides. The tibia of the fore legs are not grooved and the palpi are never acute at the tip. The beetle is of pale brown colour, the head is distinct and well–developed and large eyes and powerful trophies, the heavy biting mandibles being prominent. The beetle emerges in June and starts feeding on the tender bark of twigs. After mating, female grow elliptical nitches into the inner bark and in each deposit one to several eggs. The eggs are long, oval in shape and measures 1.5 x 0.75 mm. in size. Freshly laid eggs are whitish in colour. After 6 to 9 hours, the colour of the eggs change to brown. The number of egg varies from 5 20. The larvae normally called round headed borer is elongated, cylindrical, fleshy and white.

Grubs are legless but thoracic and abdominal segments swell into distinct bulbs. The full grown grub varies from 40–55 mm with an average of 42 mm where as breath varies from 7.8 mm with an average of 7.5 mm. The first, second and third instar grubs measures 6.3, 12.5, and 25.5 mm in length respectively. The larval period varies from 45–60 days. Larvae over winter in third to fifth instar. With the onset of the winter, generally from November to February, feeding ceases and the larvae construct galleries. Larvae commence feeding after over wintering until pupation, which occurs in cavity constructed by the pre-pupal larvae approximately 25 mm below the surface of the bark. During the following spring, larvae bore further into the sap wood and heart wood where they feed and later pupate. Pupation also occurs in larval galleries in the xylem. The duration of the pupal stage is about 2–3 weeks. The length of the pupal chamber varies from 35–40 mm long, 18–20 mm broad and 10–15 mm towards the cortex of the stem. Pupation takes place inside this chamber keeping the head upwards, adults emerge from the pupal case within the pupal cavity and remains there while their untanned cuticle hardens. The pattern of emergence of each sex is similar and the ratio of the male to female is approximately 1 : 1. Both the grubs and adults are injurious to primary tasar food plants, *T. arjuna, T. tomentosa* and *S. robusta*. Grubs are more destructive than the adults. The presence is indicated by the accumulation of granular frass and discolouration i.e. 'wet spot" caused by the sap leakage that occurs around the attack site. The infested plant oozes sticky gums as well as faecal matters, which cover the opening of the hole. A severe damage is caused by tunneling under the bark by larval stage and feeding on the bark by the adults. Larva feeds inside the main stem and due to that tunnels are formed in rummified way. These ramified tunnels reach the center of the main stem, thus causing damage to phloem and plant ultimately dies. Tunnels are filled with powdery tubing's and small frass pellets.

Control

The removal and burning of *Terminalia* twigs after pruning results in reduction of the beetle population, by way of disturbing their oviposition sites. Light traps are used to attract and kill the adult beetle. Monocrotophos 36 EC or Phosphamidon 100 EC @ 0.05 per cent soaked in cotton are applied in the holes made by the borer in the stem. This cotton gauze is inserted with the help of wire rod in the holes of infected plants. Cypermethrin 10 EC is also effective against the buprestid borers (Singh *et al.*, 1990; Singh *et al.*, 1993; Mandal *et al.*, 1989). Apart from chemical control measures some natural control agents have also been identified which is very potential in minimizing the stem borer population in natural plantation. The parasitoid consists of Szelenyiola sp. near batocerae Ferriere (Encyrtidae : Hymenoptera) and a chalcid fly *Megachalcis fumipennis* (Chalcidae : Hymenoptera). *Szenyiola* sp. is an important egg parasitoid and *M. fumipennis* is a larval parasitoid. Both the species inhabit in the same ecological situation and seek similar ecological requirements (Thangavelu and Singh, 1994). Both the species have high reproductive potential and female biased in nature. It parasitizes the eggs and grubs of these beetles and helps in minimizing the population of pest insect. Besides this, an important predators *Amblystomus guttula* Dejean (Coleoptera : Carabidae) and *Ichthyurus* sp. (Coleoptera : Cantharidae) has been reported as important predator. It is a voracious feeder and kills maximum number of host larvae. It attacks the host larvae in the plant holes made by the borer. Further, *Tryophagus putrescentiae* Schrank (Acridae : Astigmata) has been recorded to feed on the eggs of buprestid borer. Biological research on this aspect has indicated that it possesses most of the desirable attributes and therefore, may be used as a tool in the biological control of stem borer.

***Batocera lineolata* Chevrolet (Coleoptera : Cerambycidae)**

The pest is polyphagous in nature. This beetle is widely distributed in India and is a highly destructive pest of oak. The adult is generally known as longicorn beetle. The body length varies from 4.5 to 5.5 cm. The general colour of the beetle is black with grey brown and white spots. Adults emerge in April–May and the mated female oviposits singly in slits in the bark during July. Incubation period varies from 7 to 14 days. The grub feeds on the internal tissues of the plant for 4-5 months. The mature yellowish grub measures about 10 cm in length and is stout with well-defined segmentation. The grubs tunnel into the sapwood. It pupates in a chamber. The adult cuts its way out of the pupal chamber. It feeds on the bark of the young twigs and petioles. Young grub on emergence, feed first on bark and make zig-zag galleries. As they grow in size, they bore deeper inside the woody portions of the tree branch. Owing to the damage to the woody tissues, the vitality of the tree is reduced. As a result of their feeding and cutting into the wood, sap flow in that portion of the branch or the growth of the trunk becomes restricted. It gradually causes death of infected portion. The infected plants can be located from the distance due to presence of the frass that comes out of the holes in the branches or in the main trunk. Presence of more than one hole on the same plant indicates multiple damage

Control

Collect and destroy the grubs and beetles. Insert in holes cotton wicks soaked in Dichlorvos (0.1%) or Dimethoate (0.03%) and seal with mud.

Batocera titana **Thoms (Coleoptera : Cerambycidae)**

It is a large yellow or greenish yellow longicorn beetle with blackish markings and orange spots. The adult beetle varies from 5.5 cm to 7.5 cm in length. The sides of the beetle produced medially into a sharp black spine bending downwards. The adults make their appearance on the plant from July to September. The mated female deposits her eggs in cracks or in the bark. The eggs hatch into grub, which starts feeding on the bark. The development of the grub is completed in 5-6 months. The winter is passed as grub, which measures about 2.5 to 4.0 cm in length. They start feeding as soon as the weather becomes warm in the spring and continue throughout the season. They bore through the wood of the tree, cutting out large galleries, sometimes as much as 2.5 cm in diameter. When they are full grown, the larvae are about 5.0 cm long and tunnel a large pupating chamber which is plugged up at either end with wood fibres and dust. They pupate in the cell and remain there for 2 to 3 months. The life cycle of the insect takes at least one year. It is also one of the most destructive borers attacking oak trees. Infested trees have many large burrows, with opening through the bark. The entrance of these burrows is usually packed with coarse wood fibres. There is generally a discharge of sap from the opening to the burrow, which wets and discolours the bark of the tree for some distance below it. The wood of the tree is weekend, so that the branches or the main trunk breaks

Control

Collect and destroy the grubs and beetles. Insert in holes cotton wicks soaked in Dichlorvos (0.1%) or Dimethoate (0.03%) and seal with mud.

***Batocera rufomaculata* (Coleoptera : Cerambycidae)**

Stem Borer (*Batocera rufomaculata*) is widely distributed in India and attack a number of forest trees including sal. The adult beetle is large, cylindrical with long antennae usually at least half as long as body or often much longer. They are often brightly coloured and others exhibit striking cases of cryptic coloration. In most species the inner margins of the eyes are notched and antennae may arise within this notch or just opposite to it. The tibial spurs are well developed and the tarsi five segmented, the third segment being bilobed conceals the further segment. Many cerambycids stridulate either by working the hind margin of prothorax against the specialized striated area at the base of scutellum or by the hind femora rubbing against the edges of the elytra. Both the type of stridulatory organs may occur in the same insect. The grub are apodus, elongate, cylindrical and whitish and they bore into woods, stem or roots and thus prove more destructive to shade and forest trees. The grubs feed inside the stem, making tunnel upward which results in drying of branches and in severe cases death of tree. Eggs are laid either in the cracks of tree trunk or in the cavities of main branches which covered with viscous fluid. Grub pupates inside the stem and beetle emerges in July/August.

***Xylotrichus sp.* (Coleoptera : Cerambycidae)**

Xylotrichus sp. is important stem borer of muga silkworm host plant. The beetles normally emerge during April–May and October–December (Singh and Das 1996). After mating, the female oviposits about 100 eggs in 3-4 weeks in cracks and crevices of the bark of the stem. The incubation period is about 10 days. Larval period lasts for about 9 months. When full grown larvae cut an exit holes in the stem and pupate near it, inside the tunnel. The pupal period lasts for about 30 days. One generation is completed in one year. The larvae of this beetle are mainly destructive as they bore into the stem, which wilt and break easily. The young plantations of som are heavily affected and are killed and older ones give unhealthy appearance.

Control

The swabbing of main stem and thick primary branches with BHC 50% emulsion @ 1 kg mixed 50 litres of water per acre once during April-May and twice during October – January is effective.

Carpenter moth, *Zeuzera multistrigata* Moore (Lepidoptera : Zeuzeridae)

Zeuzera multistrigata is one of the stem boring species which are considered as serious pest of som and soalu plants in Assam. It prefers to attack mature plants however, seedlings have also been reported to be attacked. Survey reveals that 79 per cent soalu and 54 per cent som trees have been annually affected by this pest (Annonymous (2001). The insect is of moderately large size. The antennae are frequently bipectinate in both sexes. The frenulum is short and non functional. The moths are nocturnal fliers and lay more than 1000 eggs on the bark of the trees . The eggs hatch into larvae and enters into the branches of the main stem.

The another important species of this genus is *Z. indica* (Herrick-Schaffer). It is a medium size moth, whitish in colour having narrow wings with black spots on fore and hind wings and a long abdomen. Abdomen and thorax are densely covered with white cottony hair. Seven dark blue spots are found on the thorax. The antennae are bipectinate. The males are conspicuously smaller than the female. Adult emerges between September and November and a female lays 1200-1900 eggs approximately. The eggs are creamy to pale yellow in colour, oval in shape ranging from 0.4 to 0.6 mm in size. The egg stage lasts from 7-11 days. The newly hatched larvae are brown in colour with black head and tapering at the posterior end which measures about 1.5 to 3.0 cm. The full grown larva is creamy white in colour and measures about 6.0 to 8 cm in length. The pupa is dark brown dorsally and with light yellow abdominal shade. The moth emerges through the hole leaving the pupal skin which protrudes from the exit hole. The most damaging stage of the pest is the larval stage. The larvae bore the main trunk of the tree and feed on the internal tissues gradually. The growth of the plant is retarded and as a result, quantity and quality of the foliage gets deteriorated. They make the tunnel which is of bee hole type and run up wards from the chamber of the surface of the sap wood and penetrates to the depth of several inches in the wood. An average 6-7 holes has been observed in one branch. They prefer to oviposit in the trunk base between 1-2 feet above ground level. After emergence, the larvae start feeding on the bark and damage the internal portion of the plant. The larvae are internal feeders boring large galleries in the branches of som and soalu plants.

Control

The infected plants are uprooted immediately. Spraying of monochrotophos 0.05 per cent on the infected plant is highly effective to control the pest population

Weevils (Curculionidae)

Various weevils damage the plantations of *Terminalia arjuna* and *T. tomentosa*. The most prominent among them are *Myllocerus undecimpustulatus maculosus*, *Myllocerus viridanus*, *M. transmarinus* Hbst, *M. discolor*, *M. subfasciatus*, *Crinorrhinus nobulosus*, *Atmelonychus peregrinus*, and *Apoderus tranquebaricus* Fab, *Xanthochelus faunus* Oliv.

Ash weevil ; *Myllocerus undecimpustulatus maculosus* West, *Myllocerus viridans* Fab. (Coleoptera : Curculionidae)

The ash weevil *Myllocerus undecimpustulatus maculosus* Desbr. is a notorious pest and damages wide range of plants including Arjun (*T. arjuna)* and Asan (*T. tomentosa*) on which tasar silkworm (*Antheraea mylitta*) larvae are reared. It damages in all the stages of plant growth, the adult damage the foliage and grub damages the root system. Damage due to this pest have been recorded up to 52–75 per cent, if adequate and timely plant protection methods were not adopted. The adult beetles are minute or large size (0.3 to 5 cm in length) and coloured dull brown, grey, black and a few green. The head is more or less prolonged into a snout which varies considerably in size, shape and length and bears the mouth parts at the tip. Weevils are recognizable by their rostrum and a elbowed antennae. Weevils have peculiar habit of shamming dead. When approached the legs and abdomen are folded close to the body and the insect drop to the ground. This is a valuable defense, especially in the thick vegetation of tasar food plants, the insect falling to the soil and being extremely hard to find. The ash weevil *M. maculosus* lives and reproduces through out the year in tropical tasar region of India. They are abundant in field during winter months, October to January. During this period mating is quite common. Newly emerged adults are sexually not active and requires a maturation period. Hence mating is initiated only after 2–3 days of adult emergence. The males live 25–30 days. During the entire adult period, female mates only 4–5 times while male mates more frequently. The adult female oviposits in leaves, leaf-rolls, stems or in the soil. The eggs hatch in to larvae. The apodus grubs are phytoyphagous and have a variety of diet. They feed in the leaf mines on rolled leaves, leaf or flower buds, stems or on the roots of the plants. Some cause galls on the stem of the plants. The entire life cycle may be completed on the different parts of the plants above ground or in the soil; in others the pupa lives in the soils. The grubs pupate inside the plants part in cocoons of plant fibres or silk or in the soil in earthen cells. The grubs remain in the soil feeding on roots of seedlings and destroy stem. One grub can destroy seedlings in 40 days. The adult weevils feed on leaves from edges.

Control

Spray BHC 50 WP in 150 litres of water per acre or dust the plantation with BHC 10 per cent at the rate of 10 Kg/acre. *Dinocamus mylloceri* Wlk. has been reported to parasitize the larvae of *Myllocerus* sp.

Grey weevil ; *Atmelonychus peregrins* Olive (Coleoptera : Curculionidae)

Grey weevil, *Atmelonychus peregrinus* Oliver is an important pest which damages the primary tasar food plants. The weevils appeared during the June and the population attained its peak during August September and subsequently the population decreased with gradual fall in the temperature since middle of the October. Adults were 10–12 mm long and 4–5 mm in width, colour black with uniform grey or brownish scaling, under parts paler. Head transverse and constricted behind the eyes. Fore head with a deep central furrow. Prothorax is broader than abdomen. Elytra gradually acuminate behind. Legs black, with grey scaling the femora rugosley punctuate. At severe state of infestation, the weevils consumed the entire lamina along with mid rib. They also devoured flowering shoots wholesome leaving only the bare stem. It is imago only which is responsible for, foliage loss.

Brown beetle ; *Crinorrhinus nebulosus* Marshall (Coleoptera : Curculionidae)

Adults of *Chrinorrhinus nebulosus* cause extensive damage to Asan plantation in tasar culture. The eggs are laid singly in the soft thinner bark portion of the plant. The eggs hatch into the larvae. They are soft elongate and white. The larvae are internal feeders. Generally, larvae are white, soft, legless grubs with distinct brown head and a much wrinkled body, which is fleshy and slightly curved. Pupation occurs in the plants. The newly emerged weevils are active insects, nocturnal and feeding on the leaves or other part of the plant or plant sap. It causes maximum injury during October–November.

***Auletobius densatus* Voss (Coleoptera : Curculionidae)**

It is a polyphagous pest. The adult weevil is about 4 mm long and reddish brown in colour. The mated female oviposits about 200 eggs over a period of two months. Eggs are kidney shape and laid in a small hole in the top shoot of the host plant After hatching the grubs girdles the stem. The tiny grubs hatch in 5 to 10 days. The newly hatched grubs are white, short, without legs. The head is distinct, hard, brown or yellow in colour. The larvae feed and grow within 4 weeks, eating out the pitch of the stem and, if this becomes exhausted, descending to the soil to complete growth by feeding upon the fibrous roots. The larvae pupate either in the growing branch or in the soil among the roots. The adults emerge a little before mid summer. They feed for a short time and then go into hibernation till mid summer. There are only one generation a year. The adult weevil congregates on the top of the plant and causes extensive damage by feeding on the bud of the growing plants. The tender leaves are also eaten after feeding on the bud. The pest causes maximum damage to foliage during monsoon period.

Control

Soil application of Aldrin 5 per cent or Chlordane 5 % @ 10 kg per acre will effectively control the grubs.

***Coccotrypes integer* Eichhoff (Coleoptera : Curculionidae)**

This scolytid insect infest acorns of *Q. acutissima* and *Q. griffithii* in north eastern India. It is very small insect about 1.4 mm length and dark reddish brown in colour. Wings are well–developed and are capable of flying to a considerable distance. The adult insects mate just before the onset of spring and eggs are laid near the flower of the plant. The eggs are small, white and spherical. The newly hatched eggs are short, fat, and legless. The young grub mines into the newly forming acorn seed and remains feeding upon the acorn until full grown. The mature grub is pinkish white in colour and about 10 mm in length. The grub chews its way out leaving a small pinhole sized opening in the tough covering of the acorn and pupates outside in the silk. But, in many cases the full grown grub pupates within the hollowed out seed case. On maturing, the beetle bores a round hole through the acorn cell and escapes during June and July. There is only one generation in a year. The damage is caused by the grub, the interior of the corn seed being partially or entirely hollowed out. Therefore, under the attacks the seed is completely destroyed.

Control

Foliar spraying of Phosphamidon 0.05 per cent is effective to minimize the scolytid population on the growing oak plants.

***Calendra sculpturata* Gyllenhal (Coleoptera : Curculionidae)**

This weevil is the most serious pest to the acorns of *Q. lanuginosa*, The adult is dark red brown in colour and measures about 5 to 6.6 mm length. It has a long curved snout, which is about one half the length of the body. The body is thicker near the base of the elytra and abdomen is broadly rounded. Adult weevil emerges from the acorns in the middle of June and starts feeding on the plant. The female with her long beak like mouthparts chews a small pin hole sized opening in the young new acorns and then turns round, inserts her ovipositor into the hole, and lays over one to several eggs in the nut.. The tiny grubs hatch in 4 to 9 days. They are white, short, chunky, humpbacked grubs, without legs and with a distant, harder, brown or yellow head. The larvae feed and grow for several weeks. The mature larva is nearly 10 mm long. Usually after the acorn drops in the fall, the larva chews its way out, leaving a much bigger hole than the oviposition hole, and digs down into the earth. The pupal stage lasts for five days and the adults appear in the autumn. Some larvae may stay in the acorn throughout the winter and enter the soil in the spring. The larvae pupates in the spring in a pupal cell in the ground, and the adult emerges any time from June through August. There are one or two generations in a year. The grubs burrows and feed in the fruits and causes considerable loss. It causes one or more holes in the acorn.

Control

Collection of the infested acorns after the fall and burning is usually very effective.

Leaf roller; *Apoderus notatus* (Fabricius) (Coleoptera : Curculionidae)

It is common leaf roller of oak plants. Adult is a medium sized, reddish brown weevil with a long snout . It has the curious habit of cutting and twisting the oak leaves . It occurs in February to March and July to August. There are apparently two broods in a year. The female usually inserts her eggs in the rolled leaves. Eggs are oval yellowish, 2 mm long and hatch after 4 days. The small legless, yellowish grub feeds inside the roll and finally pupate within 10 days. The pupa, is bright yellow in colour and about 3 mm long. It is enclosed in the folded leaf in the midst of black powdery excretal matter left by the grub. The adult bores its way out of the rolls. Adult female cuts across near the base of the leaves, crossing the midrib from one margin only. The leaf is then folded longitudinally, and the tip is rolled in the rolling process. It forms a compact cylindrical mass, consisting of tightly rolled and folded leaf blade, with the egg in the center. Thus the entire leaves are damaged.

Control

Collection and killing of adult weevils and their roll from the plant reduces the pest population.

Leaf roller; *Apoderus tranquebaricus* Fab. (Coleoptera : Curculionidae)

Leaf roller *Apoderus tranquebaricus* has been reported to cause extensive damage to the leaves of tasar food plants (Singh *et al*, 1994). The weevils destroy the entire flush of the new leaves either by eating or by rolling them in the nursery or young trees. Leaf rolls are constructed by females after making certain cuts in the leaf blade and the egg is laid in the center of the roll. The entire development of the weevil occurs in the rolls which fall on the soil. The life cycle is very brief constituting of 13–15 days viz. 3 days of embryo genesis, 5 of larval and 5–8 days of pupal period where after the weevils emerge. The formation of characteristic rolls themselves cause the loss to the leaf quality and quantity.

Red pumpkin beetle; *Aulacophora foveicollis* Lucas (Coleoptera : Chrysomelidae)

It is a polyphagous insect which feeds on variety of plants. The pest has been reported to cause severe damage to *T. arjuna* and *T. tomentosa* plantation. The grubs are creamy white, with a slightly darker oval shield at the back. When fully grown they measure about 12 mm in length and 3.5 mm across mesothorax. The beetles are oblong and 5.8 mm long. Their dorsal body surface is brilliant orange red and the ventral surface is black, being clothed with white hairs. The beetles resume activity as soon as the season warms up and lay about 300 oval, yellow eggs singly or in batches of 8-9 in moist soil, near the base of the plant during their life span of 60-85 days. The incubation period lasts 6-15 days. The larval and pupal period are 13-25 and 7-17 days, respectively. The pupation takes place in the thick walled earthen chambers in the soil at a depth of about 20-25 cm. On emergence, the beetles begin to feed and breed. The life cycle is completed in 26-37 days. There are five generations from March to October. The grubs lead a subterranean life feeding on roots of the tasar food plants. The beetles are very destructive to soft and succulent leaves of *T. arjuna* and *T. tomentosa* during April-June.

Control

Deep ploughing of the soil to kill the grubs. Apply 2.5 kg Carbofuran 3G per acre 3-4 cm deep in the soil near the bases of the plants

Besides this some other chrysomelid beetle *Tricliona picea, Tricliona pulvis, Colasposoma semicostatum* Jacoby and *Chrysomela* sp. are prevalent in tasar ecosystem (Singh *et al.*, 1994). The larvae of chrysomelid are leaf skeletonizer where as adult consumes much of the leaf tissues. The grubs feed on the secondary and tertiary rootlets leading to wilting and stunted growth. Estimated loss due to sucking the chlorophyll content of matured/semi–matured leaves of tasar food plants ranges from 30–40 percent. The incidence is very high during July–August. The beetles appear during July and voraciously scrap the chlorophyll content of plants and make the leaves unsuitable for tasar silkworm. Mainly the adult that feeds extensively on the leaves causes the damage.

White grubs; *Holotrichia consanguinea* Blanchard (= *Lachnosterna consanguinea* Blanchard) (Coleoptera : Melolonthidae)

Among coleopteran defoliators May–June beetles, *Holotrichia consanguinea* is the most serious pest of mulberry tasar and muga plants. It is polyphagous pest distributed through out India. The beetles are active at night and hide in the soil during day time. The females enter the soil and deposit the eggs vertically in a row. One female lays about 50-60 eggs. The larvae emerge after 8-10 days by rupturing the egg shell in a zig–zag manner. The full-grown larva is C-shaped with three pairs of prominent thoracic legs. The abdominal segments being translucent, the intestinal contents are clearly visible. The larval period varies from 8-10 weeks. Pupation takes place in the earthen cell and pupal period varies from 12-16 days during October–November. On emergence from pupal case, the adult has a soft, creamy yellow abdomen and thin crumpled, pinkish, papery elytra which later become hard and stiff and change gradually to dark brown Adult longevity varies from 80-90 days.

Control

Soil application of BHC 10 per cent @ 100 kg. per ha mixed with equal quantity of farm yard manure during May–June, or Quinalphos 5 G @ 2.5 kg a.i/ha against grub is recommended. Spraying the host trees with 0.05 per cent monocrotophos 40 EC or 0.1 per cent lindane 20 EC is effective against adult beetles.

May–June beetle; *Anomala blanchardi* Blanchard (Coleoptera : Scarabaeidae)

Anomala blanchardi Blanchard (Coleoptera : Scarabaeidae), commonly known as May–June beetle is nocturnal in habit and defoliate large number of plant in summer. The damage is most severe on the marginal trees of the plots. The damaged leaves present a characteristic sieve-like appearance due to shoot holes. Presumably, the grubs feed on humus, roots of grasses, weeds and other forest plants. There may be one or two generation in a year. The adult emerges from the soil after first shower of rain in the evening and immediately fly to their host plant in swarms and start feeding on the foliage. The congregation of beetle starts from the top of the tree and move downwards with the result that defoliation of the leaves also takes place in the same manner. The beetles leave the host plants in the early morning next day and go back to soil. Only the adult beetles are foliage feeders but the larvae too damage considerably.

Control

Soil application of BHC 10 per cent @ 100 kg. per ha mixed with equal quantity of farm yard manure during May–June, or Quinalphos 5 G @ 2.5 kg a.i /ha against grub is recommended. Spray the host trees with 0.05 per cent monocrotophos 40 EC or 0.1 per cent lindane 20 EC to kill adult beetles (Singh *et al.*, 1992).

Defoliating insects

Defoliating insects are recognized by their feeding habits. They skeletonise the leaves consuming the cell walls and its watery cell contents. The loss of leaf tissue reduces the food making capability resulting in the weakening and stunting of the growth of host plants. A magnitude of insects from different groups specially moths, beetles and weevils are the most injurious and causes reduction in the quality and quantity of the leaves. This category includes insects feeding on the terminal parts (twigs, tips or shoots, small rolls, seeds and cones), wood of the trunk and branches. They seldom kill the trees, but they can reduce the rate of terminal growth or cause deformities. The borers attack the trees by cutting large tunnels through the tissue of oak. Mulberry and Non-mulberry silk host plants are attacked by large number of defoliators. The most important group of insects which causes maximum injury to the plants are defoliating insects (Singh and Kulshrestha, 1989, 1990). Feeding injury to the leaves by defoliators disrupt the production, utilization, regulation of growth and survival. When the defoliation is severe and continuous, death of entire plant occurs where as less frequent defoliation causes reduced growth and branch drying. Non–mulberry silk host plants are defoliated generally by caterpillars of moth, leaf beetles and chafers. In the field number of species of lymantriids are pests of tasar food plants. The common species are *Euproctis fraterna, E. lunata, E. lutifacia, Dasychira horsefieldi, Notolophus posticus* and *Orygia antiqua.*

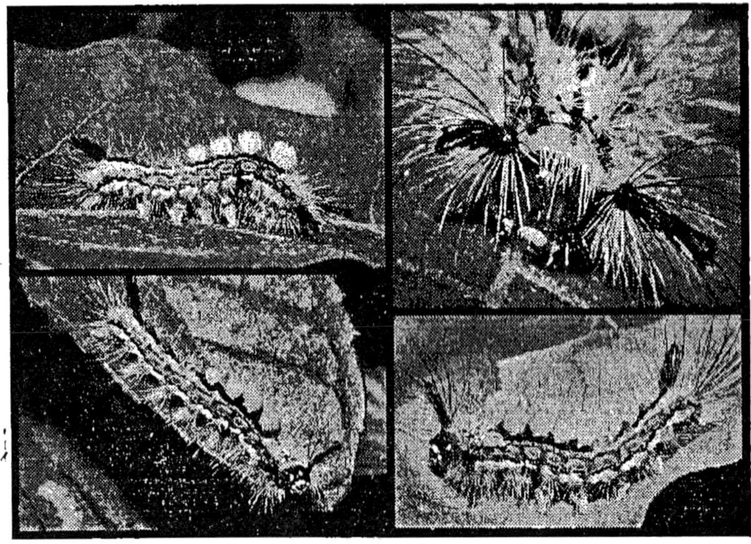

Fig 4.3 : Caterpiallrs of Vapourer Tussock moth

Vapourer Tussock moth ; *Notolophus antiqua* ((Lepidoptera : Lymantriidae)

Vapourer Tussock moth *Notolophus antiqua* is the most important defoliating pest of primary tasar food plants in tropical tasar region of India (Singh *et al.*, 1991). It is a sporadic pest and frequently occurs in large numbers that it seriously injuries the foliage of the plant. Some times the damage is so serious that the plant may be completely defoliated by the caterpillars. During severe infestation the saplings and young plants are completely consumed and the growth of the plant is checked. The moth emerges from the cocoon by making a round hole. The newly emerged moth is active and moves about waving its antennae. The moth is white to brownish in colour. The forewings are crossed by two deep brown bands and have a conspicuous white spot near the anal angle. The hind wing is comparatively paler than the forewing bending to darker near the edges. The colour of the abdomen is similar to that of the fore wings. The colour of the body is white, clothed with yellowish white hairs and resembles a grub more than a moth. The female lives in the cocoon made during the pre-pupal period. The female emerges from her cocoon and after pairing lays her eggs in masses upon the surface of the cocoon which remains attached with the leaves of primary tasar food plants. During the oviposition period, some hairy scales are attached with the eggs and mingle with a viscid secretion which is deposited along with the eggs and which, on drying, becomes hard and brittle. Each female lays 311 to 553 eggs. It takes 30 to 45 minutes to complete the oviposition. After oviposition, the female dies. The eggs are small and dorsoventrally flattened. The freshly laid eggs are white in colour. The freshly emerged larva possesses a prominent light red head and brownish black body colour with two yellow stripes on each side. All body segments bears small–setae. The young larva on the hatching has the power of spinning with a thin silk thread with which it lowers itself from the resting place when disturbed and by means of which it regains the place from where it had dropped. This power is lost as the insect develops from successive moults. The young larvae start moving in search of food. It prefers the tender leaves of the food plants. They first penetrate their head under the surface of the leaves and start feeding just below the epidermal layer more or less like leaf minor. There are five larval instars. The larval period varies from 20–40 days. The mature caterpillar is striking and hides during scorching sun beneath the leaf. The head is brilliant vermilion in colour. The body is yellow banded with white and adorned with black tipped tuffs and bundles of black colour hairs. The caterpillars pupated in cocoons on leaves of *T.arjuna* and *T. tomentosa*. During the pre-pupal period, the caterpillar gets contracted in size and changed their colour to yellowish creamy. The pupal period ranges from 9–12 days (Singh *et al.*, 1993).

Hemerocampa leucostigma **(Lepidoptera : Lymantriidae)**

It is a polyphagous pest and feeds on *Terminalia* sp. It is important pest of castor plant and som plant also. The moth emerges during night. The female is sluggish and cling to cocoon. The eggs are laid on the surface of cocoons in an irregular mass. A mated female oviposits more than 600 eggs in her life time. The eggs are creamy colour and spherical in shape. The incubation period is of seven days. The male undergoes four moults whereas female undergoes five moults. The full-grown larva has a brown head, a pair of long pencils of hair pointing forwards from the prothorax. The female larva is more robust and bigger than male and feeds voraciously on *T. arjuna* plants. The larva pupates in a transparent cocoon inside leaf fold. Pupal period is about 7 days. The wings are brown in male whereas the female moth possesses only vestigial wings. The total life cycle lasts for 27 to 33 days.

Hairy caterpillar; *Euproctis* sp .(Lepidoptera : Lymantriidae)

Euproctis is a polyphagous and sporadic pest of castor plant but it has been found to damage som plants also. There are several generations in a year. The moths emerging in March, lay large number of eggs in masses on the lower side of the leaves covered with female anal tuft of brown hairs. The eggs are flat, circular and yellow in colour. The incubation period is 5-7 days. After hatching the young larvae, feed gregariously for the first few days. Later on they feed individually. The larvae are full grown in 14-20 days. The last instar larvae make loose silken cocoons in the plant debris laying on the ground and pupate in the soil. The pupal period is about a week in summer. The egg, larval and pupal stages may last 10, 85 and 20 days respectively during the winter season. The larvae cause damage by feeding on the leaves. In case of severe infestation, there may be complete defoliation of the plants. The damaged plants remain stunted and there is a considerable reduction in the quantity and quality of leaves.

Control

Collection and killing of egg mass and caterpillars during rearing period is effective in minimizing the pest population. Spraying of malathion 0.05 per cent is quite effective in case of severe infestation. It is generally conducted keeping in view of the safe period for rearing of silkworms.

Tussock caterpillar ; *Euproctis fraterna* (Moore). (Lepidoptera : Lymantriidae)

Tussock caterpillar *Euproctis fraterna* is a sporadic pest of oak plants in India. The adult moth is stout and yellow with weavy marking on wings. Two clear dark spots are present on the apical angle and two larger ones on the cubital angle of the forewings of the female and an additional dark one near the costal margin in the forewing of the male. Females lay eggs in batches of 100 to 120 eggs on the lower surface of oak leaves and covered with buff coloured hairs from the anal tuff of the female. The egg is creamy yellow, circular, flattended and lasts for 5 to 8 days. The eggs are flat, circular and yellow and are laid in masses covered with yellow hairs on the under surface of the leaves. The incubation period varies from 6-9 days. The early instars larvae have whitish hairs and feed gregariously. There are seven larval instars, sometimes six and the larval duration vary in different seasons. It takes about 28 to 51 days. The newly hatched larva is about 2.5 mm long, slender, hairy and yellow in colour with dark tufts of hairs on the second and third abdominal segments. The full grown larva is stout, 20–25 mm long, dark reddish brown on the ventral surface; head and prothoracic shield bright orange red, thickly covered with tufts of whitish hairs and with a pair of large tufts of darker hair directed forward on either side of the head and a single similar tuff directed backwards from the anal segment. The mature larvae pupate in a thin cocoon interwoven with the larval hairs in leaf folds or up to 5 cm below the soil and pupate inside. Pupal stage lasts 12 to 18 days. The total life cycle varies from 52 to 82 days. Over wintering takes place in the larval stage. Feeding continues up to December and it is resumed for a short period before it pupates in February. The caterpillars feed gregariously on the surface tissues of oak leaves and the older leaves. Maximum damage is caused during March to August.

Control

The caterpillars and moths should be collected and killed. Dusting of 10 per cent BHC or 5 per cent Aldrin @ 10kg per acre is highly effective. Spraying of 250 ml of Methyl parathion 50 EC or

500 ml of Endosulfan 35 EC in 250 litres of water per acre is effective in containing the pest population.

Castor Hairy caterpillar ; *Euproctis lunata* (Walker) (Lepidoptera: Noctuidae)

The castor hairy caterpillar *Euproctis lunata* is widely distributed in India along with an allied species Euproctis *fraterna*, which is also a serious pest of castor plants. Apart from castor it has been reported on *Terminalia arjuna*, *T. totmentosa* and som amd soalu plants. Moth emerges in the February, lays large number of eggs in masses on the under surface of the leaves covered with the female anal tuft of brown hairs. They hatch in 5–7 days and the young larvae feed gregariously for the first few days. Later on they disperse and feed individually. The larvae are full-grown in 15–20 days after passing through six stages of their development. The last instar larvae make loose silken cocoons in the plant debris lying on the ground and pupate inside. The pupal period is about week in the summer. During winter months the egg, larval and pupal stages may last 10, 15 and 20 days respectively. There are several generations in a year of this pest. The larvae cause damage by feeding on the leaves and in case of severe infestation they may cause complete defoliation. The attacked plants remain stunned and produce very little seed.

Control

Dusting of BHC 10 per cent or 5 per cent Aldrin @ 10 kg/acre is highly effective. In case of severe infestation, spray 250 ml Methyl parathion 50 EC, or 500 ml endosulphan 35 EC in 250 litres of water per acre.

Gypsy moth; *Lymantria plumbalis* Hampson (Lepidoptera : Lymantriidae)

Various species of *Lymantria* has been reported to cause damage to Oak tasar food plants viz. *Lymantria plumbalis, Lymantria beatrx* Stoll, *Lymantria grantrix* Walker and *Lymantria mathura M*oore. However, the gypsy moth, *Lymantria plumbalis* is most prevalent in the north–eastern region of India. The female moth has scale like aborted wings and is unable to fly but walks fast where as the male moth has functional wings and can perform its normal flight. Adult females are whitish and males are dark brown.

The female moth oviposits 400 500 eggs . Eggs are covered with buff coloured hairs on trunks. The incubation period varies from 4–5 days. The first instar larvae after hatching become established in the crown of the tree and remain there until the fourth instar. Caterpillars feed during the morning and after noon period. During day hours the final instar larvae migrate towards the base of the plant and again starts feeding at night on the crown of the plant. Pupation occurs during July and adults emerge in July and August. The larvae are voracious feeders. The feeding commences from the margin of the leaves and generally the tender leaves are preferred but matured caterpillars cause damage to matured leaves. The young larvae make small holes in leaves and start feeding. Older larvae feed on leaf edges, consuming entire leaves except for the larger veins and the midrib. The entire tree may be defoliated in case of heavy attack.

Control

Control by natural enemies and the use of chemical and microbial insecticides have been used extensively in heavy infestations. Otherwise mechanical control is generally implemented.

***Phalera assimilis* (Bremer & Grey) (Lepidoptera : Notodontidae)**

The adults are medium sized, about 10 cm. across the wings, stout moths with buff colour. The forewing is having silvery grey markings on basal and inner areas. The caterpillars feed gregariously on the oak leaves during July and August, and completely defoliates the small trees. Adults emerge in the last week of June or first week of July from the diapausing pupae and live for 2 to 7 days. The adult female oviposits on the underside of the leaves in masses of 200 500 eggs. The hatched larvae are orange red in colour and start feeding at first on a single leaf with their heads all pointing towards the outer edge of the leaf. The caterpillars moult five times during the period of about 40 days. The full grown larvae (about 7.0 cm long) are black with reddish coloured head and posterior end. Body setae are long, white or pale yellow in colour. The mature caterpillars enter the ground and changed into pupa. There is pupal diapause and the whole winter season is completed as pupal stage. There is only one generation in a year. Larvae are gregarious. At first the young caterpillars skeletonise the leaf, but within a few days, they increase in size and of necessity spread over a number of leaves, sometimes all the leaves of a single twig or small branch, and begin consuming the entire leaf. During feeding, they migrate from one part of the tree to another after completely defoliating small trees, or single branch on large trees.

Control

Mechanical collection of egg masses is quite effective. Spraying with Malathion, 50 EC and Parathion, at the rate 0.05 per cent kills the caterpillars. Waiting period of 15 days should be maintained for the rearing of silkworms

Tiger moth; *Argina argus* Kollar (Lepidoptera : Arctiidae).

The adult moths are orange coloured with black spots and patches on the wings. The eggs are laid in masses on the ventral surface of leaves. The eggs hatch in 3–4 days. Newly hatched caterpillars feed voraciously on the tender leaves. They rapidly grow larger, undergo five instars and within a fortnight they are fully grown. The full grown caterpillars are 25 mm long, pale in colour with a black line along the side. There is an irregular black patch and a yellow line along the side of the body. It pupates in the soil and forms a light cocoon around its body. The moth emerges after 4 to 7 days. Total life cycle is completed in 25 to 30 days.

Control

Mechanical collection of egg masses is effective. Malathion 0.05 per cent can be sprayed to kill the caterpillar.

Castor semi looper; *Achaea janata Linn.* (Lepidoptera: Noctuidae)

This is one of the most important pests of castor in India. This is a polyphagous insect, which feeds on several ornamental and fruit plants also. The adult of *Achaea janata* is pale reddish brown moth with a long wing. The wings are decorated with broad zig–zag markings, a large pale area and dark brown patches. The fully fed larva is dark and marked with prominent blue black; yellow reddish stripes and has a pair of reddish processes and also a dorsal hump near the hind end of the body. The colour of the later instar is variable. After mating the female lays eggs on the undersurface of the leaves singly. The eggs are round and blue green in color. A female can lay up to 450 eggs during its life span. The egg size is fairly large about 1 mm in length. The larva emerges by cutting a hole in the egg shell in 2 5 days and devours it immediately. Then it starts feeding on foliage. The larva is full fed in 15 20 days and moults 4 5 times. The grown up larva prepares a loose cocoon of coarse silk and some soil particle and pupates in the soil or among fallen leaves. The pupal period varies from 10 15 days and the moth on emergence, feed on the soft and succulent leaves of castor. The pest completes 5 6 generations in a year.

Control

Application of 1 kg of DDT/BHC 50WP or 600ml of Dizinon 20 EC or 250ml Methyl parathion 25 ECin 250 litres of water per acre is effective to control the pest population.

Noctuid moth ; *Hyblaea puera* Crammer (Lepidoptera : Hyblaeidae)

Hyblaea sp. is one of the most important defoliator of oak plants in north–eastern and western region of India. The moth superficially resembles noctuids but they have prominent maxillary palps and the larva is pyraloid. It is nocturnal in habit. The wings are bell shaped and creamy brown and grey with a darker cross band. Female oviposits singly on the ventral side of young leaves. Eggs are oblong and yellowish or greenish in colour. The eggs hatch after February and the hatched larvae start feeding by rolling the leaves. The full grown larva is about 30 mm long, stout, cylindrical and, usually greenish below and dark brown or blackish above the body. Feeding is completed in 25-29 days and pupation occurs in a rolled leaf, amongst dead leaves on the ground. The adult emerges after 15-20 days. The total life cycle is completed in 30-40 days. There are 2-3 generations in a year. The larvae are voracious feeder on the leaves of oak plants. They roll the leaves and cause damage. Maximum damage is caused during July–August.

Control

Mechanical collection of egg masses is effective. Malathion 0.05 per cent can be sprayed to kill the caterpillar.

Camptoloma tigrinus **Hampson (Lepidoptera : Noctuidae)**

Camptoloma tigrinus is a common pest of oak plants in the north eastern Himalaya and Burma. The adult is of moderately large size with dark grey stripes and spots on the wings. Adults emerge in April May. Eggs are laid on the lower surface of leaves in clusters of 100–200. Eggs are covered with hairs from the anal tuft of the female. The egg is creamy yellow, circular and flattened. They hatch in about 3 to 4 days. Larvae are gregarious and live together inside a common pouch like nest. Generally, the larva hibernates during young larval stage (particularly in the third instar) inside the nest and becomes active in spring (March April). The mature final stage caterpillars measure about 9 cm. and cause extensive damage to plants. Larvae stay in the nest during day time and come out to eat leaves at night. The final instar larvae spin silken yellowish cocoons inside the nest, or sometimes they walk down to the ground and spin cocoons between the fallen leaves. It pupates inside the cocoon and the pupal period lasts for about a month. The young larvae skeletonized the leaves where as the older larvae devour all except the leaf stalk.

Control

Nests should be collected and destroyed immediately.

Slug caterpillar; *Parasa pastoralis* Butler (Lepidoptera : Limacodidae)

It is a serious pest of sporadic occurrence on oak plants. The moth is short with stout forewings, predominantly green in the middle and brownish at either end. The young larvae feed gregariously by scraping the under surface of leaves and cause extensive damage to oak plants. As they grow they get scattered and feed on the entire leaves and cause eventual drying of foliage. Eggs are laid in batches of 20 to 30 on the under surface of the leaves. The caterpillars hatch out in 6-7 days and starts feeding on the leaves. The larvae become full grown in 6 weeks and measures about 2.5 cm in length. The larva pupates in a shell like compact, elliptical cocoon having upper surface convex and lower surface flat. The cocoons are seen attached to the food plants. Pupal period is about three weeks. The life cycle is completed in about 60–70 days.

Control

The defoliators like armyworm, leaf rollers, leaf miners, bagworms, hairy caterpillars, grasshoppers, beetles and weevils are of sporadic occurrence and considered to be of minor importance. As such they do not warrant any control measures. They are polyphagous insect infesting wide variety of plants rarely assuming epidemic proportion on oak plant. The pest can be controlled by spraying Dimethoate 50 EC @ 0.05 per cent if necessary.

Slug caterpillar ; *Parasa repanda* walker (Lepidoptera : Cochlidiidae)

It is commonly known as slug caterpillar. It is gregarious caterpillar which skeletonized the complete plant . Larvae consume the epidermis, making the leaf transparent. The leaf is left with a fine network of veins. Defoliation starts in the upper crown in early summer and progress downward. By late summer, heavily infested trees may be completely defoliated. The adults are medium sized and forewings are green with big brown patches. They are usually endemic on the oaks. Adults emerged in the last week of May or first week of June from diapausing pupae and live for 2 to 7 days. Females oviposits in single rows on the lower surface of the leaf . Eggs are flat, scale like, greenish and covered by a transparent cement secreted by the female. Incubation period lasts for 3 to 5 days. The larvae are pale green, whitish or bright yellowish green above. There is a dorsal green band and short spinous tubercles on the body. Mature larva pupates in the spring. There are two to three generations per year.

Control

Mechanical control measures may be applied to control the pest population in the field. Spraying of Malathion 50 EC at the rate of 0.05 per cent is effective.

Castor slug; *Parasa lepida = Latoia lepida* Green (Lepidoptera : Cochlididae)

Slug caterpillar is a serious pest of sporadic occurrence on castor plants. The full-grown larva is flat, fleshy and greenish in color with white lines on body. It is covered with spines having red and black tips. The moth is short and stout with fore wings predominantly green in middle and brown at the ends. The female moths lay flat shinning eggs in batches of 15–35 on the undersurface of the leaves. The larvae hatch in 7 days and start feeding in clusters on young leaves. The larva passes through five instars and is full grown in 40–45 days. It then pupates in hard shell like greyish cocoons attach to the stem of castor plant or on the tree trunk of the infested trees. The pupal period is about 20–28 days. The life cycle occupies about 10 weeks. The activity of this pest is seen through out the year. The young larvae feed gregariously by scraping the undersurface of the leaves and cause drying up to the foliage. As they grow, they get scattered and feed on entire leaves and cause defoliation of the host plant. Besides this some minor pests have also been reported which causes damage to silk host plants.

Control

Dusting 5 per cent BHC or 5 per cent Aldrin @ 10 kg/acre or spraying 250 ml Methyl parathion 50 EC, or 500 ml Endosulphan 35 EC in 250 litres of water per acre.

Castor seed and capsule borer; *Dichocrocis punctiferalis* (Lepidoptera : Pyralididae)

This is the most common borer of the castor plant and distributed through out the India wherever caster is grown. The damage is caused by the caterpillars, which bore into the stem of the young plant and ultimately into the capsule. The full-grown caterpillar is reddish brown with black blotches all over the body and a pale stripe on the lateral side. It measures about 25–30 mm in length. The moths are orange yellow, with black markings on both the wings. The moths lay eggs on the leaves and other soft part of the plant. The incubation period is of 7 days. The larvae pass through 4–5 instars and full grown in 2–3 weeks. Pupation takes place inside the seed or sometimes in the frass. The pupal period is about 7 days. The life cycle is completed in 30–37 days. There is 3 generation in a year. The active period for the pest on castor is from September to March.

Control

Collecting and destroying and dusting with infested shoots and capsule with 10 per cent BHC @ 10 Kg/acre or spraying 1 Kg Cabaryl 150 WP or BHC/DDT 50 WP in 250 litres of water per acre repeats at 15 days interval.

Flea beetle; *Cleoporus lefevrei* Div (Coleoptera : Alticidae)

These are small jumping beetles, black, blue, greenish or brownish in colour. The adult is characterized by closely approximated antennae, globose fore coxae, deeply bilobed third tarsal segment, grooved tibiae and swollen hind femur. Adult is about 3.5 mm long and greenish blue in colour. This beetle jumps heavily when approached. During spring the insects become active as soon as vegetation is well started. Eggs are laid in one to three batches of 12 to 44 eggs under the bark of the oak plant. Eggs are yellow in colour and narrow. The eggs hatched into the black grubs with three pairs of small legs. Both the beetles and the grubs feed on the leaves cause damage to the oak plant. Pupation takes place inside the soil. The winter is passed as full grown beetle and there is only one generation in a year. The adults make very small holes in the green portion of the leaves, giving the whole plant a bleached appearance; growth is retarded, and the leaves wilt even during wet weather.

Control

Phosphamidon 0.05 per cent is highly effective to control the beetle population in case of severe infestation.

***Cricula trifenestra* Helf (Lepidoptera : Saturniidae)**

Cricula trifenestrata (Saturniidae:Lepidoptera), produces a beautiful golden cocoon and inhabits throughout Indonesia and elsewhere in South Asia. It is major pest of mango in Indonesia. It has also been reported to cause extensive damage to som plants in India. The caterpillars are black with white spots and hairs; the head and abdomen are bright red. They grow to be 60 mm long and the mature spin golden mesh-shaped cocoons. The full-grown caterpillar feeds on the foliage of som and soalu. The pest is active from June–September. The matured larvae wander on the leaves and branches of the food plant, and start to spin the cocoons. They make cocoons at random, single, twin and grouped.

Moth emerged at night and male emerged earlier than female. Pairing of moths started in late morning and mating period varied from 3 to 5 hours. In field conditions, eggs were deposited in clusters of 3-4 compact rows on the upper margin of the leaf, while in captivity; the eggs were laid in batches. The fecundity per female ranged from 264-410 and 226-347 eggs under the field condition and in captivity respectively. The duration of egg incubation, larva, pupa, longevity of male moths, longevity of female moths and total life cycle were 10-20, 25.78, 16, 9.4, 8.5 and 53.97 days respectively during September –October, 13.09, 35.12, 25.90, 11.4, 10.28, and 78.31 days, respectively during November- January and 13.82, 45.75, 16.40, 10.15, 7.7 and 78.07 days respectively during February April. Five generation were observed in a year. Eggs hatched mostly during morning hours and hatchability varied from 90 to 95 per cent. Among five larval instars, the first and second were highly gregarious. Caterpillar devoured the entire leaf blade except the mid rib and defoliated the plant giving a broom like appearance. Leaf infestation by pests gradually increased from July and reached the peak by December. The pest showed a decreasing trend infestation from January onwards. No incidence and infestation of the pests were observed during May–June generally. Caterpillar's lives upon the som and soalu tree and is clothed in poisonous spines and therefore, dangerous to handle. It is common belief that if the mouth touches any portion of the human body, that part will decay in leprosy; the caterpillar is accordingly feared and nothing is done to check it, though it wholly defoliates the som tree.

Control

Eggs and caterpillars should be collected and killed. Phosphamidon 0.05 per cent is effective controlling the pest. Endosulfan 0.05 % caused the highest larval mortality. Deltamethrin 0.001 per cent and azadirachtin 0.0015 per cent were found effective against the caterpillar.

Leaf minor ; *Phytomyza sp.* (Diptera : Agromyzidae)

Leaf minor *Phytomyza* sp. is a common pest of muga food plants. This pest is widely distributed in Assam and causes extensive damage by tunneling in the leaves of som and soalu plants. It is a polyphagous pest and there are several generations in a year. The pest is generally active from December–May. The rest of the period passes in the soil as pupa. The adult oviposits singly in the leaf tissue. The incubation period is 2-3 days. After hatching the larvae feed between the lower and upper epidermis by making zig zag tunnels and pupate within the galleries. Pupal period lasts for 5-6 days. One generation is completed in 13-15 days. The large number of tunnels made by the larvae interfere with photosynthesis formation which ultimately affects the quality leaf production. The silkworm avoid feeding on such type of leaves.

Control

Spray of 125 ml of phosphamidon 85 WSC or 400 ml of Dimethoate 30 EC in 250 litres of water per acre twice at 15 days interval is used. After spray 20 days waiting period is observed for brushing the worms on the treated food plants.

Bagworm; *Acanthopsyche bipar* (Lepidoptera : Psychidae)

Bagworm *Acanthopsyche bipar* is a very common species in tropical tasar region of India. The females are apterous whereas males are winged. Wings are thinly clothed with imperfect scales and exceptionally large frenulum. Antennae are bipectinate. Labial palpi are short. Some females are devoid of antennae, mouthparts and legs. Larvae inhabit cases, which exhibit great variety of shape and of materials used in their construction. Besides this rusty tussock moth *Orygia antiqua*, white marked tussock moth, *Hemerocampa leucostigma*, looper caterpillar, leaf roller caterpillar and bagworm are important defoliators. Certain tent caterpillar viz. *Malacosoma* sp. also damages primary tasar food plants. Some caterpillars have the distinctive habit of constructing small portable cases consisting of silk and leaf parts. One larvae occupies each case and only when feeding or moving they partially emerged and exposed the anterior parts of their bodies. Various species infest *T. arjuna* and *T .tomentosa* plants but Acanthopsyche moorei is very common. Further, *Eucelis critica* and *Archips conflictana* (lepidopera: tortricidae) are important leaf rollers. The larvae cause rolling on skeletonizing. Besides this certain lymantriidae, notodontidae, limacodidae and pieridae caterpillars have been found to damage leaves of non mulberry silk host plants. Among hymenopteran defoliators sawflies and ants are very common. The females of Saw fly has a saw toothed ovipositor. This structure is used for cutting slits in plant tissues in which eggs are placed. There are certain species of ants that are very injurious to young tree seedlings. Most of the damage occurs in the early spring.

Semilooper; *Plusia* sp. (Lepidoptera : Plutellidae)

Plusia sp. is a very common semilooper polyphagous insect and attack on the large number of the plants. They cause damage by biting round holes into leaves. The semiloopers bear three pairs of prolegs on the fifth, sixth and tenth abdominal segments. The larvae are known as semilooper because as they crawl they form a semiloop in their body and then they stretch straight. These larvae are normally defoliators and are pests of non mulberry silk host plants. These insects are active during winter season.

Control

Spraying of Dimethoate 0.05 per cent during infestation kills the pests.

Sap Sucking insects

Among sap sucking insects aphids, leaf hoppers, bugs, scales, white flies, thrips, mites and cicada are very common. The adults and nymphs cluster on the plant parts and sucks the sap of the leaves.

Aphids (Hemiptera : Aphididae).

The aphids are one of the major pests of som, soalu and oak tasar silk host plants. They are of minute size with winged and non-winged forms The pest is active from November to April in Northern India. Nymphal development is completed in 11–14 days for non-winged forms and in 14–19 days of winged forms, the growth being quickest in March. From the middle of November the population starts increasing progressively and is highest in the March, declining in April as the temperature warms up. There is increase of winged forms from December onwards, with a peak in March when about 90 per cent of the adults become winged. The pest multiplies most rapidly in late spring with the increase in temperature, its population declines. The aphid suck cell sap from tender leaves which ultimately affects the leaves quality The honey dew secreted by these insects becomes the source for development of various disease in the plants. Various species of *Eutrichosiphum* aphids have been reported to attack the oak plants in north western and north–eastern region of India viz..*Eutrichosiphum garhwalense, E. pyri, E. taoi, E. tapatii, E. khasyanum, E. alnicola, E. assamensis, E. neoalnicola, E. querciphaga, E. pasaniae and Greemidia (Trichosiphum) anonae*. Aphids are small, soft bodied insects with pear shaped bodies. They have a pair of cronicles called siphunculus at the posterior end of abdomen. These aphids may be either winged or wingless. They may be yellow, green, brown, or black, with darker pigmented blotches on the abdomen, and dusky bands on the wings. Oak aphids extract the sap from various soft parts of the plant viz. top shoot apices, leaves and smaller branches. They are normally gregarious and often found in the enormous numbers. Free living solitary forms may also occur as parthenogenetic adults. All aphids defecate honeydew through the anus, which drops frequently on the leaves or on articles under the tree and becomes sticky. Usually the sweet honeydew forms a very suitable media for the growth of *Chaetophoms guercifolia an* oak. fungus, and makes it to be very unsightly. Ants and other insect are very much fond of this honeydew, and the ants remain busy with these insects to collect this liquid. The ants care for the aphids and will even transport aphids to a more succulent part of the plant to ensure the flow of honeydew. it is because of this reason that honeydew producers are sometimes called 'ants cows'. These ants are popularly referred to as "aphidocolous ants and they protect the aphids from their insect enemies and sometimes provide shelter inside their hives during unfavorable conditions of environment. In controlling the aphids, therefore, it is also desirable to apply control measure for ants, such as spraying around and on the trunk of the tree. During January, approximately after two weeks of the sprouting on *Q. acutissima* and *Q. griffithii* aphids attack on the shoot. The most favorable feeding site of these aphids are shoot apices and the young foliages of oak plants.

Indian oak aphid ; *Cervaphis quercus* (Hemiptera : Aphididae)

The Indian oak aphid, *Cervaphis quercus* (1.66 - 2.08 mm long) is characterized by the yellow colour of the body and the branched projections at the posterior end of abdomen. Each branch of which bears an apical seta, except for the terminal process. Clusters of these aphids infest the underside of young leaves, leaf stalks, shoot apices, and feed along the leaves which turned yellowish and then wilted, crinkled and sometimes fell. Every leave on a tree may be curled and distorted due to heavy infestations. Though this aphid occurs throughout the year on *Quercuss acutissima*, it rapidly increased in number during February to May . This is the period during which the host plant actively produces new leaves and twigs. In the later part of the year, when the climate is very dry and the host plants ceased to grow, the aphid population declined.

Bioagent

Many coccinellid predators are known to feed on this aphid species like *Harmonia dimidiata, Harmonia eucharis, Harmonia octomaculata, Coccinella septempunctata, Coccinella transversalis, Oenopia quadripunctata* and *Menochilus sexmaculatus*. These natural enemies usually keep infestation in check. Seven species of ants are associated with this aphid (*Prenolepis melanogaster, polyrnachis levissima, Dolichoderus fuscus, Camponotus wasmanni, Crematogaster flava, Crematogaster rogenhoferi* and *Technomyrmex brunneus*) and protect them from attack of its natural enemies.

Oak leaf aphid; *Tuberculatus indicus* (Hemiptera : Aphididae)

The oak leaf aphid, *Tuberculatus indicus* generally feeds on the *Quercus acutissima* and *Q. griffithii*. Adults are pale green to yellow in colour and varies in length from 2.5 mm to 3.12 mm. It prefers to feed on the old leaves. Due to continuous feeding, the leaves become rusty brown with black powdery patches, rendering them unsuitable for silkworm rearing. The maximum population was recorded during May–July.

Bioagent

A number of Coccinellid predators consumed these aphids, viz. *Harmonia eucharis, H. octomaculata, H. dimidiata, Coccinella septempunctata, C. transversalis. Aiolocaria dodecapsilota, Oenopia kirbyi, O. quadripunctata Panialuteopustulata, Synonycha grandis and Megalocaria dilatata.* Aphidocolous ants so far identified for this species include *Polyrhachis laevissima, Ppolyrhachis dives,* and *Acantholepis capensis*, etc.

Live oak aphid; *Eutrichosiphum (*Eutrichsiphum)*pasaiane* (Hemiptera: Aphididae)

This is a polyphagous insect feeding on various plants and abundantly available through out the year. The live oak aphid, E. (Eutrichsiphum) pasaiane, is blackish in colour and infests the ventral surface of oak leaf. The injury to the leaf frequently results in the stunting of growth. The adults prefer to feed on mature leaves and its population is very high during November and December.

Bioagent

Archaphidus greenideae is the only known parasitoid of this aphid species.

Black oak aphid; ***Lachnus tropicalis*** **(Hemiptera : Aphididae)**

The black oak aphid, Lachnus tropicalis, is bigger in size (3.18 to 4.79 mm long) and black in colour. It prefers to feed on the young shoots and branches of the oak plants.

Bioagent

The natural enemies include several hymenopteran parasitodis, viz. Pauesia indica, Neopauesia manipurensis, Suberphedrus quercicola, etc. Some times it is also attacked by lady beetle, *viz.* Harmonia dimidiata. *Aphidocolous ants of this insect include* Crematogaster ebenina, Crematogaster shubunda, Dridonrmex anceps, Iridomyrmex glaber Polyrhacis halidayi, Polyrhachis laevissima, Polyrhachis rastellata, Rhoptromyrmex wroughtoni, Technomyrex elatior *and* Tetraponera nigra.

Control

It is well–known that aphid infestation has adverse effects on the chlorophyll content of oak leaves leading to ill health of the plant which in its turn lessens the nutritive food supply to the developing silkworms. Sometimes, they become vectors of many plant viruses. So, the plants should be free from aphid attack neither it would be a resident pest in sericulture. A sound knowledge of pest population is required to plan effective control measures in order to ensure protection from aphid attacks.

***Aphis craccivora* Koch = (*Aphis leguminosae* Th.) (=*A. laburni* Kalt.) (Hemiptera: Aphididae)**

The aphids are of minute size. The non winged aphid has large eyes, black cornicles and a yellowish green tip of abdomen. They measure about 2.5–2.7 mm in length. The pest is active from November to April in northern part of India. Nymphal development is completed in 11–14 days for non winged forms and in 14–19 days of winged forms, the growth being quickest in March. From the middle of November the population starts increasing progressively and is highest in March, and decline early in April as the temperature warms up. There is an increase of winged forms from December onwards, the peak period is in March wherein about 90 percent of the adults become winged. The pest multiplies most rapidly in late spring and with the increase in temperature, its population declines. The aphids suck the cell sap from the tender leaves, buds and twigs. The vigour of the plant is reduced and the quality of the leaves deteriorates. Each aphid makes several punctures, producing wounds, which leave their mark as the leaf size increases. The honeydew secreted by these insects becomes the source of development of black fungus, giving an ugly appearance to the plants.

Control

The control of aphids is practically the same for all species. In the forest no control measures are applied to control the aphid population, in spite of the serious injury that they may cause. Therefore, preventive control measures should be undertaken regularly to reduce the pest population. Recently, some biocontrol agents especially insect predators, parasitoids and pathogens have been screened against several aphids but its application at mass scale needs standardization of the technology. Though chemical control is quite effective but due to its residual effects on silkworm, it is not encouraged. In case of severe attack of aphids, spray 250 ml of Malathion 50 EC or 150 ml of Dimethoate 30 EC or 15 ml of Oxydemetonmehtyl 25 EC in 80 litres of water per acre.

Castor white fly; *Trialeurodes ricini* Misra (Hemiptera : Aleyrodidae)

Castor white fly has been reported from India and Pakistan. It is primarily a pest of castor but nymphs also feed on the leaves and stem of *Breynia rhamnoides* and *Achras zapota*. The nymphs are translucent, light yellow in colour and covered with thick waxen filaments. The adults are pale yellow with white wings covered with waxy powder. The pest is active from February to November. The female lays eggs in clusters on the under surface of the tender leaves. On emergence, the nymphs attach themselves to the leaves. After four moultings, they become full grown. During July September the life cycle is completed in 19 21 days. The nymphs suck the sap from the under surface of the leaves in large numbers and as a result of which the leaves show yellow patches. Sooty mould also develops on honeydew excreted by the pest. In young plants, severe infestation results in gradual drying of leaves and ultimately the plants die.

Control

Spraying 400 ml of Malathion 50 EC or 250 ml of Dimethoate 30 EC or Oxydemeton 25 EC or Formothion 25 EC or 75 ml Phosphamidon 100 EC in 250 litters of water per acre and repeated at 2–3 weeks interval.

THRIPS

Thrips sp. (Thysanoptera : Thripidae)

Thrips are minute insects ranging from 0.5 to 1 mm in size occurring on a wide variety of cultivated and wild plants often referred to as fringe wings. They are also known as physapoda because of the possession of a protrusible bladder - like structure at the end of the tarsus. They are known to be of considerable importance as pests of silk host plants and also due to their ability to act as vectors of some bacterial, fungal and viral diseases of plants. The body is covered by a sclerotic cuticle and the integument often shows characteristic sculpture patterns. The head is typically hypognathous produced below into a mouth cone or rostrum, which may be broadly rounded or pointed. Sutures are absent in the head capsule and the cephalic endo sclerites are not well developed, only the rudiments being present in the terebrantia but in tubulifera only one ridge is present on either side of the frons. The compound eyes are composed of varying number of ommatidia. Three ocelli are characteristic of winged species, reduced in brachypterous forms and absent in apterous forms. The antennae are 6–9 segmented, the last two or three segments forming the style in the terebrantia. The mouthcone is formed by the labrum, the labium an the maxillae and extends ventrally between the forecoxae. They are adopted for rasping and sucking and are unique because of their asymmetry, only the left mandible being present and functional, the right mandible being fused with wall of the cranium during development. The labrum is an asymmetrical, oblique triangular and bears a round socket through which the maxillary stylets and mandible pass. The maxillae forming the sidewalls of the mouthcone carry a pair of triangular palps bearing sclerites. The prothorax is well–developed, clearly separated from the head and mesothorax. The abdomen is made of the distinct segments and rudimentary eleventh. Genitalia are well developed and sexual reproduction is more common among thysanoptera. The female being larger and more abundant than males. It causes severe loss by sucking sap from the soft and succulent leaves of mulberry and non–mulberry silk host plants. There are large number of thrips which are attracted to non mulberry silk host plants.

Liothrips litseae Moulton (Thysanoptera : Thripidae)

Liothrips litseae is polyphagous pest of *Litsaea monopetala* (soalu). The damage caused to the leaves is mainly due to their feeding habits. In early stages of its infestation, thrips occur on the upper surface of the leaves and feed actively on the cell sap through an incision in epidermis made by their mandibles. The long maxillary stylets are then brought into the action, serving to puncture the deeper cell layers and sucking the fluid by using the labrum and maxillary stylets. With the increase in the population of thrips, they migrate towards leaf tips and feed voraciously damaging the epidermis severely. Heavy infestation of thrips badly damages the leaves of *L. monopetala*. Both the adult and immature stages were found infesting the leaves. Due to gregarious feeding of thrips on the tender and succulent leaves, the leaves develop small brown patches and ultimately infested portion dries up. Heavy infestation causes curling of leaves. The peak period of infestation is March–April and July–August in Assam. Adult is dark brown species with clear forewings except for a small brownish area at the base. Its head is 1.4 to 1.6 times as long as broad. All tibiae and tarsi dark except fore tarsi and the tip of the fore tibia brownish yellow; antennal segments III–V clear yellow, VI yellowish brownish in distal half and proximal half yellow in female. Tube shorter than head and setae on IX segment shorter than tube. Females are macropterous. Body dark brown including legs except fore tarsi and tip of fore tibia yellowish brown, antennal segments II to V and proximal half of the segment VI yellow, forewings transparent with small brownish basal area. Main setae yellowish brown, slightly expended distally. Life cycle is completed in five stages i.e. I instar larva, II instar larva, prepupa, pupa I and pupa II. Eggs are laid in clusters and oviposition occurs by the sides of the midrib and veins. Eggs are whitish yellow and almost hexagonal shape. Larvae are yellow and have a tube, which is shaded with greyish tinge. The total life cycle is completed in 18 to 25 days. The duration in days of different stages of the life cycle

Thrips tabaci (Thysanoptera : Thripidae)

Thrips tabaci is a common thrips abundantly available in the vicinity of som and soalu plants. The nymphs are slender, yellowish-brown and look like adults but are wing less and slightly smaller. The adults are slender yellowish-brown measures about 1 mm in length. The males are wingless whereas the females have long, narrow strap like wings, which are furnished with long hairs, along the hind margins. The pest is available through out the year but during October – November its population are very high. The female live for 20-30 days and lays 50-60 kidney shaped eggs singly in slits, which are made in the leaf tissue with its sharp ovipositor. The incubation period is 4- 9 days and the nymphs start feeding on the plant juices by lacerating the leaf tissues. They are generally congregated at the base of the leaf and prefers to feed on the lower surface of the leaf. Nymphs are full grown in 5-6 days after passing through four stages. Later on they descend to the ground and pupate in the soil at the depth of 15- 25 mm. The pre-pupal period is 1-2 days and the pupal period is 2-4 days. There are several generations in a year. The damage is done both by nymphs and adults. The pest infests the soft and succulent leaves of the muga food plants through out the year. As a result of this attack on the tender leaves, there is severe loss in the quality and quantity of the foliage of the food plants. The leaves become wrinkled and fall off and the plant gives unhealthy look. *T. tabaci* is known to act as vector of various viral diseases also.

Oak plants are mostly damaged by the three foliage feeding species of oak thrips (Singh *et al.*, 1996). Among them *Megalurothrips distalis* Karmy, *Frankliniella intonsa* (Trybom) and *Mesothrips* sp. are most common. *Megalurothrips* sp and *Frankliniella* sp. generally feed on pollen and foliage sap while the *Mesothrips* feed on the foliage sap only. Damage is done by adults as well as by nymphs. The adults are slender, brown and measures about 1mm. in length. The males are wingless whereas the female has long, narrow strap-like wings. In shape and colour nymphs resembles the adults but are wingless and slightly smaller. The life cycle consists of egg, two larval instars, prepupa, pupa and adult. The life cycle of *M. distalis* is completed in 15 days whereas *F. inthorisa* takes about 17 days. The development from egg to adult takes about 17 days in *Mesothrips* sp. *Megalurothrips* spp. and *Frankliniella* spp. reproduce sexually as well as parthenogenetically. The mated female lays about 45-48 eggs within the leaf tissues. The *Mesothrips* reproduce sexually and oviposits on the abaxial side on the young succulent leaves.

Control

Apart from the cultural and mechanical control measures, thrips are controlled by using light traps in the filed. Water pan yellow trap is highly effective for trapping of oak thrips. In the case of high infestation of this pest on the Oak plants, chemical control measures are adopted to control the pest population. Application of Malathion 50 EC @ 0.05 is highly effective against this insect. Waiting period is at least 10 days before brushing. Under tasar and muga plantation spraying of 250 ml Methyl parathion 50 EC in 250 litters of water per acre as soon as the pest appears is highly effective. Waiting period is at least 7 days before feeding of silkworm larvae.

Jassid; *Ambrosia bagatelle* Ishida (Distant) (Hemiptera : Cicadellidae)

Jassids are commonly known as leafhopper. It is one of the most destructive pest of castor, som and soalu plants. Damage to the plant is caused by the adults as well as the nymphs, both of which are very agile and move briskly forward and sideways. The winged adults jump or fly away at the slightest disturbances and are also attracted to light at night. The females lay about 15 yellowish eggs on the undersurface of the leaves which are about 3 mm long. The eggs are greenish yellow in colour during summer and acquire a reddish tinge during winter. The eggs hatch in 4 11 days and give rise to nymphs, which are weed shaped and very active. They suck cell sap from underside of the leaves and pass through the six stages of growth in 7–21 days. On transformation into the winged adults, they live for 5–7 weeks, feeding constantly on the plant juice. The winged adults jump or fly away at the slightest disturbance and are also attracted to the light at night. The pest completes seven generations in a year. Nymphs and adults remain in large numbers and suck the sap from the under surface of the leaves. While feeding, they inject the toxic saliva into the plant tissues. The leaves show symptoms of hopper burn such as yellowing coloring, bronzing and even drying of leaves. The plants become stunted and often in highly susceptible varieties it causes complete mortality of plant, if left unprotected.

Control

Spraying 100g DDT 50 WP or 500g BHC 50 WP or 350 ml Malathion or 45 ml Malathion 50EC or 240 ml Dimethoate 30EC or 300 ml Oxydemeton Methyl 25EC or 300ml Formathion 25 EC or 75 ml Phosphamidon 85 WSC in 100 litres of water per acre is followed.

Jassid; *Empoasca binotata* Pruthi (Hemiptera: Jassidae)

Empoasca binotata Pruthi (Hemiptera: Jassidae) commonly known as Jassid is an important sap sucker and leaf infesting pest of muga food plants. It is available in all the muga growing areas in North-Eastern region of India. Nymphs and adults remain in large numbers and suck the sap from undersurface of leaves and cause yellowing, curling, bronzing and even drying up of leaves. The crop becomes stunted and often in highly susceptible varieties it causes complete mortality of plants if left unprotected. The susceptible and tolerant varieties react differently to toxic saliva and based on the symptoms of "hopper burn" injury the following grades of damage for leaf hopper resistance have been recognized for evaluating resistance in muga variety.

Grade I

Entire foliage free from crinkling or curling, with no yellowing, bronzing and drying of leaves; no hampering of growth irrespective of presence or absence of adult stages of insect.

Grade II

Crinkling and curling of few leaves, mostly in the lower portion of the plant, a little yellowing of leaves; nearly hampered vegetative growth.

Grade III

Crinkling and curling of leaves all over the plant; yellowing, bronzing and browning of leaves noticeable in the middle and lower portions; plant growth noticeably hampered.

Grade IV

Extreme curling, crinkling, yellowing, bronzing, browning and drying up of leaves and progressive defoliation; plant growth remarkably stunted.

Life Cycle: - Life Cycle

The leafhopper mate early in the morning or late in the evening. The female inserts its eggs inside the leaf veins in the parenchymatous tissue. On an average a leafhopper lays about 15 eggs, but may lay as many as 30 eggs. The egg period varies from 4–11 days. There are five nymphal instars and nymphal period occupies about seven days in autumn and 21 days in winter. Longevity of the mated adults is about five weeks in summer and seven weeks in winter, and unmated adults live three months or longer. The winter broods individuals are reddish while the summer broods are greenish yellow. In North-Eastern region, there are 10 generations of the insect with considerable overlapping. Each brood occupies from 15–46 days depending upon the environmental conditions.

Control

Chrysopa cymbele is predaceous on the leaf hopper in North-East. Ants like *Camponotus* and a spider, *Distina albida* have also been reported to be predaceous on the insect. Spraying DDT 0.1% thrice at 10 days intervals may be necessary to control the pests. Spraying Carbaryl 0.1%, toxaphen, endrin etc have also been reported to be useful.

Castor Leaf Hopper; *Empoasca flavescens* (Hemiptera : Jassidae)

Plants with heavy population of nymphs show characteristic symptoms of hopper burn. The initial system is formation of yellow patches in the margins of the leaves, which is followed by the distortion of veins and leaf curling. This subsequently changes to brown. The leaves become brittle and dry. In case of severe attack the plant looses its vigour and results in poor formation of capsules. Different varieties show different degrees of resistance and susceptibility to leaf hopper attack. Life cycle: - The leafhopper is green or greenish yellow. The female inserts 15-30 eggs inside the leaf veins of the leaves and the oviposition period is 5–7 days. The incubation period is 7 8 days and 5 nymphal instars are passed in 10 days. The entire development from egg to adult occupies about 18 days. The average longevity of male and female will be 9 and 17 days respectively. The insect multiplies in large numbers during November–January and minimum population level is noticed during the hot weather and southwest monsoon periods. It is primarily a pest of castor (*Ricinus communis)*, Som (*Machilus bombycina)*, and Soalu (*Litsaea polyantha)* and tasar and mulberry plants also.

Control

Spraying of Phosphamidon, Malathion, Fenitrothion, Fenthion, Endosulphon, Parathion, BHC, DDT, Phorate, Carbaryl, etc., are applied in the field control the pest. Monocrotophos (Azodrin) has also been reported to be highly effective.

Termites (Isoptera : Termitidae)

Termite commonly known as white ants are one of the serious pests of non mulberry silk host plants (Singh *et al.* 1992). They are known to cause severe damage to sericultural and forestry crops.

Although termites are better known taxonomically than most other groups of insect, there are many gaps in our knowledge on the ecology in relation to resource management. Termites are broadly divided into two categories according to their habitat, viz. wood inhibiting termites and sub terranean termites. The former lives in their food (wood) and does not maintain any soil connection. The later obligatorily maintain soil connection in order to derive the required moisture. In sericulture, majority of termites are soil dwellers, feeding wood, humus, lichen, moss etc. The effects of the activities of subterranean termites on the soil, therefore, present a number of interesting pictures. The enormous quantity of sub-soil brought to the surface during the burrowing and mould building activities evidently contribute to fertility of the soil, as the sub-soil is known to be rich in mineral matters. In addition burrowing activities seems to increase the ratio of percolation of rain water and of aeration both the top and sub soils. This beneficial effect is rather pronounced in the laterite soils. In addition termites also play a vital role in disintegration and decomposition of fallen branches, trees, dried stems of shrubs and herbs, grasses, fallen leaves, etc. Thus enriching the soil and recycling the nutrients. On the other hand humus feeding termites which are active on the top soil, deplete the organic matter of the top soil. The rapid removal of the organic liter from the surface of the soil can thus pose considerable problem in areas deficient in humus. The grass mulch which serves to keep the soil temperature low and moisture content high is often destroyed by subterranean termites.

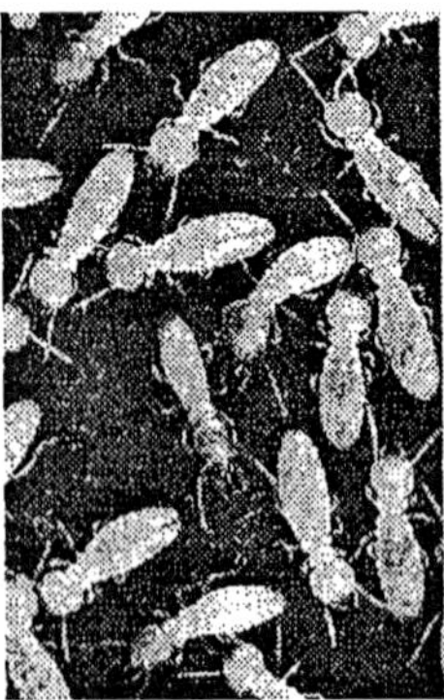

Fig. 4.4 : Termites in Tasar Plantation

Termite colony

Termites burrow into the ground, constructing a series of complex tunnels and nests with or without mound. The nest is the most remarkable structure consisting of numerous chambers and galleries. It is about 1-2 feet in diameter. The situation of nests varies with the species. The nests of many termites grow fungus garden in the center, near about the "Royal chamber". The fungus grows into a comb like structure and is fed to the royal pair and the larvae. Reproductive castes and sterile castes of subterranean termites live in the soil. Colonizing individuals are winged and are produced in large number during rainy season. The wings are meant for nuptial flight only and when they have mated, the wings usually drop off. The queen is the only perfectly developed female in the colony. She attains a much larger size. She is a phenomenon "Egg laying machine" laying one egg per second or 70,000 to 80,000 eggs in 24 hours. There is only one queen in a colony and normally she lives for 5 to 10 years. The queen is fed by the workers on the choicest food and is housed in a special area referred to as "Royal chamber" which is situated in the center of the nest where as the king develops from an unfertilized egg and becomes fully developed by consuming a superior diet. He is the father of the colony and is the constant companion of the queen, living with her in royal chamber. The king's life is much shorter, and when he dies, he is replaced by a new one. The workers develop from the fertilized eggs but remain shunted as they are reared on ordinary food. They are most abundant in the colony but smaller than the soldiers. Their mandibles are well–developed and are adopted to chew the wood. The soldiers develop from unfertilized eggs and remain comparatively underdeveloped. They are the most specialized members of the community and can be readily recognized by the large head and strongly chitinized sickle shaped mandible. The soldiers are mandibulate type and naste type. The former type of soldiers defends the colony by fighting the intruders and the later type repel them by spraying an abnoxious smelling fluid through the rostrum. The atmospheric temperature and relative humidity are two major environmental factors that influence the termites profoundly. They are extremely sensitive to changes in the atmosphere, temperature and relative humidity. Many species ascends deep into the soil in dry and hot seasons, thus escaping the extreme hit of the mid day. However, active foraging occurs in the early morning hours. The liability of human dwellings to be attacked by termite is more pronounced in damp locations than in dry locations. Every kind of aeration leads to desiccation unless the moisture contents of the surrounding area is at the saturation point or the termites have direct access to water. The ability of certain termites to survive in air with low humidity thus depends on access to water sources. To some extent water is provided as a metabolic end product of the break down of cellulose and other carbohydrates as in the case of dry wood termites. Several genera of termites do not however need free water source, as the relative humidity within their nests is constantly maintained near the saturation point. The mound building termites maintain almost constant temperatures and relative humidity notwithstanding the change in the atmosphere. The primitive termite *Archtermopsis wronghtoni* (Desn) is capable of tolerating wide fluctuation of temperature. This species occurs in high altitude of the Himalayas where temperature often rises to 37°C during the month of June and falls down to almost zero during the winter months of December February. Notwithstanding these wide fluctuations of atmospheric temperature is difficult to explain. Under laboratory conditions, *Neotermis bosei* Snyder prefers temperature of 28°C, while the optimum temperature of *Heterotermes indica* (Wasm) a subterrenean species lies between 28–32°C. *Microcerotermes beesoni* Snyder is cultured in the laboratory best at a temperature of 29°C. Much of information is available on the water relations and humidity tolerance of Indian termite. Most species prefer high humidity, almost near saturation point. Survival of subterranean termites depends on the water content of the soil which in turn is dependent on the water holding capacity of the soil. The optimum water content in the sandy soil is 2–4% and that of humus soil is 15–25%. In the humus soil the optimum feeding and survival occur in a broad range, but the same is not applicable in the case of sand which has low water holding capacity. *Heterotermes indecola* prefers 20–30% water content of the culture medium and rate of feeding is also higher in case of wood with higher moisture content (80%). However, not–withstanding the above, *H. indecola* prefers to feed near the surface soil, demonstrating a higher degree of acrotrophic behaviour. In contrast, species of *Microcerotermes* feeds maximum when moisture content of the culture medium is 15–20%.

Population Dynamics

The population dynamics in *Odontotermes obesus* varies from 4,548 to 90,961 individuals depending upon the size of the mound during non mound building months. In *Odontotermes redemanni*, proportion of workers, soldiers and nymphs are 81.33%, 14.15% and 2.42% in the fungus combs. Almost similar proportions are found in the *O. obesus*. Seasonal fluctuation in some of the termites viz. *Neotermes bosei*, *Microcerotermes beesoni* and *Nasutitermes dunensis* is well–known. In the case of *M. beesoni* the total nest population which did not include the foraging individuals at the time of collection of nests ranged from 7, 000 to 45,000 individuals depending upon the season and size of the colony. The proportion of workers, soldiers ranged from 60 to 95 % respectively. The population of immature forms was inversely proportional to worker population. The association of termite and fungi is well. Some groups of termites are obligatorily associated with fungi more than others. Termites from the families, Termopsidae and Rhinotermidae, in general, attack wood that has undergone partial decay due to fungius attack. The brown rot and white rot fungi cause decay in many species of wood, thus making it acceptable to termites. The common brown rot and white rot fungi attacking wood are *Poria monticola* (White rot), *Polyporus versicolor*, *P. hirsutus*, *P. palustris*, *Lenzites trabea*, *L. striata* etc. The white rot fungi breaks down the lignin of the wood and provides extra admissible carbohydrates to termites. Certain wood inhabiting, non decay fungi also attract termites equally vigorously and this discovery opens up the possibilities of management of termites with none decay fungi attractant coupled with slow acting toxic chemicals. The soft rot fungi and stain fungi (*Chaetomium globosum* and *Botryodiplodia theobromae*) are also attractant. The use of fungi attractants bait would be useful tool in termite survey and monitoring. It does not seem probable that t his intimate relationship between termite and fungi is symbiotic in nature. However, there is no doubt that termites are nutritionally dependent on various kinds of fungi that are associated with wood they feed on. The fungal hyphae also provide much needed protein.

Termites are better known as destroyer of wood and wood products. However, several plants of economic importance are also victims of depredation by termites. Among the sericultural plants commonly damaged by termites are, arjun, asan, sal, som and soalu plants. Mulberry plants are also attacked by this pest. The species of termites commonly observed damaging these plants are *Odontotermes obesus*, *Microtermes obesi* (Hols), *Coptotermes heimi* (Wasm).Termites have also been reported to damage various agricultural and plantation crops in the field. Termites damage to cellulose material in human dwellings and destroy building contents like carpets, furniture, documents, depositors in archives, etc. Among subterranean termites, the most important are *Heterotermes indecola*, *H. malbaricus*, *Coptotermes heimi*, *Odontotermes obsus*, *O. indicus*, *O. feae*, *O. redemanni* etc. These termites attack unprotected buildings even within three years after construction. *Odontotermes assamensis* commonly known as termite is one of the most injurious insect of som and soalu food plants in Assam. It eats the bark and leaves of the som and soalu plantation and causes heavy loss to muga sericulture industry in Assam. The som and soalu infesting termites belong to the category of wood inhabiting termites, in which the king and queen is always found in the ground, no external mounds are erected, and although the workers may carry on a large part of their activities in wood above ground. They always maintain a connection with the ground nest through either burrows or covered run ways.

Fig. 4.5 : Termites mould in rearing field

Termites in oak tasar

The oak infesting termites belong to the category of wood inhabiting termites, in which the kings and queens are always found in the ground, no external mounds are erected, and although the workers may carry on a large part of their activities in wood above ground. They always maintain a connection with the ground nest through either burrows or covered run ways. The wood inhabiting termites are social insects some what live in colonies composed of a member of castes. Each caste has a more or less specific function to perform. The primary reproductive, usually consists of king and queen are the only members of the colony that at any time have functional wings. The queen has a creamy white abdomen marked with transverse dark brown stripes. The king is much smaller than queen and is a constant companion of the queen living with her in the royal chamber. The supplementary reproductive is short winged or wingless creatures of both the sexes. In event of untimely death of the king or queen in a colony, the complementary castes replace them. Workers are sterile, usually pale in colour, slightly sclerotized and look more like nymphs than adult. Soldiers are also sterile. Head and thorax is well chitinised and the abdomen is delicate and dirty white in colour. They are four types of soldier viz. mandibulate soldiers, nasute soldiers, nasutoid soldiers and phragmotic soldiers. The king mate with the queen from time to time in the royal chambers and thus aids her in laying fertilized eggs from which the colonizing forms of workers develop. The queen is a phenomenal egg laying machine laying 70,000 to 80,000 eggs in 24 hours. The egg hatch after one week during summer and the larvae develop in soldiers or worker within 6 weeks as per the availability of the food. The reproductive forms mature in 1-2 years. There is only one queen in a colony and normally she lives for 5–10 years. The queen is fed by the worker on their choicest food and is always confined to the royal chamber with king and perform the sole function of reproduction.

The king life is much shorter than the queen, and when he dies he is replaced by a new one. The workers are developed from fertilized eggs but remain stunted as they are reared on ordinary food. They are abundant in colony but smaller in size than soldiers. Except for the reproduction and defense of the community, all other duties are performed by the workers. They take care of eggs and the young ones and remove them to safer places at the time of danger. They also tend and feed the queen, collect food and cultivate fungus food in under ground gardens. The soldiers develop from unfertilized eggs and remain comparatively under developed. The soldiers defend the colony mainly against the other insects like ants and vertebrates predators. The mandibulate types of soldiers defend the colony by fighting the intruders and nasute type soldiers repel them by spraying an obnoxious smelling fluid through the rostrum. Termites severely riddle the wood of the standing oak tree in the process of making the runways. The entire hole are found attacked and damaged in a few days. The excavations are more or less concentrate, radiating out from the pith region. Heavily infested trees often break down with gust of wind. Severe damage of heartwood by the termites adversely affects the growth and production of either wood or foliage.

Management practice

The management of termite can be conveniently grouped into two categories viz. management of wood inhabiting termites and of subterranean termites. Termites inhabiting on arjun and asan plants are primarily managed by cultural practices. The attack by *Cryptotermes* spp. to wood work in buildings can be avoided by either using treated wood or preventing swarming adults to invade wood work in houses by putting off light or screen them by wire netting. Cultural and mechanical control are applied to demolish the termites mould in the field. Burning and killing of adults in its nest is the most common method of its control. *Insitu*, treatment by toxic dusts or liquid, in general are used to eradicate existing infestations. Soil application of BHC 10 per cent @ 100 kg. per ha mixed with equal quantity of farm yard manure during May- June, or Quinalphos 5 G @ 2.5 kg a.i /ha against grub is recommended. Spray the host trees with 0.05 per cent monocrotophos 40 EC or 0.1 per cent lindane 20 EC to kill adult.

CHAPTER-5

Parasites and Predators of Silkworm

UZI FLY

Tachinids are the important parasites against several harmful pests of agricultural crops where as in sericulture, they are injurious to silkworm. Most tachinids attack the larvae of silkworm and they are commonly known as uzi fly. The best known uzi fly in sericulture is *Exorista sorbillans* (Wiedman) (= *Exorista sorbillans* Louis), *Crossocosmia sericariae* Rondani, *Ctenophorocera pavida* Meigen and *Blepharipa zebina* Walker. Indian uzi fly (*Exorista bombycis* Louis) is the key parasite of mulberry silkworm (*Bombyx mori*) (Krishnaswamy *et al*., 1964; Devaiah *et al*., 1992). Besides mulberry silkworm, uzi-flies have been reported to attack tasar silkworm (*Antheraea mylitta*), Oak tasar silkworm (*Antheraea proylei, A. pernyi* and *A. yamammai*), muga silkworm (*Antheraea assama*) and eri silkworm (*Philosamia ricini*), (Singh *et al*., 1993; Patil and Govindan, 1984). Besides these, some other lepidopteran caterpillar especially from order lymantriidae, saturnidae, and sphingidae has been reported to be an alternate host to Indian uzi fly (Kumar *et al*., 1990; Narayanswamy *et al*., 1992; Singh *et al*., 2000). Tasar uzi fly has also been reported to parasitize mulberry silkworm (Patil, 1983). Much of the information's are available on the uzi fly parasitizing mulberry and tasar silkworms (Narayanswamy, 1991; Patil and Govindan, 1984; Singh *et al.,* 1990; Singh and Thangavelu, 1993 and Thangavelu, 1991; Qu Tian et al 1990, Goodwin and Odell, 1984).

Fig. 5.1. Adult Uzi fly searching host

HISTORICAL DETAILS

Several workers have reported the taxonomic status of uzi fly. Earlier, Indian uzi fly was commonly known as *Tricholyga bombycis* (Louis). Crosskey (1976) studied the taxonomic conspectus of the tachinidae of the oriental region and reported that *T. bombycis* is synonyms of *T. sorbillans* (Wiedemann), a tachinid originally described from Canary Islands and a species having very wide host range and geographic distribution in the world. Since the genus *Thricholyga* was synonymised with *Exorista*, the species came to be known as *Exorista sorbillans* under which workers since then and even now recorded their observations. However, Gorpade (1986) after examining the uzi flies recovered form mulberry silkworm in Karnataka and also authentically identified specimens of *E. sorbillans* from Canary Islands came to the conclusion that the uzi fly from the mulberry silkworm in India is different from *E. sorbillans* and the Indian uzi fly should designated as *Exorista bombycis* (Louis). In the recent years the Indian uzi fly parasitic on the mulberry silkworm, *Bombyx mori* has been called by the name, *Exorista sorbillans*. However, recently Dr. Ian M. White, Diptera Taxonomist of the International Institute of Entomology, London has identified as *Exorista bombycis* (Louis). The absence of the uzi fly in south India on mulberry silkworm till 1980, though it was first reported from West Bengal one hundred years ago (Louis, 1880) and has been there attacking mulberry silkworm since then, and the simultaneous presence of *E. sorbillans* on a number of other hosts in Karnataka (based on identified specimens present in the collections of the Entomology Department, UAS and of CIBC, Bangalore) provide additional support to this view. Since its entry into Karnataka in 1980 (Anon., 1980), the uzi fly has spread to the entire state by 1983 (Siddappaji, 1985) and very soon in other sericultural areas of south India. Ever since this parasitoid was reported in Bengal over one hundred years ago various levels of infestation even up to 90% and in Karnataka parasitisation from 40 to 75 per cent have reported (Siddappaji and Channa Basavanna, 1981). Since infestation always results in the death of the infested silkworm, percentage of infestation means more or less the same percentage loss of silk production. Geographical distribution of tasar uzi fly is restricted to tropical tasr region of India however, it has been reported from temperate tasar and muga growing area also. It is confined especially in Jharkhand, Orissa, Chattisgarh and North Eastern region of India. Recently a large scale outbreak of its population occurred in Bastor (M. P.) and Nowrangpur (Orissa) has shaken the very root of tasar culture in north India. It has been reported to cause more than 50% loss in some of the tasar and muga growing areas.

Indian uzi fly; *Exorista sorbillans* Wiedemann (Diptera : Tachinidae)

Indian uzi fly, *Exorista sorbillans* Wiedemann, (Diptera : Tachinidae) is a serious parasite of mulberry silkworm larvae (*Bombyx mori*) in India. In recent years the Indian uzi fly parasitic on the mulberry silkworm, *Bombyx mori* has been called by the name, *Exorista sorbillans* instead of *Trycholyga bombycis*. However, recently Dr. Ian White, Diptera Taxonomist of the International Institute of Entomology, London has identified the above as *Exorista bombycis* (Louis). It is prevalent in Burma, China, Thailand,, South Korea, and some parts of India (Assam, Bengal, Jammu & Kashmir, Uttar Pradesh, Madhya Pradesh, Orissa, Andhra Pradesh, Tamil Nadu, and Karnataka). The Indian uzi fly was confined to North India especially West Bengal and later on during 1980, it was introduced into Karnataka. Maxwell Lefroy (1907) even warned that this fly pest should not get introduced into the prosperous sericultural area of Mysore. This forewarning should have been taken seriously to prevent entry and establishment of the uzi fly into the most prosperous sericulture belt of South India. This uzi fly is known to cause more than 40 per cent cocoon crop loss in some seasons in eastern region of India. Survey conducted in West Bengal reveals that the infestations range from 8-82 per cent in Murshidabad district and 20-40 per cent in Malda district. However, in Karnataka and Andhra Pradesh it has been reported up to 20-35 per cent during summer and monsoon season (Zamil *et al.*, 1991). This fly is principally a parasite on mulberry silkworms but it has been known to attack tasar silkworms also. It is the Indian counterpart of the Japanese tachinid fly, *Sturmia sericariae* (*Crossosmia sericariae*). Besides mulberry silkworm (*Bombyx mori*) it has been reported to attack on tasar silkworm (*Antheraea mylitta*), muga silkworm (*Antheraea assama*), eri silkworm (*Philosamia ricini*). Besides these, some other lepidopteran caterpillar especially from order lymantriidae, saturnidae, and sphingidae have been reported to be an alternate host of Indian uzi fly (Beeson and Chaterjee 1935). Thomson (1944) reported 44 species of lepidoptera which act as host of *E. sorbillans*.

It has also been recorded to parasitise *A. perynei* and *A. yamamai* also. Uzi fly completes five to six generations per year in Japan. The first imago appears in first half of May. It is known to pass through seven to eight generations per year in tropical zones (Krishnaswamy *et al.* 1973). Life history and biology of this Indian fly pest has been thoroughly studied and much of the information is available on the developmental stages and duration of the life cycle in various regions. Cotes (1889) reported the life cycle of uzi fly to occupy about 28 days. King (1940) reported four to five weeks under laboratory conditions. The life cycle occupied 16-23 days under Berahampore conditions of West Bengal (Sriharan *et al.*, 1971). Similar observations were made by Krishnaswamy *et al.* during 1973. Datta and Mukherjee (1978) reported 18 days in mid summer and 22 days in winter. Patil and Govindan (1984) recorded the duration of total life cycle to be 25 days. Siddappaji (1985) reported that the total life is completed in 15 days during May-June and 25 days during January–February. However, Thangavelu and Sahu (1986) reported that the fly completed total developmental period in 32 days during June-July and 58 days during November-December in

North–Eastern region of India. Narayanswamy (1991) recorded 16-25 days in southern Indian condition.

LIFE CYCLE

Life history and biology of this Indian fly pest has been thoroughly studied and much of the information is available on the developmental stages and duration of the life cycle in various regions. The life cycle is completed in 16-23 days in hot and humid region and 15-25 days in dry region of south India (Datta and Mukherjee, 1978). The female flies lay eggs preferably on the fourth and fifth instars silkworm. The maggots after hatching enter into the body of silkworm, eat away the body contents for 4-6 days and kill the host subsequently. The infested silkworm larvae can be identified by the presence of black scar developed on the skin at the point of entrance and it increased in size day by day. Uzi fly is normally active through out the year and wander at various places in search of silkworm. The host searching ability of the fly pest is very strong and after locating the host, it starts her activity for increasing her population in sericultural area. The mated female adiposity about 150 eggs in 8-10 days. Eggs are laid mostly on the dorsal surface in the inter–segmental region of the silkworm larvae. The fly prefers to oviposit on the 3rd – 5th stage silkworm larvae. The mated fly generally parasitizes 8-10 silkworms in 24 hours and inserts an average of 3-4 eggs on each silkworm.

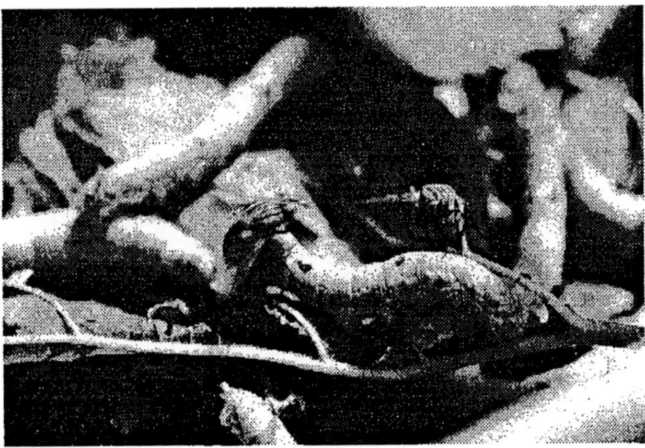

Fig. 5.2 : Uzi infected mulberry silkworm larvae

The fly generally oviposits on the dorsal surface compared to ventral and lateral region of the body. Egg is creamy white in colour and oblong in shape. It measures on an average 0.56 mm in length and 0.26 mm in width. The posterior end of the egg is rounded whereas the anterior end is little pointed. The dorsal surface is slightly convex where as the ventral aspect is flattened. The incubation period of the uzi fly on *B. mori* is 3-4 days. The eggs hatch and enter into the silkworm body. After emergence form the egg, the maggot bored into the host tissues and remained in the host until it develop fully. A small black scar appears on the skin at the point of entrance and it increases in size day by day. The scar is very prominent towards the period of maggot emergence from the host. The opening made by the maggot, while penetrating the body of the worm can be seen clearly and remains open through out. The maggot faces the spiracle towards the opening, pore through which it has entered into the body, gets the air supply and remains thereby feeding on the body contents of the host. After completion of feeding the fully-grown maggot comes out killing its host ultimately. When the infestation occurred in late fifth stage , the silkworm may be able to spin the cocoon, but fully grown maggot come out after killing its host by cutting the cocoon for pupation in the soil or crack or crevices. The maggot is fusiform, creamy whitish and measures about 1.4 to 1.7 cm in length and 0.3 to 0.5 cm in width with a pair of pharyngeal hooks at the anterior region and a pair of black spiracles at the posterior region. There are three larval instars. The detail morphology of all the three instars of maggots and the anterior and caudal spiracles of the maggot have been given by Datta and Mukherjee (1978). The larval period of the fly pest is about 4 to7 days. The mature maggot after coming out of the host crawled out and transformed into pupa in 6 to 8 h. Initially the puparium is whitish, later turned creamy, crimson-dull red, dark red and brownish black two days prior to eclosion. The puparium is barrel shaped with narrow rounded anterior and broadly rounded posterior end. The size of the puparium varies accordingly to the load of parasitization and also the host species. The pupal duration occupies 7 to 18 days. The imago breaks through the anterior end of the puparium by expanding ptilinium between compound eyes and emerges as fly.

The adult uzi fly usually emerges in the morning. The adult fly pest is blackish grey in colour with prominent black and grey stripes with four longitudinal black lines on the thoracic regions which are more clear in the male. The male is bigger than the female and varies from 11.5 to 12 mm in length whereas female measures about 10.5 to 11 mm in length. The width varies from 3.5 –3.8 mm. Lateral regions of the abdomen are covered with bristles, which are thicker in male than in the female and restricted mostly in last two abdominal segments in the latter. The longevity of the female is more in comparison to male. Male flies live for 5 to 12 days where as female live for 8-14 days. The adult uzi fly attains sexual maturity in 1 to 2 days after emergence. Mating occurs two to three times in a day. It takes about 2-3 hours in completing the mating. Flies mate on substratum however mating in flight has also been reported. The fertilized female is very active and persistent in its efforts to get at the worms, and once having reacted to silkworm rearing tray, it wanders all over it, depositing eggs indiscriminately on the worms. The pre-oviposition period varies from 30-48 hours and oviposition continues till death. The fly prefers to oviposit on dorsal region of the silkworm. Generally, third stage onwards silkworm larvae are parasitized by the fly pest however early instars worm have also been reported to be parasitized.

Fig. 5.3 : Eggs of uzi fly on mulberry silkworm

Tasar uzi fly; *Blepharipa zebina walker* (Diptera : Tachinidae)

The larvae of *Antheraea mylitta* is reared out doors on the food plants (*T. arjuna* and *T. tomentosa*) and therefore, exposed to the vagaries of uzi fly. Adult uzi fly male and female are 14 and 12 mm respectively. The female fly is smaller in size (11.45 mm) in comparison to male (13.75 mm) but very sluggish during mating, which lasted for 20–30 minutes. The body is grayish with five longitudinal depressed lines on the dorsal shield of the thorax. The abdomen has four segments with two white transverse stripes. The flies lay 150–200 eggs in a week. The maximum eggs are however laid between third and fifth day of oviposition. Eggs are laid on the inter–segmental region of the body of the silkworm. The eggs are macrotype, creamy white and measures 1.02 mm (0.75 to 1.57 mm) and 0.45 mm (0.35 to 0.45) in length and width, respectively. Generally 1–5 eggs are laid on a single larvae but during severe infestation more than 30 eggs have also been recorded. It may be due to the parasitization of single larva by more than one parasites.

After successful oviposition, fly pest rests on the plant twigs and attempts to escape. The maggot is soft with pointed head and truncated posterior end. A pair of prominent mouth hook is present on the head and posterior a pair of spiracles on deep pit. The newly hatched maggot is transparent white in color. It pierces the skin of the worm with the help of the mouth hooks, leaving a minute black spot near the hatched egg shell. The spot afterward enlarges to form a rounded balck scar. The maggot under goes three instars inside the host body in about 15–20 days. The first and second instars maggot generally remain near the scar portion. The body segmentations are not distinct in these two stages of the parasite. In the first instar the mouth hook is a simple with median hook but in second and third instars a pair of prominent hooks are developed. The second instar maggot is creamy white in colour and measure 12 mm in length and 4 mm in width. The body segmentations and spiracles are well–developed. It can move freely inside the host body actively feeding on the fat bodies and body tissues excluding the silk glands. The mature maggot wriggles out by piercing the host larval or pupal body with its pharyngeal hooks and finally kills the host. The maggots use to come out in the morning hours. They crawl in search of suitable places for pupation in the soil. If the maggots fail to find out the suitable places, it tends to pupate in the dark places. It takes 20–24 hours for pupation (Singh and Thangavelu, 1992).

The maggots passed three instars inside the host body in 20–25 days. The maggots are fusiform in shape some what acute anteriorly and rounded posterior and body segmentation is prominent. By the end of Ist stadium it is yellowish white in colour and measures about 1.5 mm in length and 1.35 mm in width. Maggots remained inside the inner surface of the host body wall. Sometimes 3-4 individuals are clamped together or floated freely in the host haemolymph. The maggots are distributed by cutaneous diffusion. Larval instars were mostly distinguished by the size and shape of buccopharyngeal apparatus. The buccopharyngeal apparatus of the first instar is 0.19 mm long and consists of a single, slightly recurved oral hook joined to a pair of dorsal and fused, ventral cornus. Second instars: The buccopharyngeal structure of the second instar is twice large that of the first, 0.38 mm long. The oral hooks are paired structures armed with paired processes and spines. It is fusiform in shape, yellowish white in colour, distinct body segment with a length of 2.75 mm. Mouth hook sharp and non serrate, posterior spiracles with two slits. After the maggot moulted to second instar it remained attached to the original trachea near the original site or crawled out of the respiratory sheath and fat body and move to the new trachea of adequate size. The buccopharyngeal hooks were considerably stronger in this instar and may also have been used if the maggot attached directly to the host cuticle. It fed on the host haemolymph of the host fat body. Third instars maggot is also fusiform in size and creamy white in colour. The body is distinctly segmented and the length varies from 1.5 2 cm. The buccopharyngeal apparatus is well developed. 25 hours after parasitization some maggots become embedded in the fat bodies of the hosts. It consisted of three segments with distal segments bearing three pairs of pointed recurved oral hooks. Th third instar maggots rasped a respiratory opening directly through the cuticle with its buccal hooks. It then turned around and pressed its posterior end firmly against the opening and retracted the posterior end, causing a suction which removed any debris or fluid clogging the hole. Suction is applied several times before the maggots pressed its spiracles firmly against this opening and resume normal feeding activity.

The pupae are oblong in shape, somewhat pointed anteriorly and rounded posteriorly. Colour reddish brown immediately following the pupation and then gradually changes to dark reddish brown. Newly pupated ones are reddish and afterwards they turn reddish brown in colour. It measures about 10 mm in length and 5 mm in width. The pupal period lasts for 10 14 days. The puparium is formed when the maggots have consumed all the available food or on account of attaining maximum size and weight or when there is reduction of humidity in the environment in which they live. When several third instar maggots come out from the host, a greater number formed viable puparia. Usually all the available food within the host body is consumed during the larval development of the *B. zebina*. The maggots then cease feeding movements. The carcass gets dried form a cream of tan colour to light brown. The maggots then enlarge the respiratory opening, or made a new opening, leave the carcass, and forms a puparium. The hardening of the last larval skin forms it. The two lateral sides are clearly marked by means of ridges. The dark brown pupal colour to slight balckish colour indicates the emergence of the adult. The adult fly escape out through the puparium by piercing off a cap shaped ptilinium and works its way through the soil beneath which it had been pupating. It moves freely and within one hour stretches its wing in vertical posture. The ptilinium decreases in size and adult can fly in one hour. The life cycle of the fly is completed in 25 35 days. The details of the morphomatric measurement are given in Table 5.1.

Table 5.1: Morphometric measurement of developmental stages of uzi fly

Stage of uzi fly	Length in (mm±SD)	Width in (mm±SD)
Egg	1.02 ±0.25	0.45±o.14
Larval instar		
I instar	1.48±0.25	0.75±0.14
II instar	5.82±0.25	1.85±0.25
II instar	13.36±2.5	3.75±0.75
III instar	12.57±0.48	6.2±0.50
Pupa	12.57±0.48	6.2±0.50
Adult		
Male	13.75±1.52	5.5±0.75
Female	11.45±1.45	8.45±1.35

Apart from *B. zebina* some other tachinids and sarcophagids have also been reported to parasitise the tasar , oak tasar and muga silkworm larvae. The identified specimens are *Carcellia sp.*, *Parasarcophaga (Liosercophaga) dus* (Thompson) and *Lioproctia annandelei* Walker. *Parasarcophaga sp.* have been reported to parasitise oak tasar silkworm also. The nature of damage and symptoms of most of the uzi fly are similar with difference in larval development period.

HOST PARASITE INTERACTION

Silkworm larvae provide food for uzi fly larval parasites and ultimately the host suffers heavy damage. After feeding on the body contents, the maggots come out of the host and crawl on the ground and pupate in the soil. However, some of the maggots which do not come out from the larvae before spinning out of the cocoon, pupate inside the cocoon. This strange pupation behaviour of uzi fly is therefore of particular significance to the organism as it involves its survival since the pupation inside the silkworm cocoon is suicidal to the uzi fly. Neither the adult uzi fly has piercing or biting type of mouth parts nor proteinase enzyme capable of dissolving sericin in order to emerge from the silkworm cocoon. Thus pupation of uzi fly inside the silkworm cocoon is a self imposed natural check on its own population. The uzi maggots emerging from the host body crawl on the soil and form a hard puparium within 24 hours. During this transition it has to survive in the environmental condition completely different from that to which it was accustomed. This abrupt and severe change is reflected in much alteration in the anatomical, morphological and physiological characteristics of the parasite. Two of the most important changes are the increase in the efficiency of reproductive system and regression of digestive system. It has been observed that temperature and humidity influence the longevity and developmental stages of uzi fly and these two factors probably play a major role in distribution of uzi fly in tropical and temperate region of India. Of course, it is likely that other ecological factors also influence the spread of Uzi fly. The Uzi fly will not be a serious problem where extremely high and low temperatures prevail. Its infestation was found to be 5 to 7 per cent during seed crop and 25 to 40 per cent in commercial crop rearing. It is because the average temperature and relative humidity is not congenial for the population build up of the fly pest population during July–August rearing season where as during the commercial crop infestation increases due to suitable temperature and relative humidity. Temperature above 30°C and below 15°C are not suitable for the fly pest population.

NATURE OF DAMAGE AND SYMPTOM

The female fly parasitizes the silkworm larvae from the third instar onwards. The gravid female oviposits directly on the inter–segmental grooves of the silkworm body usually on the dorsal and lateral sides. After hatching maggots enter into the silkworm and feed inside the body of the larvae. The infested silkworm larvae can be identified from a long distance by the presence of medium size black spot on the larval body. It is due to the entry of maggot in the larval body. The infested larvae form flimsy cocoon and ultimately the silkworm pupae die. After hatching maggots inter into the silkworm body leaving a black spot on the affected area. This black scar is an important symptom for the identification of infested worms. Maggots feed inside the body of the silkworm for 20–25 days, and the full grown maggots come out of the body of the silkworm larvae after piercing the integument. It is very interesting to note that when the infestation is less (2-5 maggots per larva), the silkworm larvae does not die and it forms a cocoon whose quality is inferior in comparison to their normal silkworm. In case the infestation is very high (25–30 maggots per worm) pupal mortality is recorded in the field itself which becomes the source of uzi fly population build up in the rearing field.

Mortality of the silkworm larvae is also dependent on the stage of infestation. When the infestation was in the early stage mortality percentage is high but if the infestation is in late stage, the infected worms are able to form the cocoons. Generally infestation has been recorded to be very high in the late stage of the silkworm. It is due to the reason that there is a strong association in the uzi fly and silkworm which attracts each other due to interference of their kairomonal activities. Fly receives a strong odour from the food plants on which the silkworm is reared and the silkworm accordingly forms a strong association which results in the heavy infestation of the silkworm larvae (Singh *et al.*, 1996). Generally fly oviposits on the third instars silkworm larvae. When the flies attack later instar larvae, they do not always lay the eggs, but if they oviposit more eggs are laid than in earlier instar worms. Oviposition has been found to be positively correlated with the body weight of the silkworm. inside the host body in about 10–15 days. During second and third crop rearing 25–40% of the infestation has been recorded at some places in the tropical and temperate, depending upon the stage of infestation and the growth of development of the maggots.

MANAGEMENT

The extensive damage to silkworm crop due to uzi menace in sericulture created an appalling situation and shaken the very root of sericulture in India. Thus the threat of notorious fly pest becomes a serious concern especially because no preventive/control measures were known to check completely the uzi infestation. Control of Indian Uzi fly and Non mulberry uzi fly has engaged the attention of several workers in the past and various approaches such as preventive measures using fly proof wire mesh, mosquito nets, ante room and skipping rearing, trapping of female uzi fly, keeping kerosene in water on the windows of the rearing houses have been suggested. But non of them have been found to be very effective due to poor and unhygienic rearing conditions in India.

The principle involved in this method is to simply exclude the female uzi fly from the silkworms. Different kinds of methods have been suggested and tried for the control of uzi fly. Fitting wire screen to the windows and doors of rearing rooms has been suggested from the beginning, as also providing an anteroom/ fore-room at the entrance of the rearing house (Mukherji, 1912; Lefroy, 1917; Jameson, 1922). Mukerji (1912) gives elaborate description of trapping the female uzi flies that rush to get into the rearing house containing worms at the right stage by providing Kerosined water placed outside in trays at the windows. All these methods of preventing the uzi from entering the rearing house may work efficiently provided there are separate rearing houses/ rooms set apart only for rearing silkworms. It is a common sight to see that in most of the rural areas of Kamataka silkworm rearing is taken up in common rooms meant for human dwelling as well as even cattle. Rearing .in thatched and tiled houses, also used extensively for rearing silkworms in Karnataka, does not permit wire-screening of windows and doors, since there will be several entry points in such homes for the uzi flies.

The other methods to exclude uzi flies from rearing trays are meant to take care of rearing situations in which wire-net fitting does not work viz. tiled/thatched houses, etc. One of the ways is to partition the portion of the room where rearing stands are placed by means of nylon netting. If necessary even the top is also provided with netting so that the rearing stands are completely enclosed with nylon net including some extra space for changing leaves, bed-cleaning, etc. (Kumar *et al.*, 1987; Siddappaji, 1985; Bhat, 1987). Covering with nylon mesh only the rearing stands individually was also tried (Siddappaji, 1985; Bhat, 1987). Though these prevention methods have provided reduction in uzi infestation, such enclosures have been reported to lead to incidence of some diseases like muscardine, flacherie and grasserie as reported in some studies (Siddappaji, 1985; Bhat, 1987). Recommending such methods to the rearers would become difficult because of these undesirable side effects, which result from rearing shelf enclosures and partitioning of rearing place with nylon net. There is yet another way to exclude the gravid females from getting access to silkworms in rearing trays which has been tested and demonstrated in villages in rearers' own rearing. This method is to provide nylon mesh covers individually to each rearing tray which provides cent per cent control if carefully done (Siddappaji, 1985; Bhat, 1987). It is also claimed that individual tray covers (nylon mesh) would not only provide 100% protection against the uzi fly infestation which is at the same time environmentally safe, simple and inexpensive, but also gives 2 to 5 kgs/ 100 dfls more yield over the untreated control, a significant increase in the weight of silk glands, cocoon, and pupa and increase in reelable filament length and fibroin content. There was improvement in renditta also (Siddappaji, 1985; Bhat, 1987, Puttaswamy, 1987). Use of pesticide is sericulture has to be restricted, as the silkworms do not tolerate even sub lethal doses of toxic compounds. Therefore, pesticides can not be used in the rearing field or houses.

Since the endoparasitoid completes almost the entire immature life period inside the body of the silkworm larva it is difficult to control through chemical means. The only stage of uzi fly which are outside the body of the silkworm is the full grown maggots which pupate in manure pits and other muddy surroundings of the rearing houses or in the rearing field in non mulberry sericulture and the free living adult. Hence Integrated Pest Management (IPM) involving population suppression by physical, mechanical, cultural, genetic, chemical and biological control strategies have to be adopted, restricting the use of pesticide to the essential minimum level. Several methods have been developed to control the fly pest population viz. nylon net on outdoor/indoor rearing, light traps and methods based on chemical control measures like uzi trap, uzicide, uzi powder and even low concentration of diflubenzuron (Kumar *et al.*, 1986 a, b). Synthetic organic pesticides have been used to control the menace of fly pest but their overuse and frequent misuse have resulted in serious worldwide problems with disastrous consequences as to the profitability of certain crops.

Further, resurgence of target pest population due to development of pesticides resistance coupled with the elimination of natural enemies and competitors, the ascent of new secondary pests that often are more injurious and harder to control than the original target pests, toxicity hazards to man, plants, domestic animals and wild life, contamination of soil, water and food chains and whole sale pollution of environment (Doutt, 1964; Van den Bosch and Haramoto, 1953) have forced the entomologists to search for other control measures. Unilateral reliance on chemical pesticides is not likely to provide solution to all our pest problems. Safer, less costly alternatives to chemical control are therefore, desirable as part of an integrated interdisciplinary approach to uzi fly management.

PHYSICAL AND MECHANICAL CONTROL

The physical method consists of attracting the flies to light in the rearing houses. The flies are not normally attracted to light which is continuously on, but when the light is put on in a dark room, some flies are attracted to the light. A sticky board hung close to the light will trap the attracted flies. Pest O flash is also effective for the control of tasar uzi fly in the rearing field. Mechanical control involving systematic collection and destruction of uzi infected silkworms and removal of maggots from the rearing bush should be adopted. The maggots that drop from the rearing bush usually crawl on the ground before reaching the suitable place for pupation, and these may be easily collected from the field and later killed by putting the pupae in hot water.

This method should be carried out without fail and can be easily done during the rearing period regularly. The method ensures prevention of uzi maggots and pupae getting into manure heaps or soil thus forming an important major source of uzi infestation to the next silkworm crop. This is a simple method, which is often neglected by the rearers. in muga and oak tasar sericulture, removal of weeds and scrapping of surface soil up to a depth of six inches in the rearing field is necessary to kill the maggots of uzi fly especially where fourth, fifth and spinning stages worms are reared. Uzi infected larvae and dead larvae should be immediately collected and destroyed. This process is very congenial during the transfer of silkworms from one to another bush in the field. In the grainage, the maggots and pupae should be collected and destroyed daily. Stray emergence of uzi fly adult should be checked in the grainage itself to avoid further spread in the rearing field. After harvesting of cocoon they should be sorted out and flimsy and uzi infested cocoons should be immediately stifled and sent to reeling sector.

CHEMICAL CONTROL

Though utilizing toxic chemicals in controlling the uzi fly pest has limited possibility because the exposed stages of the pest outside the host worm are few, and the risk of applied chemical becoming toxic to the silkworm itself is rather hard to overcome, some workers have attempted to tackle the pest by employing toxic chemicals. Sprinkling levigated China clay on spinning worms to prevent egg laying by the uzi fly (Kumar *et al.*, 1983 b; Jolly and Kumar, 1985) and feeding the worms with 1-3% bleaching powder solution treated mulberry leaves will control the uzi infestation (Sengupta *et al.*, 1980). Diflubenzuron and benzoic acid when applied on silkworms having uzi eggs on them are known to be toxic to the eggs and thus bring about control of uzi infestation (Kumar *et al.*, 1983 a; Jolly and Kumar, 1985). Since fully mature and spinning worms are least receptive to uzi eggs, application of levigated China clay is of very little utility. Diflubenzuron is well known as chitin inhibitor and has been used successfully especially against lepidopteran pest. It is very necessary to ensure safety to silkworms on which it is recommended for application (Siddappaji, 1985). Recently a combination product, Uzicide has been developed as an ovicide/oviposition deterrent against the uzi fly, and is now available in the market for use by the rearers (Anon., 1987).

BIOLOGICAL CONTROL

Biological control is desirable requirement for uzi fly control in sericulture. It is an ideal form of pest management – permanent, inexpensive, non-hazardous with its firm basis in sound ecological principles. It is the most successful and most promising alternatives of chemical control. When successful the utilization of native natural enemies is inexpensive and easy to multiply, It is non-hazardous means of reducing uzi fly populations and maintaining them often permanently below established economic thresholds level. Therefore, it is essential to know the potential biocontrol agents and its biological attributes in the native natural habitat of silkworms and its host plants.

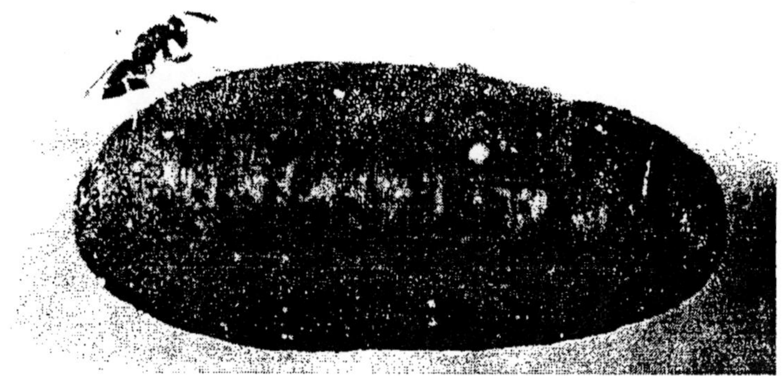

Fig. 5.4 : Nesolyx Thymus Parasitizing Uzi Fly Pupa

NATURAL ENEMIES OF UZI FLY

Both predators and parasites are animals that feed on the other animals, but a predator consumes several host individuals during its development, where as the parasite completes its development on a single host (Singh *et al.*, 2001). A parasitoid is a special kind of predator is often the same size as its host, kills the host, and requires only one host (prey) for development into a free-living adult. Several parasitoids have been screened and their parasitization potential has been tested against uzi fly pupae. *Nesolynx thymus*, *Trichopria* sp., *Exoristobia philippinensis*, *Dirhinus* sp. *Brachymeria lugubris*, *Spalangia endius*, *Pachycrepoideus veerannai*, and *Spilomicrus karnatakensis* are the important parasitoids of uzi fly (*Exorista bombycis*) (Kumar *et al.*, 1988; 1989 a, b; 1991; Samson and Ramadevi, 1985). *Nesolynx thymus* prefers to parasitize Indian uzi fly puparia where as *Trichomalopsis apanteloctena, Pediobius sp., Brachymaria lasus* and *Theronia maskaliya* generally parasitize tasar uzi fly puparia (Singh *et al.*, 1995; Singh and Thangavelu, 1995, 1996). Almost all are effective in minimizing uzi fly population but the parasitization potential of these parasitoids differ variably. There is a 10-20 percent variation in the parasitization potential of these parasitoids. Field release study has shown that even a potential parasitoid under laboratory condition does not become successful in new ecosystem, thus it needs to be standardized (Singh and Thangavelu, 1999).

DESIRABLE ATTRIBUTES OF NATURAL ENEMIES

In order to assess the usefulness of bioagents for the control of uzi fly the study of various attributes of the parasitoid has been made by the methods described by Singh and Sinha (1995). The main attribute of the effective natural enemy is searching capacity, specificity, power of increase and adaptability (Waage, 1986; Lenteren, 1983). All these attributes are of course closely related to each other and influence the population density of parasitoids in natural habitat.

HOST SEARCHING ABILITY

Searching capacity, manifested by the ability to find the host even when it is scarce, is commonly regarded as the most important attributes of effective natural enemies. The concept has been commonly expressed in terms of natural enemies area of discovery (Lenteren, 1986). A parasitoid beneficial as bioagent must successfully utilize the low-density population of the pest. *T. apanteloctena* and *N. thymus* has been found capable of attacking more than 75% of the uzi fly pupa within a six hour. It indicates its high searching capacity, i.e., the ability to find the hosts, when the host density is low. This is a more important attribute than a higher reproductive potential (Singh *et al.*, 1994). Studies made on the area of discovery of *Nesolynx thymus* revealed that it also has a high searching ability at higher host densities (Singh and Thangavelu, 1997). This behaviour enables the parasitoid to reproduce more rapidly at such densities. The search for natural enemies are preferably carried out in undistributed environments. Backyard gardens and uncultivated tracts are sometimes more likely to yield as greater varieties of species than commercial plantations. To prevent the inadvertent introduction of any undesirable organism, including hyperparasite species, all-important material should be handled under strict quarantine.

SPECIFICITY

Although some out standing successes of biological control have been achieved in several other crops through utilization of natural enemies of related species, the empirical record of biological control studies in sericulture indicate that most success have been achieved with fairly specific natural enemies viz. *N . thymus, T. apanteloctena* and *Trichopria* sp. They are attuned to the physiology, behaviour, habitat preferences, and pattern of its dispersion and phenology of uzi fly. (Kumar *et al.*, 1992). These parasitoids have degree of biological adaptation to the pupa of uzi fly and also has a greater degree of direct and rapid responsiveness to density changes in the population of the uzi fly (Singh and Thangavelu, 1999). *N. thymus* is not strictly a host specific parasitoid as it is also reported to parasitize other dipteran insect. *T. apanteloctena* is another potential parasitoid and it was observed that both the species co-exist on the same host. In the laboratory both the species were more or less equally parasitizing the uzi fly pupa. It is therefore, more likely to be highly density dependent in relation to its particular host (Huffaker, 1971) than a more generalized feeder.

On the other hand general feeders have the decided advantage of being able to survive even during period of scarcity of given host (Huffaker, 1971; DeBach, 1974). Most of the uzi fly parasitoids are oligophagous and many are truly polyphagous. However, the process involved in host finding and discrimination by the predators and parasitoids in nature are, of course, intimately associated with their searching behavior. They are seen as comprising a series of four restrictive steps: habitat selection, host findings, host acceptance and host suitability (Doutt, 1964). Habitat selection is first and sometimes most important step in sequence. Before actually searching for its host, a natural enemy may seek a certain environment, a particular type of vegetation or host plant etc. In locating its preferred habitat it may respond to various chemical and physical cues, which may be related to either the host plant or the host itself or to their interaction (Hagen *et al.*, 1976; Vinson, 1981). Chalcid parasitoids usually exhibit marked host specificity and host findings and host acceptance are heavily mediated by responses to physical and chemical cues (Townes, 1972). Host suitability is another important factors to parasitoids that oviposite to other hosts.

The success and failure of parasitization is largely dependent on the host suitability by such parasitoids. Oviposting parasitoids may inject venom that modifies the host and render it suitable for parasite development (Vinson, 1975). In most of the cases, the ability to recognize and avoid unsuitable host is highly desirable attribute of parasitoids. The relationship between the number of hosts parasitized and the host densities is found to be linear. There is a sharp increase in the number of host paraasitized as the host densities increases, but increase in the host density beyond the capacity of the parasitoid to parasitise it lead to decrease in the percentage of parasitization (Singh and Thangavelu, 1999).

POWER OF INCREASE

A natural enemy that has received host regulation in a stable environment that has reached the hypothetical steady state would only require a power of increase sufficient for replacement of the parent population in each generation. The actual power of increase of a natural enemy in the field may be affected by its fecundity and rate of development, as well as by such other factors as its searching capacity and adaptability to the condition of the particular habitat (Singh *et al*., 1995; Singh, 1998).

FITNESS AND ADAPTABILITY

To be effective in a new habitat, an introduced natural enemy should preferably be pre-adapted to it. However, the possibility of possible gradual post-colonization adaptation cannot be discounted. A well-adapted natural enemy should not require any essential requisites that are not present in the new habitat. It should be able to tolerate the prevailing climatic conditions, and should be effectively synchronized with the biology and phenology of its host in the habitat. Specific features required for synchronization may vary with the habitat. Diapauses for instance may be mandatory in one area, whereas in another area it may prove to be detrimental. In general precise synchronization is more likely to occur in a host specific natural enemy than in a general feeder. Ideally, a natural enemy should be adapted to all the habitats and niches occupied by the target pest. It should frequent all the host plant and tolerate all the climatic regimes that its host does, and should equally effective in all of them. Just as the natural enemy should be well adapted to the various natural aspect of the ecosystem, it should also be adapted to cope with man made hazards, such as pesticidal treatments. (Croft and Brown, 1975; Rosen and Huffaker, 1983). Thus the factor effecting the searching behavior of natural enemies may also determines its host specificity where as searching efficiency of a population of natural enemies is also a function of number searching, hence their power of increase and their fitness (Hassell, 1978). Just as the natural enemy should be well–adapted to the various natural aspect of the ecosystem, it should also be adapted to cope with man made hazards, such as pesticidal treatments. Resistance to various pesticides is known to have developed in the field in the number of predatory mites (Rogers and Hassell, 1974, Vinson, 1981) and further research is likely to discover additional cases among various natural enemies. Such resistant species strains should be regarded as highly desirable for biological control agents (Alphen, 1986). The tolerant doses of insecticides against the pupal parasitoids of uzi fly especially, *N. thymus, Trichopria* sp., *Spalangia* sp., and *T. apanteloctena* needs to be standardized for successful release in the field.

OTHERS ATTRIBUTES

Various other attributes have been ascribed to effective natural enemies, but they all are regarded as manifestations of the searching capacity, specificity, power of increase and adaptability. Thus, the factors effecting the searching behavior of natural enemy may also determine its host specificity, whereas searching efficiency of a population of natural enemies is also a function of the numbers searching, hence their power of increase and their fitness (Huffaker *et al.*, 1970). It does not mean that a effective natural enemy have to be superior in all these attributes. Even in outstanding biological control successes the controlling natural enemies often exhibited deficiencies in some of them. Competitive ability has often been mentioned as desirable. One should distinguish here between extrinsic or exploitable competition, involving efficiency in the exploitation of pest population by a population of natural enemies, and intrinsic or interference competition which, in parasites, involve competition between the larval forms or in a given individual host. Superiority in exploitative competition is largely determined by a natural enemy searching efficiency, whereas superiority in interference competition is a form of adaptiveness that may be less desirable than the capacity to discriminate parasitized hosts and avoid such completion altogether.

UTILIZATION

In well–design IPM program, the pest population is maintained at lower level by the action of parasites, predators and pathogens (DeBach and Hagen, 1964). There are three basic approaches to applied biological control: importation, conservation and augmentation.

IMPORTATION

Importation (classical biological control) has been used very successfully against a variety of pests (Clausen, 1978; DeBach,

1971). Screening of potential parasitoid is the most important factor in any successful biological control program. There are however, still many important pest species that are not amenable to the classical biological control approach (Nordulund, 1984). Many of these pests are found in disrupted environments such as annual row crop and green house agro ecosystems. These cropping systems are not generally conducive to the establishment of long-term stability, which is critical to the success of a classical importation program (Lewis and Nordlund, 1980). Importation including the search for, transfer, quarantine handling, colonization and establishment of exotic natural enemies, has been by far the most important and most promising approach to date, and has accounted for the great majority of outstanding success in applied biological control. It is also the least expensive method of natural enemy utilization.

However, in certain instances an already established natural enemy, whether indigenous or introduced, may show considerable promise but fall short of fulfilling its potential due to inadequacies of its own attributes or the environment. There are several potential parasitoids of uzi fly existing in the nature but its exploitation has not been done due to lack of sufficient information on the behavioural response of the parasitoids. Even well-known natural enemies of proven high efficiency still await transfer into many areas, where uzi fly is a serious problem. There are several hymenopteran parasitoids available in the nature, which can be utilized to control the uzi fly population. These unknown species are considered as potential weapons and it can be applied as biological control tools (Kerrich, 1960; Townes, 1972). Sound systematic is an essential pre-requisite to the success of biological control program. Therefore, the correct identification of pest and parasitoids are of utmost importance (Rosen, 1978).

CONSERVATION

Conservation involve permeated action to protect and preserve existing parasites, predators and pathogens; basically not taking actions that would be detrimental to natural enemies (Nordlund, 1984; Rabb, *et al.*, 1976). IPM programs that result in a reduction in pesticide use generally contribute to conservation.

Conservation and augmentation in biocontrol involves two phases, first the maintenance of existing parasitoids by avoiding harmful practices and secondly, the augmentation of parasitoids either directly releasing them in the field or by indirectly making the field environment more favorable for them. Although a number of species of eulophid parasitoids have been reported, it is rather easy to identify more effective parasitoids based on their parasitization capabilities. *Neoslynx thymus* and *Trichomalopsis apanteloctena* both multivoltine parasitoids have been identified to be such potential parasitoids. These parasitoids, which appear after the arrival of uzi fly maggots in the field (Kumar *et al.*, 1990) need to be conserved from the use of pesticides, which is very intensive on mulberry plants. Several authors have worked out the toxicity of popular insecticides against chalcids. Toxaphene, endosulfan and phosalane have been reported very safe, while fenthion and ethion are highly toxic. Conservation of parasitoids can be better served if systemic insecticides are encouraged. Another important aspect of conservation in the field is to find out the means to arrest the depressiveness of the parasitoids from field to field or away from the site of release. Waage (1986) reported the evidences for parasitoids movement and reasons thereof. Basically the movement is reported to be influenced by prevailing temperature – higher temperature induces more flight and low host density – more movement.

AUGMENTATION

Augmentation involves actions to increase the populations of beneficial effects of parasitoids, predators or pathogens (Nordlund, 1984; Rabb *et al.*, 1976; Ridgway and Vinson, 1977). There are two basic approaches to augmentation, environmental manipulation and periodic releases (Stinner and Bradley, 1989). Periodic releases can be inoculative or inundative. Inoculative releases are releases of relatively small number of biological control agents, often on seasonal basis. The control in inoculative release programme is expected to come primarily from the progeny of these agents being released. Inundative releases are releases of relatively large numbers of biological control agents, and the control is expected to come from the released agents, not necessarily from their progeny (DeBach and Hagen, 1964). Inundative releases programmes usually involve the number of releases during the season, while inoculative release programs may involve only one release. It is possible by sustained release of laboratory-reared parasitoids. Our ability to use periodic releases of the parasitoid to control uzi fly, is to rear, transport and effectively release large number of high quality biological control agents. Periodic release requires continuous release programme and thus, has commercial potential and fit IPM programmes well. The growing number of commercial suppliers of biocontrol agents is evidence of this potential (Lisansky 1990; Thomson, 1992). Adoption of biological control has had positive economic impact (Tisdell, 1990). It must be pointed out that biological control particularly augmentation and conservation does not operate in vacuum. Biological control is the most successful as part of an integrated program involving host resistance and cultural control technology. The efficacy of such natural enemies may be sometimes enhanced through various method of augmentation, or direct manipulation of their populations, such as periodic colonization, genetic improvement, or the use of semio chemicals that affect their performance. Another possible approach is through conservation, or manipulation of the environment, either by adding of asking requisites or by mitigation of various detrimental factors.

SEX RATIO

Sex ratio plays a very significant role in biological control. It helps us to understand the reproductive strategies of the parasitoids because a significant variation in the sex ratio was observed in parasitic Hymenoptera. The relative abundance of males and females commonly fluctuates when parasitic hymenoptera are propagated for several generations. The fluctuation results in part from changes in the parasitoid : host ratio at the time parasites is ovipositing.

Usually an increase in the parasitoid : host ratio reduces the per cent age of female progeny in arrhenotokous parasite species with gregarious larvae, either by reducing the percentage of fertilized eggs, i.e. female eggs that the parent females lay (Wylie, 1976), or by killing relatively more female than male parasite larvae on super parasitized host (Wilkes,1963; Benson, 1973). It has been reported that five females of *T. apanteloctena* produced relatively more female progeny from 100 hosts exposed daily than from 25 hosts ; percentage were 76.2 and 53.7 for 892 and 618 adults, respectively (Singh and Thangavelu, 1997). The results indicate that the percentage of female progeny progressively decreased as the parastioid : host ratio increased. However, in *N. thymus* it was observed that progeny production affected when 9 days old puparia were exposed to the parasitoids. The increase in the progeny production was not significant with every succeeding increase in the number of host puparia exposed to single parasitoid except when the parasitoid to host ratio was increased from 1 to 1:2 (Kumar, 1990). These findings therefore, suggest that maximum progeny production is obtained by providing 1 day old 5 host puparia to single parasitoid. Further, it has been reported that age of parasitoid had no significant influence on rate of parasitism and progeny production of *N. thymus*. But there is a direct and progressive relationship between the progeny production and increase in the number of host puparia. It is, therefore, inferred that maximum progeny can be obtained by maintaining 1: 4 or 1: 5 parasitoid to host ratio. Sex ratio is affected by several biotic and a biotic factors. Apart from this, interference among the parasitoid adults, previous parastization of the host; and differential mortality of the male and female larvae on super parasitized are the important factors in the maintenance of sex ratio population in the field (Singh, 1998). The sex ratio ranges from a slight male bias to entirely female bias (Waage, 1986). The sex ratio is affected by several other biotic and a biotic factors but the mean sex ratio among the field collected samples were always female biased (Singh *et al.*, 1995). Besides this some other basic systematic and biological information are required to be known for the potential parasitoids. Verification of primary status of natural enemy is of course, of crucial importance, but this may be done in quarantine after importation. Importation about voltanism, diapause, specificity etc. can be useful in planning; release and establishment of strategies but this too may be obtained at later stage. In any event, only reasonably well-identified primary natural enemies should be released for the control of the fly pest population in the field.

SEMIO CHEMICALS

The presence of kairomones in the IV and V instars of silkworm of mulberry and non mulberry and of a sex pheromone in the gravid female uzi fly of *Exorista sorbillans* and *Blepharipa zebina* was demonstrated by Kasturi Bai *et al.*, (1986) and singh *et al.*, (1996). Kiaromonal response of these insects has been demonstrated and attempts have been made to find out a suitable package for utilization in sericulture.

GENETIC CONTROL

Recently some attempts have been made to screen conventional chemosterilants as well as newer compounds for their sterilant effects on uzi fly. Apholate, Theotepa, Tepa are the conventional chemosterilants which have been tried against the uzi fly and are known to bring about sterility in the flies to various degree. Recently, diflubenzuron when dusted on uzi maggots and pupae has known to cause sterility in the adult flies. Quarantine measures in the pest control are intended to see that the pest does not get introduced into new area, and are mostly applied to prevent entry of a pest of foreign origin. These apply equally well to limit the distribution of a pest to the region/locality where the pest is found in a country. The Indian uzi fly was confined to north–east India and perhaps Burma and Thailand before 1980 when it was introduced to Karnataka (Anon., 1980). Maxwell Lefroy (1917) even warned that this fly pest should not get introduced into the prosperous sericultural areas of Mysore. This fore warning should have been taken seriously and thus prevented the uzi fly menace to South Indian silkworm rearers and thus a serious loss to Indian sericulture.

INTEGRATED MANAGEMENT

Though individual control methods are effective, simple environmentally safe, unilateral dependence on a single method may eventually prove unworkable due to some reason or other. Integration of two or more methods and following management principles may provide an answer to such condition. In fact integrated approach has been suggested recently, which consists of (1) spraying uzicide to kill eggs, (2) release of certain parasitoids and (3) dusting diflubenzuron or neem derivatives on maggots/ puparia to suppress reproductive efficiency of the uzi fly. In the control of Indian uzi fly nylon covers to individual rearing trays have all the attributes of an effective way of tackling the uzi fly menace. This exclusion method if rigidly followed can provide almost cent percent control of the uzi fly to the individual rearer who practices this method without any undesirable side effect. However, nylon net rearing is not economical in tasar culture but in muga culture it is profitable.

Ichneumon fly; *Xanthopimpla pedator* (Ichneumonidae : Hymenoptera)

Xanthopimpla pedator is the most common ichneumon fly in tropical and temperate region of India. It prefers to parasitize saturniid pupae of tasar, oak tasar and muga silkworm belonging to non mulberry sericulture. Besides this, some new species of ichneumon fly viz. *Xanthopimpla punctata* and *X. stemmator* are also recorded to parasitize silkworm larvae. The antennae are long and two segmented. The female fly has a prominent needle like ovipositor about 1 cm long with two identical stylets The forewing has the cross vein 2m-Cu and on hind wings anal lobes are absent but trochanters are present. The insect is yellowish with a black spot dorso-laterally on each sternum as well four black spots on the prothoracic shield. It parasitizes the late stage silkworm larvae. The adults are active on warm sunny days and frequently attacking tasar and muga silkworm larvae in rearing field during October–November. The female parasitoid uses its antennae to search for spinning worm in the tasar plantation. After locating the host, the ovipositor penetrates the spinning worm to lay a single egg. A single female parasitizes many silkworm larvae. It can lay as many as 5 to 14 eggs daily and lives for 4 to 9 days. Development of the larval parasite from egg to pupa takes place within the pupa.

Fig. 5.5 : Ichneumon fly (Xanthopimpla pedator) searching host

The maggot after hatching eats the insides of the pupa and pupates inside the remaining skin. One Ichneumon fly (male or female) from its embryo formation to adult emergence consumes fully the internal tissues of one alive pupa (male or female) inside the preserved seed cocoons in grainages. The adult emerges by making a hole in the cocoon. After pupation, the adult parasite emerges from the pupal host. The adult Ichneumon fly emerges out by piercing cocoon in the anterior end near the peduncle causing the reduction in reelability. The endoparasitisation by *Xanthopimpla* species is more at the time of cocoon formation during II and III crop in case of both bivoltine and trivoltine breeds. The killing of pupae of the preserved seed cocoons of II and III crop harvest in the grainage houses causes a great loss to the seed production (Dfls) during I crop grainage of the ensuing year. In order to minimize the parasite population, it is advisable to search biocontrol agents instead of using chemical measures. . Knowledge of biological and behavioural findings is of paramount significance for formulating the strategies required to devise efficient repressive methods. Chemical control measures can not be used during silkworm rearing period because silkworms can not tolerate even sub lethal doses of toxic compounds.

Braconid fly; *Apanteles glomeratus* (Hymenoptera : Braconidae)

Apanteles sp. is an important parasite of silkworm in sericulture. The fly is diurnal in habit. It is a small stout–bodied insects found among plantation of silk host plants. They are distinguished from the ichneumonidae in the cross vein 2m-Cu being absent in the forewing and therefore, there is only a single recurrent vein as against two recurrent veins in ichneumonids. The first sector of M+Rs which is present in the Braconidae is absent in the ichneumonidae. *Apanteles flavipes* is a very common braconid which paarasitises lrvae of muga and tasar silkworm. It prefers to parasitize early instars (first to third) tasar and muga silkworm larvae during October–November. The parasite was also reported from various other lepidopteran pests, which serve as major population limiting agents of the pest of silk host plants. Adult is a very swift flier and frequently moves from one bush to another bush for searching of the host for parasitization. The adult of *Apanteles* mated immediately after emergence. Mated females immediately even oviposited on the host larvae. The female lays egg within the body of the silkworm larvae and oviposition period lasted 1–3 seconds. Life history parameters indicated that the female laid 9–13 eggs inside a single host. The egg larval period ranged form 8–10 days. Fully fed parasitoid larvae emerged from the dead host larvae and formed yellowish white, smooth and oval cocoon immediately by the side. The pupal period was 3–5 days The duration of adult males (4–5 days) was longer than that of the female (3-4 days) The total life span of the males ranged from 14–20 days and those of the female is from 14–20 days. Adult female causes maximum damage to silkworm larvae by parasitizing. The parasitized larvae fails to form the cocoon and thereby causes extensive damage to silk industry.

Stink bug; *Eocanthecona furcellata* (Wolff.) (Hemiptera : pentatomidae)

The stink bug *Eocanthecona (=Canthecona) furcellata* (Wolff.) (Pentatomidae: Hemiptera) has been reported as a predator of several pests of agricultural crops in southeastern Asia (Cherian, 1917; Pant, 1960; Chu, 1975; Rai, 1978; Gope, 1981; Chien *et al.* 1984; Rani, 1992; Rani and Wakamura, 1993). Besides this, it has been recorded as a serious predator of tasar silkworm (*Antheraea mylitta*), oak tasar silkworm (*A. proylei* and *A. roylei)* and muga silkworm (*A. assama*) larvae causing considerable loss to silk industry in India (Jolly, 1967, Singh *et al.*, 1992a). Sen *et al.* (1971) reported the biology and population dynamics of *Canthecona furcellata* in tasar eco system. The insect completes its lifecycle in 35 to 40 days and is multivoltine in nature with 6-7 broods per year. There are five nymphal instars and the fifth moult takes place on the 19th day (12 mm) transforming into an adult which develops fully by 20th or 21st day. The adults are bronzy in colour. The mating takes place 3-4 days after the last nymphal moult. A single female lays about 200 to 300 eggs extending for about 8 days. The incubation and nymphal period lasts for 6-15 days respectively. The adults are dimorphic with thoracic marks. The females are broader than male caries mid ventrally a pink spot on the 5th sternal segment near the anterior margin. In the male the marks extend slightly to the fourth sternite in the form of a faint pink line. An adult lives for about 13 to 19 days. The survey results indicate that the predator makes its appearance in the last week of April and increases above economic injury level by the second week of May. The population increases till September and rises abruptly to its peak by the middle of October. After mid–October the population starts declining and the bug goes under hibernation between December and February (Singh *et al.*, 2000). The feeding behaviour of stinkbug is interesting. The bug approaches its prey slowly and suddenly pierces its proboscis inside the body of silkworm into the central nervous system. They generally attack the middle of the body of the prey so that the efforts of the prey to get rid of the enemy fail. After piercing, the feeding process continues for about 15 minutes at a stretch. Generally the first and second instars larvae are killed with one prick while later instars can survive 4-5 pricks. The predatory potential of the *C. furcellata* is very high. A single predator can feed and kill 130 to 225 tasar larvae of 1st to 3rd instars.

Fig. 5.6 : Adult stink bug (*Eocanthecona furcellata*) in rearing field

Several control methods (chemical, mechanical, and cultural) are known to minimize the pentatomid bug population, but farmers prefer to use synthetic insecticides due to the ease in application. Though use of synthetic insecticides has provided us with effective control of almost all major pests and predators, yet their undesirable side effects limit their continued use. However, repeated and frequent application of modern synthetic insecticide has created problems of residual toxicity, development of resistance to insecticides, pest resurgence and out break, phytotoxicity and hazards to non target species including natural enemies and other beneficial organisms, alternation in pest species population dynamics, environmental degradation and disruption of natural balance.

In view of these drawbacks associated with excessive dependency on chemical control, the concept of 'Integrated Pest Management (IPM)' which combines all possible measures into a compatible and harmonious package, has gained prominence (De

Bach, 1974). Biological control is considered as an essential component of IPM as it is economical, effective and eco-friendly. Natural enemies play an important role in suppressing pest population in the crop whenever suitable conditions prevail for their survival, development, conservation, and multiplication in any agro-ecosystem. The complexities in the behaviour and life cycle of pentatomid bugs warrant a special attention for their effective management especially in view of the changing practices in sericulture. For a sustained development in all spheres of life, sustainable sericulture has become a significant topic of discussion. Conservation of our natural resources, air, soil and water, stands at top priority and hence IPM strategies should aim at the approaches, which do not lead to degradation, depletion or pollution of nature's gifts. Thus it is high time that we may lay more emphasis on evolving eco-friendly non-chemical approaches for pest management. Biological control is one of the most important methods, which can be used to control the undesirable pests and predators. Various potential parasitoids have been screened which can be utilized as an agent of biological control in sericulture (Thangavelu and Singh, 1992). The use of bio-control agents has many advantages. Biological studies of *Psix striaticeps* revealed its importance as a promising biological control agent against the stink bug *C. furcellata* (Singh and Thangavelu, 1996, Singh *et al.*, 1992b, Singh and Sinha, 1995).

HOST LOCATION AND HABITAT SELECTION

Host location and preference using chemical cues constitute the first step leading to successful parasitization. Chemical signals have a significant role in organizing the host, pest and parasitoid complexes. Semiochemicals help natural enemies to locate and recognize their hosts through many inter and intra specific interactions. The host plant, the insect, predator or parasite, all release exudates, and these chemicals serve as mediators in host selection. These volatile chemicals affect behaviour of not only the host insect, but also the parasitoid. Studies on the enhancement of tropical tasar silk production through manipulation of host plant-insect herbivory interactions indicate that chemical constituent of host plants play a key role in effective manipulation of communication system involved in allelochemical web of triotrophic interaction in sericulture (Babu and Chauhan, 1994; Som prakash, 1996; Roger and Hassell, 1974). Similar observations on insect plant interactions were reported earlier by Bragg, 1974; Ladden, 1970; Price *et al*., 1980; and Elzen *et al*., 1983. Earlier studies were conducted on two tropic levels only, plants and their herbivores. But the recent researches on the biological control reveals that it is intimately linked with the allelochemical web of plant-pest-parasitoid/predator, resulting in a tritrophic or sometimes a tetratrophic level of interaction (Elzen, 1983; Price, 1981; Schltz, 1983; and Muller, 1983).

Our studies of allelochemical interactions among silkworm host plants and its associated herbivores and parasitoids were designed to determine how these three trophic level system operates in the light of the biological problems outlined above. Chemicals mediating parasitoid-host plant interactions are classified as kairomones, allomones, and synomones (Lewis *et al*., 1982) and appear to be key factors determining host location and governing the range of hosts attacked by parasitoids. One major task faced by a female parasitoid is location of the habitat harbouring host insect. Initially a parasitoid may seek certain environment regardless of presence or absence of the hosts (Doutt, 1964). The host however, remains only in specific locations within that environment, and a female parasitoid must locate those locations of the habitat most likely to yield their hosts. Factors attracting the parasitoid to a plant and retain it in the area have a positive selection value for the plant due to parasitoid's beneficial effects in reducing herbivore survival and fitness (Sauls *et al*., 1979). The foraging activity of the parasitoids may be influenced by a number of factors in the plants or hosts.

Plant may produce chemicals that attract and retain parasitoids (Nettles, 1979; and Vinson, 1975) or provide nutrition (Sahajahan, 1974) and thereby attracts parasitoids. A few attractants have been identified in pioneering studies of parasitoid-plant interactions (Read *et al*., 1970; Camors and Payne, 1972; Elzen, 1983; Elzen *et al*. 1984 a&b; and Lecomte and Thibout, 1984). Parasitoids are more often attracted to plants on which their host feeds (Thorpe and Caudle, 1939; Monteith, 1955 & 1964; Arthur, 1962; Manjunath,

1986 and Madden, 1970). Damaged plants may provide stimuli for increased parasitoid searching (Nishida, 1956; Bragg, 1974; and Vinson, 1975). Some parasitoids are be attracted by fruits and flowers (Nordlund and Lewis, 1976; and Voronin, 1981) or other specific parts (Nishida, 1956). Plants also may influence the general attraction of the herbivore (Muller, 1983) or kairomonal activity of the herbivore's frass (Sauls *et al.*, 1979; Nordlund and Sauls, 1981; and Elzen *et al.*, 1984b) and host reared on different plants may vary in attractiveness (Nordlund and Sauls, 1981). Retention of beneficial insect on the plant may require sources of nutrition, refuge sites and host insects (Lewis *et al.*, 1982) besides chemical cues. Our experiments on plant produced attractants for beneficial insects and effects of plant defence compounds on the quality of herbivore as reproductive resource for parasitoids focused on *Terminalia*. It was chosen because of its economic importance in sericulture and availability of many related species as well as data on *Terminalia* volatiles. Babu and Chauhan (1994) studied the volatile compounds in *T. arjuna* and *T. tomentosa* leaves and its effect on the growth and development of silkworm larvae.

REASON FOR BIOLOGICAL CONTROL

Insect populations have a tendency to fluctuate as result of their inherent characteristics as influenced by the environmental factors. The magnitude of the increase and decrease in numbers is governed by the degree of influence of various environmental factors. The rate of change in pest population is determined by the fecundity, rate of development and survival among its members. The same factor may be favourable in case of one population but may become unfavourable for another. It is thus necessary to consider the influence of various factors with reference to a particular pest population (Beine, 1984). The pest control strategy, primarily involved unilateral use of pesticides which had many shortcomings and hence the emphasis on Integrated Pest Control/ Pest management. The control of various pests and predators in sericulture is a complicated process due to the close association of silkworms and their food plants. The requirement of pesticides is high. Since the pesticides are poisonous, their use poses a threat to silkworms as well. Sometimes indiscriminate uses of hard pesticides promote speedier evolution of insect pests, affect non-target species, convert formerly innocuous species into noxious pests, and leave undesirable residues in silkworm and their host-plants. Therefore, the focus now has been shifted from chemical control to other alternative methods of pest control or adoption of an integrated approach based primarily on use of bio control application of botanicals, and safer pesticides. Such an approach may lead to reduction of residues and ancillary problems associated with pesticide application (Huffaker *et al.*, 1971). Pesticides disrupt interaction between phytophagous insects and their natural enemies, the essential ecological processes operating in nature that contribute to the regulation of insect population.

Whenever this interaction between phytophagous and entomophagous insects is disrupted, the population of phytophages, increase tremendously and they attain pest status because they become free form the constraints imposed by the entomophages. The realization that conventional pesticides could cause problem came up from the idea that it might even kill the beneficial insects controlling the pests bringing an imbalance leading to a surge in the pest population due to lack of their bio-control agents. Biological control strategies are under utilized and it cannot be a total alternative for chemical control, judging from the fact that many of the successful causes of the pest control achieved through biological control predated the era of agrochemicals (Flint and Van den Bosch, 1981). Being forgotten so long and dominated by the use of agrochemicals, biological control is now identified and considered as an alternative method of insect control, together with other non-conventional or plant based control measures (Upadhyay *et al.* 1997 and 1998). Biological control refers to management and regulation of natural biotic forces to suppress pest populations to a level below the economic injury. Identification of many naturally occurring predators, parasitoids and pathogens prevalent in sericulture ecosystem is the first step towards their conservation, augmentation and manipulation. Classical biological control program is generally less pursued in developing countries like India, where there exists a need to thoroughly explore and evaluate their native natural enemies, which may be potential biological control agents. Fortunately, in India, the native natural enemies have a little probability of extermination caused by synthetic pesticides due to lack of sufficient pesticides or the poor quality of pesticides. It is, therefore, obvious that the conditions prevailing in India can lead to more conducive environment for implementation of the biological control programme by using native or introduced bio control agents (Davis, 1967; and Napompeth, 1987). The current revival of interest in biological control is also driven by a change from pest control approaches to maximize productivity to the long-term sustainability of the sericulture ecosystems.

The biological control of the pests tends to be a long lasting, often can be implemented at little direct cost to producers and consumers. For these reasons, biological control is considered a corner stone for many Integrated Pest Management (IPM) programmes. The philosophy of modern insect pest management is based on the management of entire pest population, unlike in the past where separate techniques are employed for each individual pest. In IPM, emphasis is laid on the use of combination of methods, aimed at providing cheap and dependable long-term approach with reduced side effects. The philosophy of modern IPM programme is thus compatible with the philosophy of biological control; indeed biological control has been the core around which IPM has been developed. The reason for this is that natural enemies constitute the major pest control factors those can be manipulated. The parasitoids can be utilized in three major ways. (i) Imposition of exotic species and their establishment in new habitat, (ii) augmentation of established species through direct manipulation of their population by mass population in insectaria and periodic colonization and (iii) their conservation through deliberate manipulation of environmental factors to enhance their activity (DeBach, 1974; Kumar and Mukerji, 1996)

Fig. 5.7 : An egg parasitoid emerging from egg mass of Canthecona furcellata

APPLICABILITY OF BIOLOGICAL CONTROL

In biological control parasitoids were favoured over predators because they were most host specific, better adapted and synchronized in interrelationships, have lower food requirement per individual thereby maintaining a balance with their host species at their lower host densities and their larvae do not need to search for food (Van Leteren, 1986). The parasitoids have been used more frequently than predators in the biological control programmes and about 80% of all biological control programmes are because of them (Hokkanen, 1985). Bio control agents are now managing more than 300 agricultural and urban insect pests in more than 100 countries. Indeed, the research and development efforts on biological control are very meagre as compared to control by synthetic pesticides. Very little is being invested on the biological control programmes. The basic reason for the relative neglect is that the biological control is widely perceived as unreliable. Biological control has enormous untapped potential capable of being exploited much.

NATIVE NATURAL PARASITOID

The search for effective biological control agents against bugs of the super family Pentatomidae (Heteroptera) has been focused on their egg parasitoids and in particular, on the species of the subfamily Telenominae (Hymenoptera: Scelionidae) (Joronin, 1981; Wylie, 1982). *Psix straiticeps* Dodd. (Scelionidae : Hymenoptera) parasitizes egg mass of stink bug *C. furcellata*, a major predator of early stage tasar silkworm larvae (Thangavelu and Singh, 1992). Biological and systematic research on those wasps has focused on the large genera *Telenomus* sp. and *Trissolcus* sp. However, in tasar culture, *Psix striaticeps*, *Telenomus* and *Trissolcus* are abundantly available which are highly effective against pentatomid bug and play an important role in the biological control of some economically important predators (Chua,1979; Singh *et al.*, 2000; Singh, 2002; Wilson 1960; Okuda and Yeargan, 1988 a&b, Yeargon, 1982). *Psix* is most diverse in the Ethiopian oriental and Australian regions. Its distribution also extends into southern reaches of the pale artic region. The genus is abundantly found in south-western Asia, central Australia and the savannas of Africa. *Psix*, however, is not entirely restricted to these biomass; species have also been collected from the humid forests of South-western India, South–east Asia, Taiwan and both northern and eastern Australia.

The few host records available indicate that the species are egg parasitoids of bugs in the family Pentatomidae, Scentallenidae and Coreidae. Biological information on this genus is scarce. The scelionidae is the largest family of Proctodontoidea, with more than 70 genera distributed among three sub families and more than 20 tribes. *Telenomus* is cosmopolitan, eurytopic genus, with more than 500 nominal species. Several species show some degree of specificity, primly parasitizing only the eggs of host species from which they are reared (Bosuqe and Rabinovich, 1979). Parasitization by a single species of *Telenomus* of insect in more than one order is uncommon (Johnson and Masner, 1985). Biological control in the field has been achieved in some parts of tropical tasar region of India, because their parasitoids have great potential in managing their populations in spite of certain limitations (Singh and Thangavelu, 1991). Navasero and Oatman (1989) reported similar observation for *Telenomus solitus*. The parasitoids become established in 20 out of 25 attempts. The eggs of *C. furcellata* (0-24 hrs. old) on the eggs card were exposed for one hour to fertilized *P. striaticeps* female confined inside oviposition unit. The development of the parasitoid was recorded by mounting parasitized host eggs in Hoyer, s medium on glass microscope slides: every 2-4 hours on first day; every 6 hours in second day; and every 24 h in thereafter until adult parasitoids emerged from the unmounted eggs. The immature parasitoid stage within the host eggs were examined and measured, using a compound light microscope with an eyepiece micrometer.

LIFE HISTORY

Age-specific survival and fecundity of *P. striaticeps* on *C. furcellata* has been reported (Singh *et al.*, 1995). At 25±2°C, the development of *P. striaticeps* from egg to adult took 12 days. The total duration of immature stages was 10.75 days. The first mortality within the cohort occurred on the fourth day and increased thereafter. The number of females produced by one female *P. striaticeps* ranged from 35 to 52. Egg laying continues for about 7 days and the parasitoid lays maximum number of eggs on the 3rd day. The population increased with an infinitesimal rate (r_m) of 0.275 and finite rate (l) of 1.32 per female per day. One generation is completed in 13.25 days (T). The weekly multiplication of the population was 6.85 times. The mean sex ratio (male : female) was 1:4.5 (Singh *et al.* 1995). *P. striaticeps* is a solitary parasitoid but high incidence of super-parasitism was observed under laboratory condition. Two or three individuals emerge from the super-parasitized eggs of *C. furcellata* but with reduced body size (Singh and Thangavelu, 1996). The fecundity of scelionid parasitoids in general was reported to be 30-85. Similar observation was made by Braman and Yeargan (1989) in *T. euschisti*, a common wasp parasitizing stink bug.

SEX RATIO

The knowledge of sex ratio of parasitoids play a very significant role in biological control (King, 1961). It helps us to understand the reproductive strategies of the parasitoids. A significant variation in sex ratio was observed in parasitic hymenoptera. Fisher (1930) postulated an equilibrium ratio of equal numbers of male and female offspring for randomly mating parasitic species. Hamilton (1967) was the first to show that optimal sex ratios could vary from 0.5 if the assumption of panmixis (= random mating) was relaxed. He constructed the 'local mate competition' (LMC) model for a situation where female placed offspring in discrete patches of resource, and those offspring mated randomly within their patch before female offspring disperse to colonize elsewhere. This situation is found mainly among small gregarious and often parasitic species (Waage and Lane, 1984). The LMC model permits the production of significantly higher number of females than males, each batch of offspring containing at least one male. The males emerge earlier than females and are polygynous. The sex ratio ranges from a slight male bias to entirely female broods (Waage, 1986). Sex ratio of the stink bug parasitoid, *P. striaticeps*, was found to be female biased (Singh *et al.*, 1994). The predominance of females over males was observed in each sample of *P. striaticeps* collected in adults form or larvae parasitizing egg masses of *C. furcellata*. The mean proportion of females was 75%, which deviated significantly (P= 0.001, X2 test) from an expected sex ratio of 50:50. About 95 percent of the females were mated, which exhibited local mate finding mechanism. Males emerged earlier, remained at the place of emergence to mate with subsequently emerging females. The sex ratio varied during different season.

PARASITIZATION POTENTIAL

The parasitization efficiency of the parasitoid determines its potentiality against the pest in the target areas. In a population, the number of hosts parasitized per time unit depends upon the number of parasitoids present and the ability of individual parasitoids to locate and parasitize variable number of hosts. Usually *C. furcellata* oviposites in mass on the under surface of the leaves of plants and hence their parasitoids also distributed in patches or groups. The overall parasitization rate depends on the individual performance within a patch of certain host densities (Hassell, 1986). Thus a parasitoid responds to increased host density by increasing the number of hosts each individual destroys (functional response) or by increasing their own numbers (numerical response) while the host destruction per parasitoid remains same (Solomon, 1949). Numerical response is one of the two essential ingredients in modelling any prey-predator or host parasitic interactions, which are descriptions of the entomophages density responses (Hassel, 1986). This response is usually of more interest than the functional response because it is more often responsible for suppressing the host population although it has attracted much less modelling effort (Huffier and Kennett, 1969; Hassell, 1978; Hassell and May, 1974). A rapid and strong numerical response is the most important attributes of a successful agent for pest control (Coppel and Mertins, 1977). The study of numerical interactions between parasitoids and host population also provides the data for calculating the number of parasitoids needed to regulate the estimated pest population (Knipling and Gilmore, 1971). Numerical response was described for few parasitoids (Griffiths, 1977; Yamada, 1987; Hassell, 1971 & 1978; Stinner, 1976; and Chua, 1979). Excessive release of the parasitoids in large numbers at a time shall not ensure proportional multiplication of the parasitoid (Alphen and Vet, 1986). This was due to non-availability of sufficient number of hosts.

HOST DISCRIMINATION

Host discrimination ability among parasitic hymenoptera had not attracted the attention of biologists in past. Host discrimination ability is a desirable attribute of potential parasitoids and plays a significant role in their establishment and utility in biological pest management. 'Discrimination' in this context means preference or specificity to a particular species of host. Host discrimination ability of the parasitoid has been reported as one reason for superparasitism (Lenteren and Bakker, 1978; Lenterren, 1981; Singh and Thangavelu, 1994). The ability to reject parasitized hosts is a desirable feature of successful bio-control agents, as it provides an opportunity to infest and destroy more number of host species within the environment, thereby preventing the wastage of energy, potential eggs and searching time, which help in the speedier establishment of the parasitoid. Earlier findings showed that *P. striaticeps*, was a potential egg parasitoid of *C. furcellata* (Singh *et al.*, 1992b and 1995; Singh and Thangavelu, 1994; and Singh and Sinha, 1995). It has been reported that *P. striaticeps* can discriminate between parasitized and healthy host, a desirable trait also exhibited by some other parasitoid species (Vinson, 1975 and 1981). Even then, this cannot completely prevent super parasitism in *P. striaticeps*. Two to four male individuals emerge from a single super parasitized egg where they show reduced body size and mating efficiency. The pre-attack time is longer in in-experienced parasitoids (those which have never come in contact with the host) than the experienced ones. The inexperienced females however superparasitize significantly more hosts than the experienced females (Singh and Thangavelu, 1996). The experienced females reject heavily parasitized egg masses.

MASS PROPAGATION

Standardization of mass multiplication of bio-control agents forms the pre-requisite for using any parasitoid for pest control. Alternate hosts in place of target organisms for multiplication of parasitoids in the laboratory forms a better option since we can avoid multiplication of harmful organisms which can form a great risk even with the slightest negligence or mistake. Secondly, the bio-control agents should be as readily available as chemical pesticides without which, biological control remains a subject of academic interest with no practical role (Manjunath, 1986). Probably less than 2% of the known species of pentatomid egg parasitoids have been reared in the laboratory (Singh and Jalali, 1991). This mainly includes species of *Telenomus, Microphanurus, Aporophlebus, Trissolcus,* and *Psix,* which have been used in the biological control of various economically important pests or parasitoids. The culture environment, diet provided to the free-living adult, the size of the host required for the parasitic stages, the capability of the strain, etc., are the main factors determining the success of the mass culture programme. Of many factors that can influence the outcome of the rearing programme, the nutritional requirement of the parasitoid is among the most important one (Hughes, *et al.*, 1992). Bosque and Robinnovich (1979) mass reared *Telenomus fariai* on their natural hosts. Yeargan (1982) developed methods for mass rearing, storage and release of *Telenomus podisi* and *Trissolcus eusachisti*. Davis (1967) described this progress in the biological control of the southern green sting bug, *Nazera viridula* variety *smargdula*. Biological control in the present days is regarded as most effective method of pest control in a very stable ecosystem. But its application needs a thorough understanding of the bio-control agent and its environment and a careful handling. It is also regarded as the most important component of Integrated Pest Management, but a caution is needed since the chemical pesticides can equally damage the bio-control agents, if not planned properly. Further, a great caution is needed to balance its ecosystem without disturbing beyond certain limits, lest it can become pest of other useful organisms. In silkworm ecosystems use of bio-control agents stand a clear edge over chemical pesticides, since chemicals can cause equal damage to the silkworm, the most wanted.

Praying Mantis ; *Hierodula bipapilla* (Dictyoptera : Mantidae)

The carnivorous praying mantis occurs throughout the tropics. Mantises are commonly of large size and include no insect less than half an inch while some attain to four and even six inches. In appearance these insects are extremely striking. The majority of Indian Mantidae belongs to genera wide spread over Indo-Malayan region (Lefroy, 1933). Wood–Manson catalogued the Mantidae and Kirby,s catalogue was issued by the British Museum (Cat. Of Orth., pt. I.). 82 species was reported. *Hierodula westwoodi*, Sss. and *H. coarctata*, Westw. are the robust green insect seen upon the host plants of non mulberry silkworm. *Hierodula bipapilla* is most common predator against silkworm larvae. It is a long slender robust species of green colour. It is abundantly available in tasar, oak tasar and muga silkworm host plants on which the silkworm rearing is conducted. Its raptorial and long forelegs identify the insect. The head is deflexed and the prothorax is elongate. The tibia works in opposition to the femur like the blades of a scissors, and both are wholly or partially spinned. The antennae are filiform and conspicuous. The head is elongate and the compound eyes are large, the head very mobile and the insect have curious habit of turning the head to look intelligently to locate the prey. The mouthparts are biting type but the mandibles are not elongate as in other predaceous insects, since the forelegs capture the prey and the jaws are solely for mastication. The prothorax is long, nearly half the length of the body, and this is apparently an adaptation to secure great mobility for the fore legs and head. The forewings are moderate size, thickened; coloured and covering the large folded hind wings, which are hyaline and often coloured. The abdomen is expanded. The female deposits her eggs in a characteristic large egg case, known as oothecae, which are attached to the plants. Each ootheca, about 2.5 cm long, contains about 24 egg chambers. The egg case is made of gummy matter secreted by the female, which comes out as a frothy mass, and sets hard in short times; taking a firm position on the plant, with head down and the tip of the abdomen touching the plant, she exudes a mass of frothy gum and with the end of the abdomen works it into the shape characteristic; as soon as the base is formed and some amount of gum used, eggs are deposited in the midst of the gum.

Fig 5.8 : Praying mantis in tasar silkworm rearing field

The emission of eggs and gum continues, the eggs in the middle, the gum round, until the whole egg mass is built up, layer by layer, when she finishes it off with gum and the whole hardens to a watertight object firmly secured to the plant. The eggs are in regular rows inside the egg case and the whole mass will last through the winter on the plant. The young mantis emerge from the egg almost simultaneously and are small active insects often dark coloured with a general resemblance to an ant. All are predaceous at all times of their life; the food of the full-grown insect is large living insects, which are caught when they come within reach of the waiting mantis. The length of life cycle is not known. Hibernation appears to take place chiefly in egg stage. Generally egg masses are laid in early number and hatch in early March. The insect is active through out the year and breed and multiplies extensively. The predatory capacity of *H. bipapilla* is extremely high. It can prey on even final stage larvae of tasar and muga silkworm.

Removal of egg mass from the rearing field and killing of adult is the most ideal method to minimize the predator population in non-mulberry sericulture. The egg masses are mostly attacked by the parasitic Chalcids. The female oviposits on the newly laid eggs in the egg masses with the help of with long ovipositor. Besides this *Podagrion* sp. (Torymidae) has been screened as an important parasitoid in oothecae of mantids . There are 13 species known from the Indian subcontinent (Suba Rao & Hayt, 1986, Oriental insects, 20: 1 –430 pp. 270 – 271).

Reduviid bug ; *Sycanus collaris* (Heteroptera : Reduviidae)

Sycanus collaris (Heteroptera : Reduviidae) commonly known as Assassin bug is an important predator of non mulberry silkworm larvae. It occurs throughout the tropical and temperate tasar region, with the maximum development in moist tropical areas. Adult bug is very striking due to red and black form with the sides of the abdomen dilated. Both adult and nymphs are predaceous upon the early stage silkworm larvae of tasar, oak tasar and muga silkworm. The strong curved beak is characteristic of assassin bug. It has an elongated narrow head, the portion behind the eyes being neck like. The rostrum is three segmented, usually curved and its tip fits in a prosternal groove. The antennae are simple and of moderate length filiform. The prothorax is distinct and well developed. Metathoracic scent glands are absent. The abdomen is broad at the middle and the margin of the segment beyond the wings are exposed. The wings lie flat on the abdomen. The legs are long, formed of quick running; femora and tibiae are often spinned. The eggs are laid in the clusters on the silkworm host plants. The eggs hatch within 4 to 5 days of oviposition. There are five nymphal instars. The adult is black and about 2.5 cm long. Te rostrum consists of three segments, and the antennae are filiform. The head is long and conical. The prothorax is narrow. The nymphs are similar in form and colouring to the adult, the wings developed during the later instars. Both nymphs and adults are found on plants on which the silkworm rearing is conducted. They fed on the body fluid of the silkworms, which they pierce with the beak and suck out, while the spinned forelegs hold them. It causes painful bite, due to injection of fluid at the moment of puncture. The predatory potential of the bug is very high. It may inflict injury to 30–40 silkworm larvae in a day. Early instars larvae are more likely to be attacked by this predator.

WASPS (VESPOIDEA)

Wasps are yellow or red insects with black markings. Adult wasps usually have a long slender petiole and the forewings posses a very long discoid cell. At rest the wings are usually folded laterally. Anal lobes are absent in hind wings. The conical sessile abdomen in the female has at its apex the ovipositor. The wasp belongs to family Vespinae, Polistinae and Polybiinae are commonly available in non mulberry sericulture areas. Some of them are social wasps and the rest are solitary wasps. In the social wasps, there are males, queens and workers. Queens and workers have very effective sting. Adults pick up the tiny brushed larvae of the silkworm and causes extensive damage especially during commercial crop rearing of tasar silkworm. The most common Indian species is *Vespa orientalis* (vespidae). The nest is made up of number of circular combs contains numerous hexagonal paper cells. The whole nest is enclosed in a papery envelope. The nets are found in the open attached to the branches of trees.

The larvae fed upon another insects, decayed or fresh meat and fish. The most common Indian species are *Vespa orientalis, V. tropica, V. ducalis, V. basalis, V. cincta* etc. Among Polistinae, *Polistes hebraeus* Fabr. and *P. stigmata* are most common Indian species. These are reddish or brown, slender and elongate insects with a spindle-shaped abdomen. The nest comprises of a single more or less circular, horizontal comb of paper cells. The nest is attached by a stalk to the support. The cells containing larvae are open but they are sealed once they pupate. Polybiians are chiefly tropical wasps.The habitat selection of various wasp species is interesting. Most of the time they make hanging nests in the verandahs of grainages and in rearing huts. The papery nests of these social wasps are constructed out of chewed foliage or wood. In the temperate region the colony last for a season. The queen over winters and commences formation of new colony in September. Initially in her first brood it raises the workers in the nest newly constructed or sometimes in a nest built previously. Thereafter, it devotes to egg laying. Some times the fertilized females hide away in cracks and chinks in verandahs, roofs, trees etc. during winter season. Here they remain all the cold weather, emerging as the warm weather commences in March. The female builds the nest, lays eggs in it and reared the young till they emerge to help her.; larger nests are then made and these may last until later in the year or new nests may be begun by single females at any time. Wasps cause maximum damage during early instars rearing. They pick up the tiny caterpillars from the host plants and killed them.

Ants; *Oecophylla smaragdina* Fabricious (Hymenoptera : Formicidae)

Almost all ants come under the family formicidae, which is characterized by the presence of the detached nodes, which is formed by the basal one or two segments of the abdomen. The head bears antennae, which have a long basal joint (scape) and a number of short joints (flagellum). The mouthparts are small, the mandibles often of peculiar form. The thorax is much modified in different species. The legs are long and most species can run actively. The abdomen is distinct in the female. Ants are social, insect living in communities in which there is considerable amount of specialization of forms to serve the purposes of useful division of labour; the nest commonly consists of males and females, with various forms of workers. Commonly there are two or three forms of workers, the soldiers with large head and mandibles, and the workers major and minor, with more normal structure. A nest may consists of greater or smaller aggregation of individuals and there are few species which share the light shunning habits of termites, most nesting in soil, trees etc. but working in the light. In general the ants are scavengers, the workers bringing to the nest the food for the whole community. This food consists of dead insects, any available nutritious animal matter, the sap of plants, any nutritious vegetable matter that can be obtained; in this sense ants are excellent scavengers and they are practically everywhere in the open, they serve an extremely useful function. The life history is known in a general way but not in detail; the eggs are laid by the female and tended by the workers in the nest; the larva is a white helpless grub without legs, which is fed by the workers and is itself in capable of exertion. These larvae and pupae are found in galleries in the nests, and one may often see the nest being moved, the workers are carrying the little white larvae and pupae. In some the pupa is free, in others a silken cocoon which the larva itself prepares.

The incidence of *Oecophylla smaragdina* (Hymenoptera : Formicidae) has been reported on mulberry plants (Tikadher and

Thangavelu, 2003). It is popularly known as red ant construct nests by sewing leaves together using silken threads so that the leaves is a mixture of silk, live and dead leaves. The ants are mostly aerial and respond to the vibration of the nest by furious activity and attack filed workers. In mulberry, about 20-25 nests of different sizes can be formed depending on the size of the plant. *Oecophylla smaragdina* is the familiar red tree ant of India., which makes large nests in trees, often enclosing silkworm larvae in covering of webbed leaves. There is no definite sting, the poisons being ejected from the orifice at the apex of the abdomen. Pedicel with one joint. The green females are found yearly in June starting fresh nests on the plants, and these nests can be easily observed from the commencement. The workers are very active and fierce, collecting all manner of dead insects and even living insects which are inactive; caterpillars are attacked, cut up and carried off to the nest in pieces. A colony will have many small depots on one tree, each consisting of a number of leaves webbed together and containing a colony of Coccids or a store of dead dry insects. If one is opened and patiently watched, it will be seen that the workers draw the leaves together by their mandibles and legs, while others form inside, web them together with silk produced from a larva held in the jaws.

This is a really extraordinary sight and may be seen at any time. *O. smargdina* also induces infestations of coccids that it attends which subsequently facilitate tukra disease in mulberry, protecting scale insects producing honey dew which induces shooty mould on mulberry leaves. It is an arboreal ant, seldom seen on the ground, living in the aerial nests constructed between leaves amongst the twigs of the trees. The silk used to sew the leaves together is produced by the larvae, which are held in the mandibles of the worker ants. There may be several thousands ants in one nest system.*O. smargdina* prefers environmental temperatures between 26 –34°C and relative humidity between 62–92 % (Yang, 1982). Lokkers (1986) reported that a combination of high temperature and high rainfall is necessary to support *O. smargdina* populations. The red ants folds the leaf, reduces photosynthesis, which culminates in less leaf yield. The branches show retardation in growth and drying of leaf, which can not be used for silkworm rearing. Moreover, as the time passes after infestation, more branches, leaves and young shoots are affected resulting in reduced leaf yield. Economic threshold of the fire ant is about 15–20 % in terms of dry leaf, retarded growth of the shoot and formation of nest around other branches of infected plants

CONTROL

Removal and destruction of plant part after annual pruning will be effective in reducing pest incidence. Destruction of crop residues and chopping of stems harboring eggs and larvae reduces the pest incidence in the subsequent season.. Uniform pruning was observed to reduce the incidence of these ants. Destroying of weeds and other alternative hosts by ploughing the filed after crop harvest also reduces colonization by fire ants. Chemical control by spray of dieldrin or malathion (0.2%), at two weeks interval is effective and the mulberry leaves can be used for silkworm rearing after three weeks. Mulberry plantations, which are regularly pruned, are not affected by *O. smaragdina*.

DERMESTID BEETLE (COLEOPTERA : DERMESTIDAE)

Various pests are known to attack and cause severe damage to the stores silk cocoons rendering them unfit for reeling (Veer *et al.,* 1996; Gowda and Devaiah, 1985), Thyagrajan and Govindaiah (1987) recorded a loss of about 20 per cent of the silk production due to dermestid beetles. Dermestid are small beetles characterized by the presence of tarsi with five-joints, antennae short with a club, and received under the prothorax in a cavity. Head is retractile. The beetles are often clothed with fine hair or scales. The head in some bears a median ocellus. The apical joint of the antennae in the males may become enlarged. Dermestid larva is a elongate, cylindrical, tapering behind; the prothorax is large, the hind end bears two dorsal hooks and a ventral anal tube. Each segment has a dorsal plate, behind which is an erect row of long hairs and a backwardly directed row of stiff hairs, there are longer hairs on the sides, and a third row on prothorax. Biology and behaviour of some of the free-living dermestid beetle, which is found under the bark of the tree, has been reported by Lefroy (1935). The larvae are predaceous or feed upon dried animal matter.

The free-living larvae are found under the bark of trees and in similar situations where there is a quantity of insect larvae on which they can feed. House–hold species feed upon skins, horns, wool and similar dried animal matter. The larvae are characterized by the development of tufts of long hairs, which in some cases reaches an extraordinary development, especially in the predaceous free-living species. Some larvae are provided with small terminal and lateral tufts of hair, capable of being moved and extended. These larvae eat into their food, making holes in skins or horns and completing their metamorphosis there. The length of the life- history is not known but it can be very greatly extended in every stage, if food is scarce. It is known that the eggs are capable of remaining unhatched for long periods. The larvae will remain starve for long and that the pupal stage may be a very long one. The pupa is commonly found almost wholly enveloped by the larval skin, which is not shed, but only splits along the dorsum.

Several household species of dermestid beetles are reported and these are cosmopolitan insects spread by commerce. The most common recorded species are *Dermestes cadaverinus*, F., and *Dermestes vulpinus*, F. The larvae of *D. vulpinus*, feeds upon the cocoons of silkworms, is common in India. It is curiously fond of these cocoons eating through the silk to reach the pupa within on which it feeds. Cleghorn mentions it as a destructive insect to silk in India. The cocoons should be quickly reeled off to avoid loss. Silkworm cocoons (containing pupae) must be packed properly so that the beetle cannot get access to them neither the cocoon on arrival in the market will probably be infested and partly spoiled. Three pests viz. *Dermestes undulatus*, *Tribolium freemani* and *Trogoderma* sp. are reported to attack stored silk cocoons of *Bombyx mori* (Khan and Nighat, 1989; Sahaf, 1997). *Trogoderma* sp. also cause damage to silk cocoons as it makes circular openings of cocoons by chiseling the cocoon shell to feed on the pupa inside (Sahaf and Munshi, 2003).

Aethriostoma undulata Motsch

It is a very common dermestid beetle in mulberry field. Its larva is broad, with short hairs, with no tubes or hooks. The beetle has been reported to be predaceous upon the other insects there or to feed on their dead bodies. The larva of *Attagenus* is similar but the segments are completely hardened above and each segments fits over the next; there are no hooks or anal tube, and each segment is clothed in scales, with also a row of hairs which extends on to the sides; the hind ends bear a bundles of hairs. *A. gloriosae* Fabr. Probably occurs in India. The larva of *Anthrenus, Tiresians, Trogoderma*, are provided also with bundles of long hairs on the posterior segments, these hairs being moveable and erect, often of peculiar form; in *Anthrenus* the bundles are on the three posterior segments. *A. vorax*, Wat., is known to attack skins and horns in India, as well as woollen clothes and the bristles used in making brushes, and is constantly reported as destructive. Apart from this some other dermestids are also reported which effect silkworm cocoons viz. *Dermestes frishchi, Dermestes tesselatocollis, Dermestes coarctatus, Trogoderma versicolor, Anthrenus verbasi L, Anthrenus pipinellae Fab., Attagenus piceus (Oliver), Attagenus Japonious Reitter.*

Dermestes cadverinus **Fab**

Among all dermestid beetles *Dermestes cadverinus* Fab. is very serious and frequently available in sericulture. The adult is oval elongated; dark brown with club shaped antennae and is about 1 cm in length. The larvae have their upper surface covered with hairs. The larvae are reddish-brown and spindle shaped. They moult 5-7 times and attain a length of about 1.5 cm. Eggs are elongated, oval, milky white in colour and measure about 2 mm in length. The eggs hatch in a week and young larvae starts feeding on dried animal matter. The insect prefers to live in darker places. It moults 5-6 times in months and become pupa. The adult, after emergence, mates and lays egg, which develops to become the adult of second generation. Although the insect generally passes the winter in the adult stage, since the time of metamorphosis is not fixed, both larval stage and the pupal stage may be encountered in winter. The smell of stifled cocoons and the dried pupae inside attract the larvae and adults. They bore into the cocoon and eat the dried pupae and sometimes the eggs and damaged cocoons become unfit for reeling. Rarely the young larvae attack living silkworms.

***Dermestes vulpinus* Fab.**

It is one of the most common species. The larvae are destructive to dead animal matters. Adults prefer to live under humid condition. Adults feed on the pollen and nectar of the flowers. Eggs are laid in batches of two or three. Total number of eggs varies from 198-845. Incubation period varies from two to four days. The full-grown larva is dark brown with a medium yellow stripe dorsally and is densely clothed with hairs. They moult 7 to 14 times and the larval period varies from 25-60 days depending upon condition of temperature, humidity and amount of kind of the food. It pupates in the last larval skin and the adult emerges in about 5 to 8 days. The number of generation in a year may varies from three to six depending upon the temperature and humidity conditions. The larvae of Dermestes *vulpinus* are similar to the larvae of D. vorax but it has only one generation in a year. *Dermestes frishchi* dwells in humid places and develops better at high temperature; it eats away pupae and moths.

Anthrenus verbasi

Adult beetles are 2-3 mm in length, convex, slightly elongated and brownish with yellow white scales on a black background. Eggs are small measures about 0.5 mm and milky white in colour. The larvae are grey brown. It feeds on the silkworm cocoon and causes extensive damage.

***Trogoderma* sp.**

The adult *Trogoderma* sp. is dark brown, oval in shape measuring 2.5–3.5 mm in length and 1–1.7 mm broad at the widest point. The body is hard and compact densely covered with recumbent hairs. Head is heavily sclerotized and is of prognathus type. Antennae capitate type and 11 segmented. Pronotum is composed of single sclerite, elytra are dark brown with lighter brown patches towards the basal region. The tarsi of thoracic legs are five segmented with distal and proximal segments slightly longer than males and lay eggs on dead pupae. The eggs hatch within 6–7 days into larvae, which are slender, and eleven segmented. The newly emerged larvae was yellowish in colour and has brown head. The full grown larva was brownish in colour measuring 6.5–7 mm in length, covered with long and short, brownish hairs that form a tail like structure at the posterior end. Each thoracic segment of the larva bears three segmented legs which has sharp claws. The larval period lasted for 45–50 days, after which it pupated following adults within 6–8 days. The total development takes about four to six weeks depending upon temperature and humidity. There are about 12 generations in a year.

CHAPTER-6

PRINCIPLES OF PEST MANAGEMENT

All stages of silkworm and its host plants–root, stem, foliage, shoot and terminal leaders—are vulnerable to attack by pests. Pest damage can range from slight damage that has no effect on the leaf yield to severe damage that stunts or kills the plants. Silkworm food plant pests include insects and mites, diseases, weeds, vertebrates, and nematodes. Managing host plant pests effectively should be based on thorough consideration of ecological and economic factors. The pest, its biology, and the type of damage are some of the factors that determine which control strategies and methods, if any, should be used. Pest management decisions largely determine the degree and amount of pesticide used. Ultimately, pest management decisions represent a compromise between the value of the product, the extent of the pest damage, the relative effectiveness and cost of the control measures, and the impact on the environment. Insect control methods can be largely classified into cultural control, physical control, mechanical control, genetic control, biological control, and chemical control. The most suitable principle in sericulture is Integrated Pest Management (IPM). The goal of IPM is to use all appropriate tools and tactics to prevent economically important pest damage without disrupting the environment. Information gathering and decision making are used to design and carry out a combination of measures for managing pest problems. IPM is the best approach to manage pests of silkworm and its host plants. Monitoring forests and newly established economic plantations will help to detect problems early, while there is still time to take action.

The information gathered through monitoring is a key element in any IPM program. During monitoring, it is better to examine the center of the area as well as the margins. Competition levels among plants are recorded. Details of the weeds present in the location are also recorded. Succession of other insects is recorded. It is better to check a representative sample of trees for signs and symptoms of insect and disease problems. All the parts of the tree, from top to bottom and from branch tips to trunk should be checked. Depending on the pest, the use of traps or microscopic examination may improve the information gathered by visual examination. Each and every information should be recorded carefully. The destructive forms of many insect pests are generally most active from April through August, but infection by many disease organisms is more dependent on weather conditions than on calendar date. Scouting and monitoring for all pests and pest problems must be done regularly and frequently to avoid surprises. Weather plays an important part in the development of most insect and disease pests. Pest forecasting is done based on the daily weather conditions (high and low temperatures, humidity, and the amount of rain) and succession of insects. Identification of pests and the diagnosis of pest damage are key elements of IPM. Therefore bionomics and life table characteristics of the pest should be known to find out appropriate control measures. Apart from this some basic information viz. type of pest, its stage, size, and extent of damage occurred should be recorded. Further, how much damage is likely to occur if no control measures are taken, should be known. Does the pest or damage require immediate attention, or can control measures be postponed until the plant leaves are near harvest? All these information will throw light to find out suitable control strategies. Certain tools are useful in carrying out an IPM program. A hand lens is essential for magnifying disease signs, insects, and weed plant characteristics. If pests are in the tops of trees, binoculars may be beneficial. Pruning shears and a pocket knife are needed when probing for insects or disease or collecting weed specimens. Field guides, Extension bulletins, or other references with pictures and biological information on silkworm and its host plants diseases will help with identification. Plastic bags, vials, and containers are important items for identification. It is important to know where to find help in diagnosing pest problems. Threshold level is defined as the point at which the pest or its damage becomes unacceptable. It should be determined for every major pests. The threshold level may be related to the health, or economic value of the tree crop. Once the threshold level has been reached, it is essential to determine what type of control procedure is needed. This decision will be based on the size of the pest population, the kind of damage the pest is causing, and the control measures that are available. It is also very important to consider the cost effectiveness of potential controls. Therefore, it is better to carefully weigh the cost of control, the value of the tree, and the impact of the pest damage on the value of the tree.

MANAGEMENT STRATEGY

Generally management options are very different for forest plantation than for other economic plantations. The following are some examples of management strategies. In situations where the pest does not damage the crop value or the crop value is so low it is not cost effective to apply a control measure, no action is needed. Cultural, Mechanical, chemcial, Biological management are the important componenet of suitable pest managemejnt system. Results of management strategies should be evaluated properly. Cultural management manipulates the environment to make it more favourable for the plant and less favourable for the pest. Cultural controls such as good site selection, planting resistant varieties, or selective pruning make it less likely that the pest will survive, colonize, grow, or reproduce. Cultural management can be very effective in preventing pests from building to unacceptable levels. Some measures exclude or remove the pest from the habitat. Mechanical traps, screens, fences, and nets can remove the pest or prevent access by the pest. Tillage and mowing are used to mechanically manage weeds. Biological controls include the beneficial predators, parasites and pathogens that kill pests. There are many more known natural enemies of insect pests than there are natural enemies of silkworm parasites and predators. Biological control is generally aimed at non-native introduced pests. Ladybugs, lacewings and certain mites are common predators of insects. Some tiny wasps and some fly species are parasites of insects.

Many beneficial parasites are host specific and do not control a wide range of pests. Parasites and predators are often very effective at keeping insect pests at low levels. For example, aphids, scales, and mites rarely build to damaging levels in forests because their populations are controlled by predators and parasites. Insects are also affected by a variety of bacterial, fungal, and viral diseases that affect only insects. Biological control organisms are very sensitive to pesticides. Pesticide applications to control a pest may have the unwanted side effect of wiping out part of the natural predator and parasite population along with the pest. This, in turn, may cause a population explosion of a different pest in the void left by the predators and parasites. Results of management strategies should be evaluated properly. It is very important to determine the effectiveness of management and control tactics. This information will determine whether any follow-up treatment is needed and will improve management strategies for next year. Its effect on pre-treatment and post-treatment pest activity should be compared. This is where a pest management logbook will become invaluable. Detail observations about the incidence where pests first showed up and what kinds of natural enemies are associated should be recorded. Further the record on where and when specific treatments were applied, and what the results were obtained should be recorded with full care. Sound IPM practices pay off both economically and environmentally.

LEGAL AND REGULATORY CONTROL

Regulatory actions are often employed to prevent immigration of foreign pests or to prevent the dispersal of established pests. Such actions are termed as legal methods of pest management. Legislation, regulation and voluntary codes are used at national and international levels to ensure good practice in pest management, to govern the safe use and availability of the pest control products, the quality of food and other products that require treatment against pests and the introduction of exotic organisms across national boundaries. Such legislation takes many forms and has a major impact on farming practices. The legislation passed by the governments can be one of two types, it is either passed as direct acts, orders, or regulations or it is enabling legislation that authorizes other bodies to issue rules, orders or directives. In most cases the legislation produced by the governments will not intrude on the day-to-day practice of IPM but in specific cases it can have a significant and major impact on general approach taken. Legal control tactics, therefore, include all forms of legislation and regulation that might prevent establishment or reduce spread of an insect population. The Plant Quarantine Act of 1912 was the first legal action taken in the United States to prevent the introduction of pests from foreign countries. This law, and others that followed it, established a network of inspection stations at major ports of entry and gave the federal government authority to organize border quarantines, to inspect all agricultural products, and to restrict entry of any infested goods.

Today, these inspections stations operate under the jurisdiction of the Animal and Plant Health Inspection Service (APHIS), a branch of the U.S. Department of Agriculture. APHIS operates about 85 inspection facilities, employs 400-500 inspectors, and intercepts over 20,000 potential pest introductions in an average year. It also operates the Plant Quarantine Training Centre where inspectors from all over the world are trained to spot problems before they cross international borders. These agents, now representing nearly 40 countries, cooperate in efforts to limit the spread of indigenous pests under an international agreement that requires inspection and certification of nearly all agricultural commodities prior to export. Federal import/export laws have been enacted to establish standards for a wide range of agricultural products. Most of the fruits and vegetables imported to the U.S., for example, must receive some form of fumigation, heat treatment, controlled atmosphere storage, or radiation exposure before they are released from quarantine.

These post-harvest treatments are designed to kill whatever pest species may be present without affecting taste or quality of the produce. Despite the best efforts of APHIS inspectors, some pests still slip into the U.S. and become established (e.g., cereal leaf beetle), or move naturally across international borders (e.g., imported fire ants and Africanized honey bees). When this happens, APHIS cooperates with state and local governments to establish domestic quarantines for limiting the movement of these pests within our borders. Whenever appropriate, eradication programmes may be launched together with quarantine actions in an effort to destroy the entire pest population while it is still small. Inspections and quarantines are also operated by some state and local government agencies. The California Border Patrol, for example, is notorious for its interceptions of uncertified agricultural products at the state line. The rigid laws that prohibit private citizens from bringing a bag of oranges into southern California are generally given credit for excluding numerous pest species that might have caused millions of dollars in losses. The price tag for these prevention efforts is only a fraction of the estimated $350 million Californians spent to clean up their most recent infestation of the Mediterranean fruit fly (*Ceratitis capitata*).

INSECTS UNDER QUARANTINE

Since 1915, nearly 300 pest species have been under some sort of domestic quarantine. The success of these programmes has been erratic. In some cases (e.g., screwworm flies, khapra beetles, and parletoria date scale), eradication programs have managed to eliminate the pests. But for every noteworthy success, there has been a complete failure (witness the cereal leaf beetle, the gypsy moth, and the European corn borer).

In general, eradication efforts are most successful when the pest population is relatively small, does not disperse rapidly, and is highly susceptible to control tactics. Even if a newly established pest population cannot be eradicated, it can often be slowed down a bit by regulating the movement of its host plants or animals, prohibiting sales of infested products, and inspecting all shipments of affected goods. This type of containment strategy usually gives growers and producers a short reprieve before invasion of the pest and buys a little more time for researchers to find better methods of control. Licensing and certification are important regulatory tools used in some commodities to ensure that infested or contaminated material is not sold commercially or used as breeding stock. Maintaining disease-free mother plants in strawberries, for example, can often prevent the spread of insect-borne disease during the growing season. Certified seed or plant material is produced under strict guidelines established by commodity organizations in their own interests. These products may be more expensive initially, but eventual savings in pest control usually provide a good return on the investment.

LEGISLATION IN INDIA

The history of pest legislation in Indian is very interesting. The first Act in this country was passed in 1906 under the Sea Custom Act of 1878 to stop the entry of Mexican cotton boll weevil. It was followed by the present destructive Insects and Pests Act No II passed on 3rd February 1914 at the instigation of Bombay Chambers of Commerce. Thus, provision were made for preventing the entry of foreign insect pests, like the American cotton boll weevil (*Anthonomus grandis* Boh.) and others that might be harbored in agricultural products. The entry of such plant material as sugarcane and cotton were banned totally. Plants and seeds, in general, could be imported under certificate of health through notified ports where they could be inspected by qualified staff, fumigated properly and kept in quarantine before being released. There, they could be examined for presence of eggs of pests or for the presence of bacteria, fungi and other pathogens, which are normally not killed by fumigation. The import of useful insects such as parasites was permitted only to scientific institutions. According to various amendments of the Government of India Act, 1914, provisions were made for adopting control measures against local and exotic pests in centrally administered territories and States after first obtaining sanction from the President of India or Governor. Before control measures could be adopted by states, it was considered necessary to (1) declare the organism to be injurious, (2) place the infested area under quarantine, (3) ensure that preventive and remedial measures were prescribed. In this act, provisions were also made for the State Governments to pass their own legislation for adopting remedial measures. Thus, the East Punjab Agricultural Pests, Diseases and Noxious Weeds Acts were passed in 1949. Other States have passed similar legislation.

According to notification No. 1581, dated the 1st October, 1931, under the Destructive Insect Pests Act, 1914, provisions were made to restrict the imports of cotton from America and the West Indies and to bring air freight within the scope of Act. The import of cotton was allowed, provided it was fumigated and disinfested at the port of entry. The import of plant material by air was prohibited except under strict quarantine for scientific purposes, with prior permission of the Government of India. Similarly fruits and vegetables from Afghanistan were allowed to be imported, provided they were accompanied with certificates of health issued by the Competent Authority. The import of seed by air was exempted, but the entry of some of the seeds, viz. cotton, Egyptian clover, flax, rubber, coffee and sun flower was banned. The import of potato seed from infested areas infested with golden nematode and wart disease was prohibited. Under this Act, notifications were also issued from time to time for inspection, disinfections or destruction of plant material to be transported by road or rail within the country. This check the further spread of pests and diseases that might be present in certain plant parts. According to the provision of East Punjab Agricultural Pests, Diseases and Noxious Weeds Act 1949 the State Government could enforce, when necessary, control measures for the eradication of pests, diseases or weeds. Some times free pesticides were supplied for the control of pests, the success was greater.

The Department of Agriculture and the community of the Development Organization have carried out anti-rat campaigns. An integrated legislation was made to control the locust population in India. It is the credit of the Government of Punjab and Haryana and to the rural farmers that, since 1945, the locust has not been allowed to breed and not a single swarm originated form these areas, although the conditions have been favourable for its breeding. It is only natural to believe that in spite of best quarantine measures to restrict the import of new pests from abroad or to restrict the spread of organism from one part of the country to another, new problems will continue to appear and therefore, notifications or amendments to the 1914 Acts will be needed from time to time. The East Punjab Act of 1949 and other similar Acts of various States also need amendments to make Government effective in their implementation. The undulated sale of pesticides and weedicides under proper levels, giving instruction for their safe usage, also needed the attention of the Government. In the absence of any legislation, certain pesticides were brought under purview of the Poison Act of 1919, and of the Drugs Act of 1940. However, in view of the fast expanding pesticide industry and the wide scale use of these highly poisonous chemicals, more specific provisions were needed to safe–guard public interests and even human life. Consequently, a special legislation to regulate the manufacture, transport, handing, sale and use of pesticides was envisaged and a new Bill–namely the Insecticide Bill was passed by the Indian Parliament in September 1968. Under this Act, provisions have made for compulsory registration of pesticides and for establishment of pesticides laboratories to carry out various functions under the legislation.

QUARANTINE PROCEDURE

Quarantine is required to protect an importing country from the entry of noxious exotic organisms. Introduced biological control agents present a risk due to several reasons. Spores, or other infective stages may contaminate cultures. The silkworms may be infected with natural enemies–parasitoids, parasites and diseases The material may be being imported for study in quarantine before risk, safety and other factors are understood fully. Quarantine is essential for most importations of biological organisms, so that release of imported cultures of certified purity may be permitted, however, even these should be subject to careful inspection before liberation. The precautions required in a quarantine facility for processing silkworms will depend on the level of risk. Since quarantine facilities are costly to build and operate it is sensible to consider the purpose of the facility before making financial commitments.

Further, it must be emphasized that the effectiveness of a quarantine facility is entirely dependent on its staff; thus, funding will also be required for training good quality staff who will maintain the strict procedures required. The likelihood of survival of any organism should it escape should be of prime consideration for the location levels of containment and levels of security desired. Separate Quarantine Testing Laboratories that are fully equipped should be established for the purpose. The quarantine testing should be done scientifically adopting centrifugal method for detecting the pebrine disease in the consignments coming into our country and for export. The equipments and other facilities should include precision incubator with RH facility, tissue homogenizer, mixie, centrifuges, binocular microscopes, cyclomixer, etc. and glassware, plastic wares and chemicals.

In silkworm grainages, most insects move towards light and can be discouraged from escaping through entry points by constructing darkened double entry door systems and by fitting light traps which operate within the double entry vestibule when the doors are shut. Such arrangements are recommended for preventing the escape of flying insects. Tight fitting seals are added to doors to prevent the escape of crawling individuals. Drains should be designed so that insects cannot escape. In high security facilities effluent will be sterilized and any insects escaping down drains will be destroyed. When the drains empty into the public sewer or a septic tank it is important to include a water trap, designed so that it is not possible for the water in it to evaporate or to fit a filter with sufficiently fine mesh to prevent the escape of insects. A footbath containing a disinfectant is required to sterilize the shoes of personnel on exit from the facility.

PLANT QUARANTINE

Plant quarantine through legislative measures restricts the movement of plants and planting materials between countries and within country so as to exclude introduction of new pests/diseases/ weeds. Destructive insects and Pests Act of 1914, the Plants, Fruits and Seeds (Regulation of import into India) Order of 1989 (PFS Order) and New Policy on Seed development (NPSD) 1988 have been described, besides regulations governing import of biocontrol agents and edible fungi. Integrated Pest Management envisages plant quarantine as first and foremost method of pest management options in addition to being environmental safe and economical. The first quarantine is a legally enforced procedure through which plant parasites are intercepted at points of entry into geographical and political subdivisions. The second certification of propagative parts of plants to be free of pathogens is principally a cooperative procedure in which groups of agriculturists work with advisory personnel of the region produce pathogen free seed and other propagative parts of plants for sale to farmers (Roberts, 1991, Rohwer, 1991). Plant Quarantine is an endeavor, by legislative measures and through the policing of these measures, to exclude from a defined geographic area such as countries, state or region, pest or disease of plants exotic to that particular geographic area and assessed to be potentially serious to agricultural, horticultural or forestry industries (Rohwer, 1991). With the advent of speedy and sophisticated transport system, the problems of spread of dangerous pests and diseases across national boundaries have enhanced manifold. Coupled with this, the vast increase in quantum of import of seeds, propagating materials by national and international institutions for evaluation, research as well as multilocational trials for sustaining the green revolution and yellow revolution increased the problems of quarantine screening and treatment. Each country or region has its own collection of insects, fungi, bacteria, viruses, nematodes, weeds etc., which originate there and as the plants, its seeds, fruits or any other vegetative propagative materials are moved from one country to the other, there is a distinct possibility of entrance of associated -pests and diseases to new location from its original native habitation. It is always not possible to predict the behavioral performance of the introduced pests/diseases in to its new location/environment and sometimes the introduced pests/diseases may prove to be more serious in the new location. Hence, plant and quarantine seeks to be a filter against exotic pests and diseases thus ensuring minimum risks of introduction or spread of these pests and diseases in the new location (Mathys and Baker, 1980).

QUARANTINE REGULATIONS

Keeping in view the apprehension of lurking disaster as a sequel to the introduction of exotic pests and diseases, the movement of plants/plant materials is regulated by prescribing the relevant rules/regulations with the sole aim of precluding or at least hindering the entry of the exotic pests and diseases into the country. The first Plant Quarantine legislation in the world was enacted in Netherlands in 1977 prohibiting the importation of coffee plants and seeds from Ceylon. In U.S.A. the Plant Quarantine work may be said to have started in 1891, which was followed by Australia by adopting the Plant Quarantine regulations in 1909. The purpose of the Plant Quarantine Regulations is as follows:

- To prevent the introduction of plant pests/diseases not already known to occur/widespread in the country.
- To control pests and diseases already introduced in the country by restricting their spread and by endeavouring to eradicate them.
- To provide facilities and services for the import/export of plants/plant materials.

The Plant Quarantine Regulations the world over aim at enforcing legal provisions on the movement of plants/plant materials between countries (Foreign Quarantine) and between states within the country (Domestic Quarantine) to check the infiltration/spread of exotic pests/diseases. The International Conference on Phytopathology held in Rome in 1914 recommended several quarantine measures to preclude the spread of pests and diseases across national boundaries. In consonance with those recommendations the Government of India enacted Destructive Insects and Pests Act, 1914 (II of 1914) (DIP Act, 1914). The salient features of this Act are as under (Anonymous, 1941):

(i) The Act authorizes the Central Government to prohibit or regulate the import into India or any part thereof any specific place therein, of any article or class of articles.

(ii) It authorizes the officers of the customs at any port to operate and enforce, as if the Customs Act of 1968.

(iii) It authorizes Central Government to prohibit or regulate from a State or the transport from one State to another State of any article likely to cause infection to any crop or of insects generally or class of insects.

(iv) It authorizes State Government to make rules for detention, inspection or destruction of any insect or class of insects in respect of which Central Government have issued notifications and, for regulating the powers and duties of officers.

(v) It provides penalties for persons who contravene the domestic quarantine rules.

(vi) It provides protection of officials from any suits, prosecution or for anything done in good faith.

The plants/plant materials are imported into the country as per regulations enshrined under "The Plants, Fruits, and Seeds (Regulation of Import into India) Order, 1989 (PFS Order) issued under DIP Act aiming at preventing the introduction of exotic pests/ diseases/weeds into the country. Plant Quarantine comprises of two aspects, namely, Foreign Quarantine and Domestic Quarantine which are narrated hereunder:

DOMESTIC QUARANTINE

The Domestic quarantine aims to control/contain further spread of the exotic pests/diseases already introduced in the country. In order to assess the crop losses due to insect pests and diseases, surveys are conducted at different intervals for assessing the extent of infection/infestation, the notifications are issued from time to time under Sections 4A and 40 of the DIP Act, 1914. Some of the important existing domestic quarantine measures already in force in the country to prevent diseases are given below:

(i) To prevent the spread of Banana bunchy top virus, complete prohibition of transport of banana planting material from the states of Assam, Kerala, Orissa, Tamil Nadu and West Bengal to any other state.

(ii) To prevent the spread of Apple scab disease caused by *Venturia inaequalis*, transport of planting materials of apple is prohibited from Jammu and Kashmir, Himachal Pradesh.

(iii) To prevent the spread of potato wart caused by *Synchytrium endobioticum* movement of potato grown in W. Bengal is prohibited to any other state.

(iv) To prevent the spread of codling moth (*Carpocapsa pomenella*), movement of apple and walnut plants including fruits are prohibited from Ladakh and Kargil districts of Jammu and Kashmir.

(v) To prevent the spread of potato cyst nematode (Gobodera rostochiensis, G. pallida), movement of potato from Tamil Nadu and UT is prohibited except for table potato accompanied with special permit issued by Plant Protection Adviser to Government of India.

(vi) To prevent the spread of sanjose scale (*Aspidiotus/ Ouadraspidiotus pernicious*), transport of host material is prohibited unless accompanied with the certificate of freedom of the pest from the Directors of Agriculture.

FOREIGN QUARANTINE

The Foreign Quarantine deals with the movement of plants/ plant materials between the countries in order to ward off the risk of the infiltration of exotic pests and diseases into the countries. In India, the goal of Foreign Quarantine is realized by observance of PFS Order, 1989 (Anonymous, 1989). This Order prescribes regulations for the import of plants/plant materials. The observance of these regulations helps the country to be free from the exotic pests and diseases. There are general and specific conditions for import of plants and plant materials

GENERAL CONDITIONS FOR IMPORT

All consignments of plants, fruits and seeds (hereinafter referred to as 'consignments') shall be imported into India subject to the following conditions, namely.

(1) No consignment shall be imported into India without a valid permit issued under clause (3).

(2) (a) All applications for a permit to import consignments by land, air or sea shall be sent in triplicate at least one month in advance to the Competent Authority, and application for the import of seeds, fruits and Plants for consumption shall be made in form 'A' and that for the import of seeds and plants for sowing of planting shall be made in form 'B',

(b) A fee of Rs, 50 shall be payable along with the application for the import of seeds, fruits and plants for consumption and Rs.100 for application for the import of seeds and plants for sowing or planting and the fee shall be payable in the form of Demand Draft payable to the Competent Authority having jurisdiction.

(3) The Competent Authority shall issue permit in Form "C" for import of seeds and fruits for consumption and in Form "O" for import of seeds and plants for sowing or planting, if he is satisfied that the applicant meets all the necessary conditions;

(a) The issue of permit may be refused or withheld by the Competent Authority after giving reasonable notice to the applicant and for reasons to be recorded in writing;

(b) The import permit issued under this clause shall be valid for a period of six months provided that the Competent Authority may, on request, extend the period of validity for a further period of six months, for reasons to be recorded in writing;

(4) The Competent Authority shall forward to importer an orange and green colour tag specified in form "E", in the case of permits issued for import of seeds and plants for sowing or planting so as to facilitate the identification of consignments at the time of their arrival at the land customs station or port of entry.

(5) (a) All the consignments for consumption, sowing and propagation or planting shall be imported into India only through entry points notified by the Central Government from time to time in this behalf, provided that all consignments of dry fruits, fresh fruits and vegetables for consumption, imported from Afghanistan, Pakistan and West Asian countries by land shall be imported only through Attari-Wagha Border.

(b) All consignments of plants and seeds for sowing and propagation or planting shall be imported into

India through land customs station, seaport, airport at Amritsar, Bombay, Calcutta, Delhi and Madras and such other entry points as may be specifically notified by the Central Government from time to time.

(6) (a) The consignment, on arrival, at an entry point, shall be inspected by the Plant Protection Adviser or any other officer duly authorized by him in this behalf, in accordance with the guidelines issued by the Plant Protection Adviser from time to time.

(b) The Plant Protection Adviser or the officer authorized by him may, after inspection, fumigation, disinfection or disinfestation, as may be considered necessary by him, accord quarantine clearance for the entry of a consignment into India or require, in public interest, destruction of the consignment or return of the same to the country of origin;

(c) Where fumigation or disinfestations or disinfections is, considered necessary in respect of a consignment of plants, seeds and fruits of more than 1000 cubic meter in volume, the importer shall on his own or at his cost through an agency approved by the Plant Protection Adviser arrange for the fumigation, disinfestations or disinfections of the consignment, under the supervision of an officer duly authorized by the Plant Protection Adviser in that behalf;

(7) It shall be the responsibility of the importer,

(a) to bring the consignments to the concerned Plant Quarantine and Fumigation Station, or to places of inspection, fumigation or treatment as directed by the Plant Protection Adviser or the officer duly authorized by him;

(b) to open, repack and load into or unload from the fumigation chamber arid seal the consignments; and

(c) to remove them after inspection and treatment, according to the directions issued by the Plant Protection Adviser or an officer duly authorized by him.

(8) The consignments intended for other countries shall be allowed transit through or transshipment at air or sea ports or land customs stations, provided they are packed in such a manner as will not permit spillage of any soil or material or escape of any pest, and subject also to the condition that they are not opened in any place in India.

(9) No consignment shall be imported unless accompanied by an official Phytosanitary Certificate issued by the authorized officer of the country of origin of the consignment: Provided that cut flowers, garlands, bouquets, fruits and vegetables weighing less than two kilograms imported for personal consumption may be allowed to be imported without a Phytosanitary Certificate or an Import permit.

(10) Consignments for import should be packed in the packaging material envisaged as in clause 2(g) of this Order. No consignment wherein hay or straw or any material of plant origin is used for packaging or as a part of packaging materials shall be allowed to be imported.

(11) Import of soil, earth, compost, sand, plant debris along with plants, fruits or seeds shall not be permitted except under the following conditions,

(a) The consignments of soil, earth, clay and similar material for any microbiological, soil-mechanics or mineralogical investigations and peat for horticultural purposes may be permitted through specified air or sea ports or land custom station, on applications made for that purpose;

(b) The application for the purpose referred to in (i) above shall be made to the plant Protection

Adviser, at least one month in advance in form "F""

(c) The Plant Protection Adviser may, after scrutiny of the application, and if satisfied of the purpose, for which such consignment is being imported, issue special permit in Form "G".

(d) The consignments shall be inspected, fumigated disinfected, on arrival, by the Plant Protection Adviser or any other officer duly authorized by him in this behalf.

(12) The importer of the consignments or his agent shall pay to the Plant Protection Adviser or any other officer duly authorized by him in this behalf, the fees prescribed in Schedule III to meet the cost of inspection, fumigation disinfestations, disinfection before the release of the consignments.

POST-ENTRY QUARANTINE

(A) Plants and seeds, which require Post-entry Quarantine as laid down in Schedule II of this Order, shall be grown in Post-entry Quarantine facilities, approved and certified by the Designated Inspection Authority, to conform to the conditions laid down by the Plant Protection Adviser. The period for which, and the conditions under which, the plants and seeds shall be grown in such facilities shall be specified in the permit granted under clause 3.

(B) The Post-entry Quarantine facilities shall be established and provided by the importer or his agent at his own cost and these shall be ready for use at the time of arrival of the consignment in India. The importer shall obtain a certificate from the Designated Inspection Authority who, after inspection of the Post-entry Quarantine facilities, shall certify that such Post-entry Quarantine facilities have been duly established and provided in accordance with the guidelines of the Plant

Protection Adviser. The importer shall produce this certificate before the Officer- in-Charge of the Quarantine Station at the entry point, at the time of arrival of the consignment.

(C) (i) The Officer-in-Charge of the Quarantine Station, if after inspection of the consignment is satisfied shall accord quarantine clearance with Post-entry Quarantine condition on the production, by an importer of a certificate from the Designated Inspection Authority as envisaged in clause 7, with the stipulation that the plants shall be grown in such Post-entry Quarantine facility for the period specified in the import permit.

(ii) After according to quarantine clearance with Post-entry Quarantine, conditions to the consignments of plants and seeds requiring Post- entry Quarantine, the Officer-in-Charge of the Quarantine Station at the entry point shall inform the Designated Inspection Authority, having jurisdiction over the Post-entry quarantine facility, of their arrival at the location where such plants would be grown by the importer.

(D) The importer shall inform in advance the Designated Inspection Authority having jurisdiction, about the time of planting of such material.

(E) The importer shall permit to the Designated Inspection Authority complete access to the Post-entry Quarantine facility for the inspection of plants and shall, at all times, abide by his instructions concerning the plants in the Post-entry Quarantine.

(F) The Designated Inspection Authority shall inspect the plants grown in the Post-entry Quarantine facility of the importer for the detection of the incidence of pests and diseases and observance of general terms and conditions governing the approval of the Post-entry Quarantine. Such inspections shall be at the time of planting and at such intervals as may be considered necessary by the Designated Inspection Authority in accordance with the guidelines issued by the Plant Protection Adviser.

(G) (i) The Designated Inspection Authority shall permit the release of plants from Post-entry Quarantine, if they are found to be free from pests and diseases for the period specified in the permit for importation.

(ii) Where the plants in the Post-entry Quarantine are found to be affected by pests and diseases during the specified period,

(a) The Designated Inspection Authority shall order the destruction or return to the country of origin of the affected consignment of whole or a part of the plant population in the Post-entry Quarantine if the pest or disease is exotic, and

(b) The Designated Inspection Authority shall advise the importer about the curative measures to be taken to the extent necessary, if the pest or disease is not exotic and permit the release of the affected population from the Post-entry Quarantine I only after curative measures have been observed to be ; successful, Otherwise, the plants shall be ordered to be destroyed.

(iii) Where destruction of any plant population is ordered by the Designated Inspection Authority, the importer shall destroy the, same in the prescribed manner under the supervision of Designated Inspection Authority.

(H) The importer shall be liable to pay the prescribed fee for inspection of plants in the Post-entry Quarantine facility as laid down in Schedule III.

CHAPTER-7

Cultural, Mechanical and Bio Ecological Practices

Prior to the advent of chemical pesticides, humanity relied primarily on cultural and mechanical methods of pest management. With the development of pesticides, the relative impact of various cultural and mechanical practices on pest populations was often overlooked. With the gaining of public interest in environmental issues, cultural and mechanical pest management practices started regaining its due attention. Cultural control is one of several general approaches to insect pest management. Simple modifications of a pest's environment or habitat often prove to be effective methods of pest control. As a group, these tactics are usually known as cultural control practices because they frequently involve variations of standard horticultural, silvicultural, or sericultural practices. Since these control tactics usually modify the relationships between a pest population and its natural environment, they are also known, less commonly, as ecological control methods. Simplicity and low cost are the primary advantages of cultural control tactics, and disadvantages are few as long as these tactics are compatible with a farmer's other management objectives (high yields, mechanization, etc.). Unfortunately, there are still a wide variety of insect pests that cannot be suppressed by cultural methods alone. Cultural control involves modification of standard farm practices to avoid pests or to make the environment less favourable for them. Crop rotation, tillage practices, regular irrigation, stimulating plant growth with fertilizers, pruning and thinning, time of sowing and harvesting, destruction of crop residues, barriers, hedge rows, traps, and other forms of environmental modifications all influence the incidence of pest problems. Crop resistance to pests, which may be considered a cultural practice, remains a key factor in many pest management programs.

Cultural control of insect pests is affected by the manipulation of the environment in such a way as to render it unfavourable to the pest. This is achieved using variety of techniques. Many are considered old and traditional but serve to reduce the chance of insect colonization, multiplication and dispersal. Cultural control includes techniques such as rotation, hermetic storage systems, intercropping, mixed cropping, manipulation of crop sowing dates and management of weeds and field margins. Cultural control is a largely a prophylactic measure and sometimes a control measure too. Coaker (1987) considered cultural control as the first-ditch defence around which to other control options lie. On its own, it is unlikely to reduce pest infestation to desirable levels, but within the framework of an insect management system it can be used to reduce pest infestation to desirable levels of infestation below economic threshold. Van Lenteren (1987) noted that in study of environmental manipulations for the purpose of improving natural enemy efficacy the only area that had received attention appeared to be behaviour modifying chemicals, a typical example of the emphasis placed on the development of the products. Of all the control techniques cultural control is rarely dependent on a particular product and as a consequence, offers very few opportunities for capital return. Hence, cultural control has attracted little interest and little funding for research. Cultural control is largely dependent on techniques and practices and since it is difficult to sell some one idea or, an approach it is left to the universities and Government research institutes rather than the industry to carry out the necessary research. Here though the cultural control has suffered in that the techniques on their own are rarely capable of exerting the required levels of control and there is no glamour or prestige associated with control options that only partially solve problems. The panacea mentality has not really been subsumed by the concept of insect pest management and cultural control representative an important component in the development of a coherent, holistic approach to pest management and warrants a great deal more attention than it is currently receiving.

There are several types of cultural controls with abundant number of examples. Every time a grower rotates a crop or tills the land; weed, insect, nematode, slug and pathogens are affected. It is helpful to consider all of the available options while developing overall pest management strategy. Crop rotation is one of the oldest and most effective cultural control strategies. Growing the same crop year after year in the same field gives the pests and pathogens a fair opportunity to multiply and establish to the extant they can cause serious damage to the crop. Crop rotation keeps the host plants off the pests and pathogens and can thus break this cycle by subjecting the pests and pathogens to starvation. However this limited to those which have no alternate host-plants in the near vicinity. As a rule, crop rotations are most likely to be practical and effective when they are used against pests attacking annual or biennial crops and having a narrow host range, cannot move easily from one field to another, and are present before the crop is planted. Crop rotation replaces a crop that is susceptible to a serious pest with another crop that is not susceptible, on a rotating basis. For example, root knot larvae can be starved out by following legume with one to two years of a non-host crop such as soybeans, peas, or okra. Rotation schemes have also proven successful for controlling pests in pasturelands.

Intercropping (also known as mixed cropping) is another way to reduce pest populations by increasing environmental diversity. In some cases, intercropping lowers the overall attractiveness of the environment, when host and non-host plants are mixed together in a single planting while in other cases, intercropping may concentrate the pest in a smaller, more manageable area so that it can be destroyed easily. Vegetable crops intercropped with tasar or muga host plants serve as trap crop for several defoliating insects. Legumes sometimes serve as trap crops to reduce populations of leafhopper, *Emposca flavescens* and rootworms (*Diabrotica* spp.) in mulberry gardens. In some crops, it is possible to create discontinuity in the pest's food supply simply by altering the time of year for planting or harvesting. This strategy, often known as phenological asynchrony, allows farmers to manage their crop so that it remains "out of phase" with pest populations. *Terminalia arjuna*, for example, can escape injury from gall insect (*Trioza fletcheri minor*) if it is pruned in early autumn and the leaves are fully utilized. In the tasar silkworm rearing, farmers pre-pone brushing of silkworms on sal plants to protect their crop from larval mortality due to the toxicity of over mature leaves. Since trees become more susceptible to beetle outbreaks, as they grow older, good management dictates that silkworm rearing should be conducted in time fully utilizing the leaves. Further, harvest timing is often the most practical method available to farmers for controlling uzi fly infestations in non-mulberry sericulture.

Sanitation refers to keeping the area clean and devoid of plant material that may harbour pests. Sanitation is the basis for pest management in most livestock, public health and food establishment pest management programmes. Sanitation is also applied for crop protection where removal of sources of pest is a critical factor. Examples include removal of weeds that may harbour mites, aphids, or whiteflies from chowki garden; destruction of crop residues such as dead woods and dry leaves that may be over wintering sites for pests; cleaning of farm equipment that can spread pests from field to field; and removal and management of manure that provides breeding sites for flies. The planting of several crops in close proximity is called polyculture (as opposed to monoculture), a practice that makes it difficult for pests to find their favoured host crop and also promotes favourable habitat for beneficial natural enemies. Many field crops, such as peas, soybeans, carrot, and vegetables can be grown in parallel strips, a practice called strip cropping, which again creates the habitat diversity favourable for natural control. Pruning and pollarding is one of the best techniques to minimize the pest population. Some pests are normally carried from the old crop to the new one. Proper pruning of the undesirable portions of arjun (*Terminalia* spp) plants keeps the gall insects, stem borer and leaf minor under check. In mulberry, pruning, training and crop scheduling helps to a great extant in controlling the pests of the shoot system.Trap cropping is a system where a pest's preferred food crop is cultivated within or adjacent to the crop to be protected; the insects are attracted to the trap crop that is then destroyed. For example, mealy bugs and weevils concentrate on *Hibiscus* spp, which is destroyed along with the pests.Selection of pest resistant or tolerant or avoiding crop varieties is one way of crop protection. Some varieties have physical and chemical properties that repel or tolerate or kill pests. Many crop varieties grown today are resistant to one or the other pest. Breeding for pest resistance is tedious and time consuming, requiring many generations of plant hybridisation and selection. However, non-conventional methods like biotechnology have shown a short cut to this goal.

Mechanical and physical control measures involve the use of force or physiological factors of the environment with or without the aid of special equipment. The physical control measures give immediate tangible results and are generally popular and convincing to the farmers, even though they are time consuming, laborious and are often applied when much damage has already occurred. Physical controls are methods that physically keep insect pests from reaching their hosts. Hand picking, use of hand nets, beating and hooking, mechanical exclusion which consists the use of devices by which the insects are physically prevented from reaching the crops and the produce are the most ancient methods employed by man and is still being used to minimize pest population. Handpicking and destroying the pests can be used for gregarious or large or brightly coloured foliage feeders such as chafer beetle, hairy caterpillars and grubs of stem borers. Mechanical control methods can be rapid and effective, but many are mostly suited for small acute pest problems, and are popular with sericulture farmers. Importantly, mechanical controls have relatively little impact on natural enemies and other non-target organisms, and are therefore well–suited for use with biological control in an integrated pest management approach. Deep ploughing exposes many soil insects to desiccation or predation by birds and flood irrigation suffocates the insects and their life stages hiding in water to death. Shaking plants will dislodge many pests. For example, melolonthid or cuculio beetles can be removed from arjun and asan trees by diligently banging tree limbs with a padded stick and collecting the adult weevils on a white sheet as they fall out of the trees. A strong spray of water will dislodge aphids and mites from chawki garden, and houseplants. Fly swatters and mousetraps are forms of mechanical control.

Various types of traps viz. uzi fly traps, cricket traps, light traps, air suction traps, electric traps have been devised for collecting and killing different types of insects. Stink bug can be trapped under cardboard traps wrapped around arjun trees; the bands are removed and destroyed. Barriers include window screens for keeping pests out of buildings. It is the best method to minimize grainage pests. Deep trenching prevents the soil-dwelling grubs from infested plots to the neighbouring plots. Repellents, confusants, and irritants are not usually toxic to insects, but interfere with their normal behaviour and thereby keep the insects away avoiding damage. Wide use of synthetic sex pheromones may attract the insects searching for mate to a wrong place making them unable to mate and produce offspring; 'codling moth control' is one such product commercially available. Using insect pheromones in this manner is called mating disruption, a practice that works best in large commercial plantings where it is less likely that mated females will move into the planting from outside the treated area. Many of these types of behaviour inducing chemicals break down or wash away quickly, and must be reapplied frequently, used in an enclosed area, or formulated to release slowly over a long period.

APPROACHES AND OBJECTIVES

Understanding the underlying processes and the mechanisms of cultural control is of prime importance to know the insect-host interactions and hence are of immense interest to ecologists, for instance, the role of insect dispersal in the colonization and exploitation of crop plants. The interest in the ecological principles has produced some of the more illuminating work on the processes involved in cultural control. Knowledge and understanding of these processes will enable entomologists to predict more accurately the potential value of similar control techniques in other systems. Research into cultural control revolves around the evaluation of various combinations of practices and techniques for specific crop systems. There have been little or no efforts to identify the processes that account for the differences between treatments. While there is a need for both approaches, it would be difficult, and often impossible, to predict the value of new techniques and practices using the latter approach. Often greater emphasis is laid on limited evaluation of treatments than evaluating the processes. As a result, cultural control is often considered to include a mismatch of techniques without cohesive relations. Too little has been done to allow the development of an integrated philosophy with a greater knowledge of the host and pest biology and the underlying ecological principles to obtain a more rational/holistic view of the subject. The objective of cultural control is relatively straightforward, being reduction in the initial colonization, reproduction, survival and dispersal from the system.

CONDITION OF THE HOST

The ability of the host to withstand pest infestation may be due to as much to the general state or condition as to any inherent resistance. A healthy and fit animal and plant can have a greater tolerance of pest attack than an unhealthy one. The immunological response of silkworms can be impaired by any other factors, which can indirectly affect their susceptibility to pest attack. The immunological response of silkworm to *Nosema* sp. is affected among other things by stress, photoperiod and infestation with other parasites while the uzi grub *Blepharipa zebina* can be impaired by vitamin deficiency. In insect plant systems such factors can be influenced by the use of cultural control techniques. Crop plants provided with adequate nutrition and water may prove to be more attractive to pests, which may increase to even greater levels and hence, the use of fertilizers and water must be balanced against the increased size of pest population (Van Emden *et al.*, 1969). Coaker (1987) states that irrigation, mulching, manuring and fertilization are practices that in general promoting rapid growth and shorten the time the susceptible plant stage is available for attack, providing the crop with greater tolerance and opportunities to compensate for insect damage. However, such practices can only change the physiology of the plant as food for insect pest species, making it lusher there by enhancing pest survival (Coaker, 1987).

The value of these and other cultural practices must be at present being assessed on an individual crop/pest basis since there is little current information on which general conclusion can be based. Pest outbreaks may occur because host crops suffer environmental stress (Brodbeck and Strong, 1987; Mattson and

Haack, 1987), the most devastating of which can be drought caused by high temperatures and decreased water availability. Under these conditions, virtually every plant process is affected. Water stress induces changes in trichome size and density, their thickness and waxiness (as means of increasing reflectance and decreasing transpiration water loss); the plants are generally warmer due to less transpiration cooling (Mattson and Haack, 1987); and have increased levels of soluble carbohydrates and amino acids in the leaves (Wheatly *et al.*, 1989). Depending on the pest, these changes may or may not influence their ability to survive, and increase reproduction and development rates. In the water stressed mulberry plantation, it has been shown that threat to the leaf miner was acute on the plants suffering from severe stress and high leaf surface temperature; whereas the cicadellid, *Empoasca* sp. tended to concentrate where there was no drought and leaf temperature were low (Wheatly *et al.*, 1989). Thus, in situations where irrigation was introduced to reduce the effects of drought, it would be expected to have differential effects on the two pest species. Further detailed studies on the influence of irrigation and fertilization on the condition of the host crops and the levels of pest infestation are required.

BIOECOLOGICAL METHODS

Under bioecological methods of pest control male sterilization technique, use of attractants, repellents, hormones, antifeedants and antimetabolites, resistance of plant to insects are most common approaches to minimize the pest population. The male sterilization technique has been successfully employed in recent years in the control of some pests in United States. The pioneering work of Knipling in this direction has led to the effective control of screw worm (*Cochliomya hominivorax*), the melon fly (*Dacus cucurbitae*), the Mediterranean fruit fly (*Ceratitis capitata*) etc. It involves the artificial rearing and releasing of sterilized males, which are other wise physically, healthy to compete with natural males in the field. The females are thus deprived of the opportunity to produce viable eggs. Consequently, certain proportion of the expected population does not appear and the pest is suppressed in numbers. The sterility caused in male is irreversible. The procedure involves mass rearing, sterilization and release of males, which competes against their own kind for mating. Sterilization of both sexes of the pest population was also found to be useful. Reproductive capacity of the individual is reduced due to sensitivity of insects to radiation. Insects are irradiated by exposure to cobalt 60 irradiation. Radioactive isotopes are also applied on insects by spraying the insects with a solution of a radioactive compound. The common radioactive isotopes used for labelling studies with insects are carbon 14, phosphorus 32, iodine 131, bromine 82, arsenic 76 and sulphur 36 etc. This technique can be used in controlling several parasites of silkworms.

For effective control, the males and females should mix thoroughly over a considerable area. Chemosterilants like apholate, Tepa, Metepa, Thiotepa, etc., act as sterilants on contact with the body or leg pads. Morphogenetic agents like Diflubenzuron are effectively used in controlling *Exorista sorbillans*. It is necessary that sterilized insects are not affected in their vigour, longevity, behaviour and mating competitiveness. This provides an ideal situation in cases where females mate only once in their life. Use of semiochemicals to modify the behaviour of the pest population is a recent approach of pest management. Signalling chemicals that an organism can detect in its environment and may affect the organism's behaviour or physiology is called semiochemicals. Those that act between the members of same species are called pheromones, and those that act between species are called allelochemics. The allelochemics may be allomones, which favour the emitter, or kairomones, which favour the receiver. Many allelochemics act as both allomones and kairomones and such chemicals are called synomones. The insect sex pheromones are the best known to produce a wide variety of allelochemics that influence plant insect relationships. The semiochemicals act as repellents, feeding or oviposition deterrents, antibiosis factors (allomones); and attractants, oviposition excitants and feeding stimulants (kairomones).

ATTRACTANTS

The use of attractants in insect pest management is precise, specific and ecologically sound. Most of them are natural products but certain synthetic attractants are also known. Analogues of certain natural products have also been systematically synthesized. Of the several natural compounds isolated, the pheromones of silkworm (*Bombyx mori*), Gypsy moth (*Prothetria dispar*) and the pink ballworm (*Pectinomorpha gossypiella*) are well–known. The pheromones of the honeybee (*Apis mellifica*), which suppress sex differentiation among workers, have been known to be 9 keto trans 2 decenoic acid. Of the synthetic compounds, gyplure, related to natural compound of the gypsy moth has become important in insect control work. Scolytids or bark beetles have also been known to produce pheromones, which though produced only by one sex, causes an aggregation response in both sexes. The male secretes the pheromone into the hindgut, which gets incorporated into the faecal pellets, and the pheromones released through them attract flying males and females towards the galleries. Pheromones have been successfully used in insect control and demonstrated a positive anemotactic orientation from distances of over a mile and attracting them towards the trap. The attracted males are rendered immobile by coating the trap with sticky material or destroying them by means of insecticides or by sterilization by exposing them to the chemosterilant, tepa. The release of synthetic pheromones is timed matching with the natural release by their females. Synthetic attractants are capable of attracting both sexes. In silkworm rearing field uzi trap is used to control uzi fly population. During rearing period Citronella oil bait is also used to capture various predators. Attempts are on to develop plant-originated baits for use in insect control.

ANTI-FEEDANTS OR FEEDING DETERRENTS

The anti-feedants are the chemicals, which inhibit or deter the feeding of insects due to their presence on the natural food of the species concerned. Anti-feedants act as gustatory repellents, inhibiting the gustatory or taste receptors. It is a type of feeding deterrents, which deprives an insect from continued feeding on the host; death at the end is due to starvation. Compounds like Eulan CN and Mitin FF are being used for protection of fabrics against insect and their larvae. Further, some compounds viz. Triazines compound 24055 organotins stenous chloride, carbamates, baygon, botanical extract like pyrethrum and some miscellaneous copper, stearate, phosphor, cycocel are known to have antifeedant activities. Anti-feedants are nontoxic to many insects but inhibits the surface feeders from feeding and caterpillars, beetles, weevils have been controlled particularly when they happen to be surface feeders. Recently anti-feedants have been identified in the plant extracts of *Coccus trilobus, Clerodendron trichotomum* and *Lindera triloba.*

REPELLENTS

Insect repellents are the substances whose stimuli elicit avoiding reactions. They lead the insects away from the source. They are typically behavioural responses arising through the stimulation of chemoreceptors, olfactory or gustatory receptors. The naturally occurring feeding repellents are effective while the artificial chemicals were not successful. Repellents are highly effective indoors compared to outdoors. Repellents have been used successfully against crawling insects, termites, fabric eating insects and blood sucking insects. Application of Citronella oil and diphenylamine repels the uzi fly from laying eggs on silkworms.

METABOLITES

Many higher plants produce biologically active natural products known as metabolites that affect the growth and development of other organisms. These compounds are considered defensive substances useful to the plants in discouraging or preventing attack form many pests and such compounds are called allelochemics. It is an info-chemical that mediates interaction between two individuals belonging to different species. Plants have a number of chemical constituents known as secondary metabolites, since they do not seem to have any specific function to plant physiology. To circumvent the effect of plant defensive substance, insects have developed detoxification mechanisms through the action of Mixed Functional Responses (MFO), glutathione transfer and a host of other enzymes, followed by active excretion. Insects avoid ingested toxins through strategies such as (i) preventing the toxins from reaching the target sites (pharmacological strategy) (ii) metabolizing the toxicant into less toxic or non toxic forms and (iii) through target site insensitivity. Excretion, adsorption and sequestration are the principal modes of overcoming the toxic effects. The ongoing mechanism has led to the evolution of several physiotypes or chemotypes in plants and many biotypes in insects. Ellegic acid is a toxic compound present in the leaves of *T. arjuna* on which silkworms feed. Tasar silkworms through either of these mechanisms detoxify it.

RESISTANCE OF PLANTS TO INSECT

Large-scale use of pesticides in the field has manifested a wide range of side effects. These adverse effects include hazardous effects on non target organisms, pest out breaks, resurgence of pest populations and development of insect strains resistant to insecticides. This has necessitated the exploration for alternative methods of pest control. The use of insect resistant varieties offers one such promising possibility. The use of insect resistant varieties provides insect control at no additional cost and is compatible with other methods of pest control. All the green plants possess the essential amino acids, sugars, lipids, vitamins, minerals, etc. However, the physiological condition may vary from plant to plant and also with the age of the plants. The secondary substances vary from species to species qualitatively. For this reason, all insects do not attack the same plant and the same insect does not attack all the plants. The selection of host plant is primarily the result of an increase or decrease in the thresholds of feeding stimuli, which include the gustatory, tactile, visual, olfactory and chemotactic ones.

Some insects feed on diverse variety of food; while in majority of species there is restriction of host plants. Strict host plant specificity or monophagy is rare, feeding on large variety of plants (polyphagy) is not infrequent, but the most common is the choice of a few related plants, or to some extent unrelated plants (oliphagy). The ability of the insect to choose a particular variety under field condition is known as selection. The ability of the plant to suppress the reproductive potential of the insect is known as antibiosis, promoting production of a lesser number of offspring. The capacity of some plants to support large population of insects without showing any side effects is known as tolerance. Selection, antibiosis and tolerance are operative factors in plants and are taken into account in experimentally building up resistance to insect attack. Groups of secondary compounds or a mixture of several closely related compounds of the same group produced by the plants behaved either as chemical attractants or as repellents. The accepted hypothesis is that plants first synthesized all secondary compounds affecting insects or herbivores as a general defence against feeding.

An insect species could evolve the means to detoxify the xerobiotic, which may act as a repellent or as a toxicant, thus enabling the insect to utilize the plant. To the successful insect, this compound, henceforth, become a valuable signal to guide to the host plant. In fact, such close association of the compound with the insect, makes it an essential feeding stimulant. The host selection mechanism thus largely depend upon the type and concentration of secondary substances and very little on the physical appearance, physiological conditions and nutritive value of the plant. It will be logical to conclude that a chemical affecting insect feeding can primarily be a plant's defence mechanism to which insect can adopt to a varying degree ranging from mere host guide to an essential feeding stimulant.

CHAPTER-8

CHEMICAL CONTROL

Chemical control is the use of chemicals to kill pests or to inhibit their feeding, mating, or other essential behaviours. The chemicals used in chemical control can be natural products, synthesized mimics of natural products, or completely synthetic materials. Insecticides and miticides include many types of commercially available toxins, some naturally derived, others synthesized that are used for killing insects and mites. Chemical controls, particularly synthetic organic insecticides, have been developed for nearly every insect pest. They are widely used in industrialized nations for several reasons: they are highly effective — one product often controls several different pests; there is relatively low cost for product or labor; and generally their effects are predictable and reliable.

Chemical insecticides have allowed management of larger acreages by fewer individuals because of the reduced labor needed for physical and mechanical controls. Besides their use in agriculture and sericulture, chemical insecticides has been very important in the battle against disease-carrying insects. Of the various methods of insect control, chemical control brings about the prompt and conspicuous relief from the attack of the pests. Even after adoption of all approaches to control the pest population, sometimes pest population continues to increase and cause extensive damage to silkworm and its host plants Therefore, in order to kill the pest very quickly, chemical control method is adopted.

The commonly available chemicals are used to control the pests. Various chlorinated, organophophate , carbamate and plant originated insecticides have been tested against the major pests of silkworms and their food plants. These tested insecticides are used in case of severe attack of the pest. This is only the best method to knock down the population of pest in the field. Conventional insecticides are among the most popular chemical control agents because they are readily available, rapid acting, and highly reliable. A single application may control several different pest species and usually forms a persistent residue that continues to kill insects for hours or even days after application. Because of their convenience and effectiveness, insecticides quickly became standard practice for pest control during the 1960's and 1970's. Overuse, misuse, and abuse of these chemicals have led to widespread criticism of chemical control and, in a few cases, resulted in long-term environmental consequences.

HISTORICAL ASPECTS

References to insect pests are found in the Bible. The prophet Joel vividly described the ravages brought on by the desert locust and the approaches to insect management. Earlier references to locust invasions and burning as a control measure were recorded during the Shang Kingdom in China (ca. 1520 -1030 BC) (Harpaz, 1973). Around 2000 BC, Sinuhe, observed the effects of flea-borne bubonic plague in Syria. The whole of Syria became a huge open grave due to death of more than 90 per cent of its population. Most of the sufferers were war soldiers who were attacked by dreadful fleas. This devastation led to the discovery of insecticides. Documentation of the ancient use of chemical controls for insects appears in Homer's writings before 1000 BC, where sulfur was identified as an insecticide. Pliny (AD 79) recommended the use of arsenic as an insecticide (Cremlyn, 1978). The Chinese recognized the use of mercury and arsenic for body louse control (AD 100-200), and white arsenic was used to protect rice plants from insect attack as early as AD 400 (Konishi and lto 1973). Fortunately, the chemical tools for insect control have evolved rapidly in the last 100 years, starting with inorganic agents such as lime sulfur and arsenicals, and augmented by natural products such as pyrethrum, rotenone, and nicotine, whose agricultural uses were beginning to be established by 1870 (Shepard, 1951; Cremlyn, 1978). At the time of their introduction, such material must have seemed to represent a marvelous shift in the balance between man and insects, providing for the first time reliable tools to combat important insect pests. On the other hand such chemicals have serious disadvantages in characteristics such as potency, flexibility of use, variability in composition, limited range of insects that could be controlled, etc. In case of arsenicals, environmental persistence and hazards to the applicator and consumer of treated produce were all evident and led to the first notable public concern regarding pesticide hazards and to the detection of residues which exceeded established tolerance limits for arsenic in foodstuffs (Shepard, 1951).

The first recorded use of a synthetic organic insecticide, dinitro-o-cresol, occurred in 1892, and by 1930s a range of such compounds have been discovered and had found limited use (Cremlyn, 1978). From 1920 onwards, the increasing use of insecticides as tool for insect control led to their growing predominance as an approach to insect control; and it is claimed with such justification, to the neglect of the studies on pest biology and ecology and of alternative methods of insect control (Smith 1978). A series of dramatic discoveries during late 1930s provided new synthetic insecticides of unprecedented power and range of activity that further shifted the emphasis towards the chemical approach to insect control. Krueger (1932), while synthesizing organ phosphorus esters, noted that they caused a series of unexpected physiological effects including respiratory distresses, disorientation, and visual disturbances. So started a long series of investigation, particularly by Schrader that led to development of the organophosphates both as insecticides and as nerve warfare agents (Fest and Schmidt, 1973, Eto, 1974). At the same time, the studies of Paul Muller of Geigy Co., Switzerland, on moth proofing agents led to the discovery of DDT, the first of series of organochlorine insecticides, which were developed during the 1940s. The interesting history of development and impact of DDT during the Second World War has been well-documented (Price, 1970). The impetus provided by these discoveries encouraged further efforts to find synthetic insecticides that generally met with success and led to a constant stream of new products over the following 20 years. However, many of these newer compounds were not without serious faults. The organophosphates in many cases shared the familial trait of high mammalian toxicity with the nerve gases. Even today this group is responsible for the majority of accidental deaths attributable to pesticides. In 1971 the use of several of the most toxic organophosphates was prohibited in Japan because of their hazardous nature (Anonymous, 1982). During the same period the impact of chemical and biological persistence of organochlorines especially DDT were brought to the notice of all concerned. Rachel Carson (1962) in her book, 'Silent Spring' reported that the accumulation of DDT in the biota had serious toxic effects on many species. This aroused a lot of public concern and speeded up political action. As a result, during the 1970s, the organochlorines were subjected to regulation and their use has been phased out one by one in the US, Europe and Japan. However, some of these compounds remain in commerce in some developed nations, and their utilization remains high in many developing nations due to their low price and high effectiveness (Wigglesworth, 1976; Wright, 1976).

The use of organochlorines was reviewed and the vacuum left by their loss was filled by the enhanced use of the organophosphates and the related cholinesterase-inhibiting carbamates, which were introduced during the 1950s (Kuhr, and Dorough, 1976). However, the high intensity of use the pesticides provided by these discoveries encouraged use of much more potent insecticides after the Second World War, which greatly increased the selection pressure on insect pests, and rapidly led to severe insecticide-resistance problems (Georghiou, 1982; Georghiou, and Saito, 1982). This led to the need to find alternative insecticides. In this strategy there is a necessary to formulate a continual stream of new insecticides. The introduction of new compounds has slowed considerably during the 1970s (Goring, 1977; Braunholtz, 1981), which precludes such an approach. Many pests are known to have developed resistance to a wide range of insecticides. Such populations can be insensitive to many new materials, or may be able to quickly become insensitive. At present resistance to insecticides remains a considerable problem, though not universal, with few remedies available. However, an effective means to manage this problem must be found if insecticides are to continue as major weapons in future.

Another consequence of the intensive use of pesticides is the creation of new pests due to destruction of the natural enemies that normally keep them in check. For the same reason, resurgence of damaging population levels of a primary pest has sometimes been observed after an initial insecticide application (Metcalf and Luckmann, 1975; Pimentel, *et al.*, 1980). The need for increased and more frequent application of pesticides to combat these pesticide-induced problems has been disastrous. A well-known example is the abandoning of cotton cultivation in the lower Rio Grande valley of southern Texas and north–eastern Mexico in the late 1960s, as attempts to control the induced secondary pest, *Heliothis virescens,* failed because of high insecticide resistance. Eventually growers had to resort to making 20 or more insecticide applications per year without saving their crop (Bottrell and Adkisson, 1977). Both the farmer and the pest control strategy became bankrupt. A similar example with cotton in Peru has been described in detail by Doutt and Smith, (1971). Although such cases are rare, they do illustrate vividly how chemical control can have severe repercussions, particularly when used indiscriminately and without regard to underlying ecological considerations. All the revised pest management systems now emphasize the preservation of the environment and the natural control agents.

PESTICIDE RESEARCH AND DEVELOPMENT

The large-scale pesticide industry dates back to the end of World War II with the commercial introduction of the phenoxyacetic acid, selective herbicides, synthetic organochlorine and organophosphorus insecticides. It has been estimated that $ 32.6 million worth pesticides were exported from the major pesticide manufacturing countries in 1949 (U.S.A. - $ 28 million, U.K. - $ 3 million, France - $ 1 million and Switzerland - $ 0.6 million). By 1965 the exports from United Kingdom expanded $10 million and Western Germany emerged as a major pesticide exporting country. The total value of all pesticides applied to crops all over the world in 1949 was approximately $ 200 million, which clearly emphasizes the point that most of the pesticides are used in manufacturing countries and not exported. The relatively low proportion exported may be partly due to the fact that many under developed countries import chemical intermediates from industrialized countries and then carry out the final stages of production themselves in which case the chemical intermediates do not count as pesticide imports. Developing and underdeveloped countries like to have political control of pesticides manufacture. In India, benzene hexachloride is the most widely used insecticide because it can be produced simply in crude form. The annual production of this compound in India in 1965 was around 30,000 tonnes which is about four times to that produced in United States of America though it represented only 20% of the total American production of all organochlorine insecticides (Crowdy, 1977).

The pesticide reviews of American Department of Agriculture gave the value of American pesticide production in 1965 as about $200 million and a reasonable estimate of total world pesticide production would be around $ 600 million. British pesticides production was nearly $ 23 million, including $12 million worth herbicides, $ 6 million of insecticides, $ 3.6 million of fungicides, and $1 million of miscellaneous pesticides. The pesticides manufacturing companies were actively engaged in basic discovery, development, evaluation and manufacture of modem pesticides. Research and Development was actively associated with screening, determination of general biological characteristics, analytical and metabolic studies, basic and applied toxicology, chemical synthesis, formulation studies, etc. The evaluation of biological performance and assessment of possible hazards under the intended conditions of use are studied in different locations though it caused heavy loss to human beings.

CLASSIFICATION

Pesticides are chemicals designed to combat various pests attacking crop plants or animals of human interest. Most important among them are insecticides and fungicides. There are also rodenticides, (for the control of rodents) nematicides (to kill nematodes), molluscides (to kill snails), and acaricides (to kill mites). In most of the places, pesticides are an integral part of plant protection system and are applied for the purpose of protecting plants from injury by insects and disease causing organisms. There have been a number of evidences of pest resistance, pest resurgence, ground water and food chain contamination, risk to workers, and pests continue to remain as pests in spite of repeated use of pesticides. Still the demand for pesticides has been growing throughout the world. In sericulture, it is reported that pests and diseases cause loss of more than 25 % of leaf yield thus indirectly affecting the cocoon production (Singh *et al.*, 2000). Insecticides contain one or more active ingredients that serve as toxicants (poisons). In their purest form (technical grade), these chemicals may be too toxic, too unstable, or too volatile to be handled or applied safely.

Therefore, technical grade insecticide is always mixed with other compounds, known as adjuvants. In order to improve the performance, safety, or handling characteristics of a commercial product, these mixtures (technical grade insecticide plus adjuvants) are known as formulations. Almost anything could be an adjuvant: pumice, ground walnut shells, buffalo gourd root powder, vegetable oil, etc. These compounds are usually listed on the label as "inert ingredients", but they are certainly not inactive. Many adjuvants are proprietary products, protected by patents and closely guarded as industrial secrets. They may represent 90-95% of the total volume of a commercial formulation. An insecticide can be manufactured in a variety of formulations, each tailored for specific applications. Some of the more common formulations are dusts, granules, wettable powders, emulsifiable concentrates and solutions; concentrate liquids, fumigants, mixed formulations, etc.

DUST

Dusts are dry powders usually formulated with inert particles of ash, chalk, talc, or clay. They are designed to be sprinkled or blown onto target surfaces. The powder sticks to feet, legs, and other body parts of passing insects, and the toxicant is eventually ingested when the insect cleans itself. Insecticides that act as stomach poisons are often formulated as dusts. These formulations are not always suitable for outdoor applications because they have a tendency to drift (blow away with the wind). Insecticides, which are required to be used in dry form, are mixed with impregnated organic flours such as walnut-shell flour, soyabean and wood-bark or pulverised minerals such as sulphur, diatomite, tripolite, lime, gypsum, talc, pyro-phyllite, or clays such as bentonites, kaolins, attapulgite and volcanic ash. The concentration of toxicant present in the dust formulations ranges from 0.1 to 25 per cent. They are ground to fine sizes ranging from I to 40 microns and most dust particles will pass through a 325-mesh screen. In general, the toxicity of a dust formulation increases as the particle size decreases. Some insecticidal compounds are inactivated by alkali and, therefore, the quality of the finished dust formulation largely depends on the properties of the carriers used. Compatibility, particle size, abrasiveness, absorbability, specific gravity, wettability (if used as a water dispersion) and cost are some important basic points to be considered in the selection of dusts. Dusts are easy to handle. Best results are usually obtained if applied early in the morning when the plants are wet with dew. Even slight winds may cause drift of dust to non-target fields leading to undesirable problems.

GRANULAR INSECTICIDES

Granules (G) or pellets (P) are coarse particles (e.g., clay, ground corn cobs, or walnut shells) can serve as a carrier for certain types of insecticides. The toxicant slowly leaches out of the carrier, minimizing its movement within the ecosystem and maximizing its active life (persistence). Granular formulations are commonly used for controlling soil-dwelling insects and in systemic insecticides that are applied around the base of plants. Since granular formulations do not blow or drift, they are considered relatively safe from the standpoint of accidental human exposure. Insecticides are formulated as granules consisting of inert material with the toxicants absorbed on to them. The granules vary from 0.25 to 2.38 mm in diameter and the range of particle size is designated by a two-figured mesh classification such as 30/60 which means that virtually all of the granules will pass through a standard 30 mesh sieve (30 openings per linear inch) while only a negligible quantity will pass through a standard 60 mesh sieve. Some granule sizes are 20/40, 30/60, 40/60, etc. The formulations commonly contain 2 to 10 per cent toxicant applied by solvent impregnation to highly absorptive clays such as bentonites and attapulgite and diatomaceous earths of proper particle size. Granules are broadcast either on to the plant or soil and no water is required for their application. Granular particles being heavier, there is very little drift, which prevents undesirable contamination to adjacent crops and undue loss of insecticide. The toxic material in the granular formulation is released over a long period than does a spray deposit. Formulations of fine grain size can be used to carry systemic insecticides, which dissolve in the due drops on the leaves and later absorbed by the plant. Such formulations are available for the control of sucking pests like leafhoppers, plant hoppers, aphids, thrips, etc. Heavier formulations are used in case of soil fumigants, which are convenient to be placed in holes made in the soil with greater ease.

Such formulations are used to control soil-borne diseases like root-knot, root-rot, etc. Baits also come in the form of large granules or globules. These are used for burrowing pests in the soil like grubs, rodents, etc. Granular application may sometimes cause scorching if the chemical is concentrated in a smaller volume of carrier. The efficacy of the granular formulation applied to soil is dependent on many factors such as dosage, type of formulation, soil type, moisture conditions, method of application, rate of release of the toxicant, type of crop plant, rate of respiration and growth of the target insect, stage of the crop plant, weather conditions, etc.

WETTABLE POWDER

Wettable powders (WP) or Soluble powders (SP) are dry formulations that are mixed in water to form homogeneous spray solutions. A soluble powder dissolves completely in water, whereas a wettable powder contains an emulsifier that produces a uniform suspension (colloid). Many foliar insecticides are formulated as wettable (or soluble) powders because they are easy to transport and generally have a long shelf life. Wettable powders contain 15 to 95 per cent toxicant blended with a dust carrier such as attapulgite and a dispersing agent. Common formulations in use contain 25, 40 or 50 per cent toxicant. They are diluted or suspended in water and used as sprays. Adding 1 or 2 per cent surfactants improves the efficiency of the product.

SOLUTIONS

Many of the synthetic organic insecticides are not soluble in water but most of them are soluble in organic solvents; the common organic solvents used in insecticide formulations being amyl acetate, carbon tetrachloride, cyclohexanone, dibutylphthalate, ethylene dichloride, kerosene, refined kerosene, monochlorobenzene, petroleum naphtha or Stoddard's solvent, pine oil, technical di- and tri-methylnaphthalenes and xylene. Solvency, toxicity to plants and animals, fire hazard, compatibility, odour and cost are some factors to be taken into consideration in choosing a solvent. The solvents also possess, in many cases, some insecticidal properties of their own. At high concentrations many solvents are toxic.

EMULSIFIABLE CONCENTRATES

Emulsifiable concentrates (EC) are liquid formulations contain the toxicant(s) and an emulsifier dissolved in organic solvent. The concentrate is diluted with a large volume of water to produce the final spray mixture. Some foliar insecticides are formulated as emulsifiable concentrates. Unlike most wettable powders, they do not leave a visible residue on fruits and vegetables. Organic solvents in the mixture, however, may injure sensitive plants. Emulsifiable concentrates are also commonly used in sprays for urban and industrial pests. This is the most common and versatile formulation, which contains the toxicant, a solvent for the toxicant and an emulsifying agent. Incorporation of a small amount of an emulsifier (surface-active material) into the mixture ensures emulsification at desired stability, wetting and spreading characteristics. When mixed with water, the concentrate forms an emulsion of oil-in-water type. When sprayed, the solvent evaporates quickly leaving a deposit of toxicant from which the water also evaporates. The classes of emulsifying agents of importance in insecticide formulation and application are alkaline soaps, organic amines, sulfates of long-chain alcohols, sulfonated aliphatic esters and amides, mixed aliphatic-aromatic sulfonates, non-ionic types (ethers, alcohols, and esters of polyhydric alcohols and long chain fatty acids) and natural materials such as resins, proteins, carbohydrates, lipids, alginates and saponins. Sulfonated aliphatic esters and amides are excellent wetting and spreading agents and in wettable powders they are of use as dispersing agents. The agents act as stickers and wetters. Spreading and sticking agents are found in finely divided bentonite, other clays, flours, etc.

POISON BAITS

Poison baits consist of relatively small quantities of toxicants combined with food material attractive to pest. The toxicant in the bait may be of plant origin or a synthesised chemical. Poison baiting is adopted in places where spray application is generally not feasible as for the control of household pests and insects, which live underground and damage crops. It is also useful for control of biting and chewing insects infesting crops over a large area. For the control of rats, crabs and slugs also poison baiting is adopted. Poison baits are used in dry carriers for the control of armyworms, crickets, cutworms, earwigs, grasshoppers and locusts, white grubs, wireworms, etc. Poison baits are also used in liquid carriers for the control of ants, houseflies, sucking moths, root maggot, flies, etc. The poison bait sprays for flies and moths include toxicants like malathion with composite protein or yeast hydrolysate. Further, the adhesiveness of certain formulations in the form of emulsions and wettable powders is used in the form of stickers. These are intended for specific uses. Apart from this, waxes containing insecticide are sometimes used on floors. Shampoos intended for use on pet may contain an insecticide.

FUMIGANTS

Fumigants are released in the vapor state (as gases) and enter the insect's body through its tracheal system. Fumigants are most effective in storage bins, ship holds and even in soil where the gas can be confined. It is highly effective in an enclosed area such as a greenhouse, a warehouse, or a grainage. Fumigants are most often formulated as liquids under pressure and quite often they are mixtures of two or more gases. The liquid fumigants are held in cans or tanks. In some cases, the gas is required to be produced at the place to be fumigated. The advantage with fumigants is that the gas is able to penetrate easily to places not accessible to other formulations and is readily dispersed. However, great care in handling is required as the vapours are often pungent and toxic and de-flavours the treated crop. Aerosols (A) — Insecticides that are formulated together with a solvent may be pre-packaged in pressurized spray cans or sold unpressurized for use in special fogging machines. Spray cans are relatively expensive (per pound of active ingredient) but they are convenient, easy to store, and have a long shelf life. Commercial foggers are typically used indoors (e.g., greenhouses and grain ages) or for control of biting flies in community-wide pest control operations. Insecticide-fertiliser mixtures are also formulated by adding granular insecticides to chemical fertilisers or by spreading insecticide directly on to the fertiliser. This basically reduces the manpower requirement in application of the same. Formulations or mixtures containing two or more active ingredients are used in practice and the response to such combinations is referred to as "joint action". Joint toxic action is classified into 'independent joint action', 'similar joint action', 'synergistic action' and 'antagonistic action'. In case of 'independent Joint' action two active ingredients work independently with different physiological modes of toxic action. Combination of methyl parathion and sesamex is used to control housefly is an example of independent joint action.

Similar joint action is defined as combination of two active ingredients applied separately or jointly to produce the same response or toxic action. Combination of BHC and DDT is an example of 'similar joint action'. In 'synergistic action' the mixture of active ingredients increase the toxicity of the other, and the total response or increased toxicity produced will be greater than the sum of toxicities of the components of the mixture used separately or the additive effect of the components. Mixed formulation of methyl parathion and DDT (1:3), carbaryl and BHC (1:1) and methyl parathion and carbaryl (1:1) was frequently used in India during seventies. The advantages of such mixed formulation are that the mixture is more effective due to synergistic action and saves on the cost. It is believed that mixed formulation having independent joint action may tend to prevent development of insecticide resistance in pest species. It also helps in controlling resistant strains of insects. Sometimes there is 'antagonistic action' in which one component may reduce the activity of the other and therefore should be avoided. Mixtures of aldrin and parathion, or thionophosphates and sesamex come under this group.

ULTRA LOW-VOLUME CONCENTRATES (ULV)

Ultra low-Volume Concentrates (ULV) are highly concentrated formulations (more than 8 pounds of active ingredient per gallon) are designed to be used in specialized spray equipment that atomizes the concentrate droplets. ULV spray equipment is used by most aerial applicators (airplane sprayers) that treat forested lands or large agricultural acreages. Despite their many advantages, conventional insecticides are not ideal pest control agents. Indeed, one of their greatest strengths, broad-spectrum activity, is also one of their greatest weaknesses. While it is certainly an advantage to control multiple pest species with a single chemical treatment, the non-specificity of most conventional insecticides poses a serious threat to non-target organisms in the environment.

High mortality among natural enemies can have an enduring impact on the ecological balance of any community. In the absence of biocontrol agents, more insecticide applications may be the only recourse available to stop pest resurgence. Once we step onto this "insecticide treadmill", it can be very difficult to get off. The non-target effects of insecticides are not limited to natural enemies. Most of the organochlorine compounds (e.g., DDT) have been banned from use in the United States because of their indirect effects on reproduction in birds of prey (e.g., eagles, ospreys, condors, etc.). These environmentally stable compounds are highly soluble in lipids, making it possible for them to accumulate in the body fat of non-target organisms. Predators, particularly those at the top of a food chain, amass pesticide concentrations many times greater than anywhere else in the environment. This process, known as bioaccumulation (or biomagnifications), was responsible for high levels of DDT and related compounds in birds of prey during the 1960's and 1970's. These pesticides did not injure the birds themselves, but caused thinning of their eggshells and high rates of breakage during incubation. Most organochlorine insecticides were banned during the 1970's and 1980's and many of the threatened bird species are now recovering from the brink of extinction.

RESISTANCE TO INSECTICIDES

In some cases, sub-lethal concentrations of an insecticide can stimulate rather than suppress the growth of a pest population. This phenomenon, known as homoligosis, has been observed in a number of pest species, including two spotted spider mites (*Tetranychus urticae*), western corn rootworms (*Diabrotica virgifera*), and brown plant hoppers (*Nilaparvata lugens*). Low doses of pesticide seem to improve the nutritional quality of host plants, thereby increasing a pest's reproductive potential or decreasing its time of development and finally, there is the problem of resistance to insecticides. Insects are among the most adaptable organisms on the face of the earth. For the past 400 million years they have managed to survive by adjusting to changes in their environment, so it should come as no surprise that they can also adapt to chemical pesticides. Resistance has increased exponentially since the late 1940's, and today, there are over 500 pest species that exhibit some level of resistance to at least one type of insecticide. Insects may become resistant to insecticides in several ways.

Biochemical resistance usually involves changes in the metabolic pathways that insects normally use to break down plant defenses and other environmental toxins. This detoxification is facilitated by enzymes (esterases, hydrolases, transferases, and oxidases) that change the chemical structure of toxicants before they cause physical harm. Physiological resistance involves functional changes in basic life processes that alter the way toxicants interact with the body.

Some German cockroaches (*Blatella germanica*), for example, have become resistant to carbaryl (a carbamate insecticide) as a result of genetic changes in the permeability of their cuticle. Behavioral resistance may occur as a result of any innate change in behavior that reduces an insect's probability of encountering a toxicant. In some parts of Panama, the mosquitoes that vector malaria (*Anopheles albimanus*) have become hypersensitive to certain insecticides. They manage to escape lethal doses of insecticide by avoiding enclosed areas and refusing to land on treated surfaces. When an insect population becomes resistant to one insecticide, it may also become less susceptible to other toxicants in the same chemical family. This phenomenon, known as class resistance, is a common problem in all major groups of insecticides (organochlorines, organophosphates, carbamates, and synthetic pyrethroids). When an insecticide loses its effectiveness because of pest resistance, users typically replace it with another compound from a different chemical group. But resistance to one group of compounds does not prevent subsequent development of resistance to compounds in other chemical groups. Over the years, some pests have developed multiple resistances to nearly every insecticide that has ever been used to fight them. Colorado potato beetles, *Leptinotarsa decemlineata*, are probably the best example of a population that has developed multiple resistances. It may never be possible to prevent insects from developing resistance, but there are ways to manage these chemicals that will prolong their useful life.

CHEMICAL NATURE

Chemical types include: (1) elements, such as sulphur, phosphorus, antimony, arsenic , barium, boron, copper, fluorine, mercury, selenium, thallium and zinc; (2) inorganic compounds such as lead arsenate, sodium fluorosilicate, zinc phosphide, paris green; (3) organic compounds viz. (a) a compound of plant origin such as pyrethrum, nicotine, rotenone; (b) animal and mineral oil, such as fish oil, diesel oil, (c) synthetic organic compounds such as DDT, malathion and carbaryl; and (4) poisonous gases such as hydrogen cyanide, ethylene dichloride, carbon tetrachloride, methyl bromide, phosphine etc. used as fumigants. The above-mentioned compounds could be rearranged based on their mode of entry into insect body and lethal action on the pest.

MODE OF ENTRY AND ACTION

Most of the insecticides are either systemic or contact in action. Systemic insecticides are basically stomach poisons, which are absorbed by the plant system and enter the body of the pest as it feeds on the plant parts. The other is contact insecticide, which are absorbed by the insect through the skin on coming into contact with the cuticle. Systemic insecticides are a special type of stomach poison. These compounds are absorbed by the tissues of a plant (or animal) without ill effects. Insect pests ingest the insecticide when they feed on the treated organism. Systemic insecticides are sometimes included in the diets of domestic animals to protect them from internal parasites (e.g., cattle grubs and other bot flies). Plant systemic can be incorporated into the soil around ornamentals or bedding plants. The insecticides are absorbed by the roots and translocated to leaves, stems, and flowers. If the insect that feeds on a treated plant doesn't acquire a lethal dose of insecticide, it may at least be deterred from further feeding. Although systemic insecticides are commonly applied to plants, but they are not suitable immediately for silkworm rearing because the insecticide remains in the leaves even after its maturity. They are generally used against the insects with biting and chewing type or sucking type mouthparts. The host plant sprayed with the insecticide absorbs the systemic insecticides and the same enters the plant system through the transport system. On consumption of the plant tissues or sap, the pest gets poisoned and killed. The effectiveness of an insecticide usually depends on when and where the pest encounters it. Most insecticides are absorbed directly through an insect's exoskeleton. These compounds are known as contact poisons because they are effective on contact. Contact poisons gain entry to the body through vulnerable sites on its body. Contact poisons are applied as sprays or dusts either directly on the insect body or to the strategic places which are liable to come in contact with the body of the insects. They kill the insect by clogging spiracles and respiratory system or by entering through the cuticle into the blood and acting as nerve or general tissue poisons.

It has been shown that the insect cuticle possesses very high absorption properties, so that the lethal doses applied externally is almost the same as the quantity required to kill by injecting it into the body (Cremlyn, 1978). The contact insecticides are highly liphophilic and are readily absorbed by the lipids present in the epicuticle of the insect exoskeleton. When applied, the toxicants spread quickly over the entire surface of the body of the insect, possibly in the wax monolayer and apparently absorbed by the body of the insect through surface membranes or the bases of setae. The residual film of a contact insecticide may kill an insect by its action on sensory organs present in the tarsi of the legs. Though it was widely held that insecticides penetrate through the integument and carried to the target organ, the central nervous system by the haemolymph. Recent investigations have shown that the toxicant does not penetrate into the haemolymph in significant quantities and reaches the site of action in the integument of the tracheal system. There is a wide range of contact poisons available in the market.

INORGANIC INSECTICIDES

Among inorganic compounds sulphur, sodium fluoride and arsenicals are very common.In an essential insecticides, two points, viz. the percentage of total arsenic present and the proportion of water soluble arsenic have significance. Ideal arsenical insecticides should have very high water insoluble soluble arsenic content, which should otherwise be readily soluble in the digestive juices of the insect. If the arsenic content is water soluble, it can enter the foliage and cause "burning" injury to plants. The arsenates are more stable and safer to use on plants than arsenates and hence the latter are used in poison baits and not on the plants, as they are phytotoxic. In insect arsenic poisoning causes disintegration of mid gut epithelial cells with vacuolized cytoplasm and clubbed chromatin of the nuclei. The symptoms of poisoning are regurgitation, hoper and quiescence. Death of insect is primarily due to inhibition of respiratory enzymes. Among arsenate, lead arsenate and calcium arsenate are the most common insecticide against the control of castor looper, red beetle, black beetle and grasshoppers. Apart from this there are several arsenicals viz.white arsenic, sodium arsenate, paris green, copper arsenate, magnesium arsenate, zinc arsenate which are commonly used against various pests. In sericulture use of Bordeaux mixture is frequently used to check the infection of microbes on the pruned plant of tasar food plants Sodium selenate is used to control aphids and mites. Zinc phosphide is applied to control rodents in grainages and field.

ORGANIC COMPOUNDS

Among organic compounds, hydrocarbon oils, animal origin, plant origin and synthetic organic compounds are very common. Hydrocarbon oils contain the petroleum or mineral oils and tar oils. The mineral oils are the petroleum oils derived from sedimentary rocks. Complex solution of hundreds of hydrocarbons is present in the petroleum oils and kerosene, gasoline, lubricating oils; asphalt, tar and many other mixtures are derived from them. In their natural state oils are highly phytotoxic, but under certain conditions, if applied with emulsion they prove to be safe for use on the plants. Oils are used as solvent or carriers for insecticides. Oils are relatively cheap and easy to mix with good spreading capacity and low toxicity to animals and insects have not developed any resistance to them. However, they exhibit relatively low toxicity to most insects and instability in storage. Toxicants derived from plants have been from long time and still used as arrow –tips poisons and fish poison. Plants products are used in many ways in insect control and among those nicotine, pyrethrum and rotenone are well–known. Earlier nicotine was used to control soft bodied insects having sucking type of mouthparts.

The main source of nicotine are the two species, *Nicotina tobacum* and *Nicotina rustica* . Nicotine sulphate is much less toxic to warm blooded animals and is quite stable. Pyrethrum is another important toxic compound to most insects and acts as a contact insecticide with a quick knock-down effect. It is relatively harmless to mammals, and has no phytotoxic effects. Pyrethrum is extracted from flowers of *Chrysanthemum cineralifolium*. It is formulated as dust, spray and aerosol. Rotenone is an extract of derris plants. Now a days several leguminous plants are used for extraction of rotenone but roots of *Derris* and *Lonchocarpus* is used for commercial production. Rotenone acts as contact as well as stomach poison. Before the discovery of synthetic insecticides, it was widely used in several crops. Owing to its low mammalian toxicity, it is particularly useful in killing external parasites. It is available as dust, wettable powder, emulcifiable concentrate and aerosol. Another important plant products are neem derivatives. Neem *Azadirachta indica* is known to contain diverse array of biologically active principles, of which azadirachtin (tetranortripenoid) is one of the best known derivatives. It has antifeedant, anti ovipositional, growth disrupting and fecundity reducing properties on different insects. In recent years many neem seed products like oil, cake crude extracts and purified fractions have been tested against variety of insect species. Neem products are most suitable for inclusion in Integrated Pest Management programme in sericulture.

SYNTHETIC PYRETHROIDS

Natural pyrethrins were effective against a variety of store grain pests but were unstable in light and very expensive. Therefore, synthetic pyrethroids were synthesized, which exhibit high activity against insects, low mammalian toxicity, greatly increased stability, and effectiveness at very low doses, rapid action and degradation to innocuous residues. The activities of pyrethroids to insects, mammals and other groups depend upon optical and geometrical configuration of their acidic and alcoholic components. Their morbidity in the soil is very small. The mammalian toxicity of the pyrethoids is generally lower than that of other class of insecticides. These are more effective as contact insecticides and to a lesser extent as stomach poisons. These can be used for the control of a wide range of pests even when used at very low dose and being biodegradable leaves no residue to accumulate in the biological systems. In sericulture cypermethrin and Fenvalerate is commonly used to control the various pests of tasar food plants. Cypermethrin is sold under the trade name Ripcord, Cymbush, Cyperkill and is available as emulcifiable concentrate. It is applied at the rate of 20 g a.i. per acre in low concentration in the field. Permethrin is sold under the trade names Permasect, as emulcifiable concentrate. Fenvelerate is sold under the trade name Sumicidin and Fenval as emulcifiable concentrate. These synthetic pyrethroids are highly effective against the stem borer particularly *Aeoslesthes holosericea* or *Psiloptera fastuosa* and several defoliating insects on *T. arjuna* and *T. tomentosa* plants.

CHLORINATED HYDROCARBONS

The chlorinated hydrocarbon compounds have carbon, hydrogen and chlorine as their basic molecular constituents and some compounds may have oxygen and sulphur also. DDT, BHC, toxaphene, chlordane, heptachlore, aldrin, endrin and endosulphan (cyclodene group) are the most common and widely used insecticides. DDT is also sold under the trade name chlorophenothene, teffidex and is available in various formulations including emulcifiable concentrate, wettable powder and dust. It is applied at the rate of 250-100 g a.i. per acre in concentration varying from 0.1-0-2.5 percent on practically all plants. It is effective against jassids, white flies, leaf miners and beetles. It persists for quite a long period on the plants, in the soil and it is also accumulated in the body fat of birds and fishes. Owing to its repeated use, even though some of the insects have developed resistance, it still continues to be a leading protective insecticide for out door use.

ORGANOPHOSPHORUS INSECTICIDES

These insecticides are characterized by the molecule having one or more rarely two atoms of phosphorus and are usually derivatives of phosphates (Po4). They mainly act as contact and stomach poisons and some are systemic, some slowly give out vapour which have fumigant action. The phosphorus esters are their biological activity to inhibition of enzyme cholinesterase and this is determined by the magnitude of the electrophilic character of the P atom, the strength of the P-X bond and the satiric nature of the substituents. It has been shown that the phosphate esters (P-0) are more active as the P is much more electrophonic than in the phosphorothionate esters (P-S), the later, being the stable is activated by the by oxidation to the corresponding p-0 compound in the animal body. The P-X bond of the toxicant is broken during incubation process due to its biomolecular reaction with enzyme cholinesterase. In the highly toxic compounds strong electron withdrawing substituents such as P-NO_2 and CH_3S- may be found . Similarly the alkyl and alkox Y substituents of phosphates and phosphonates also influence the electrophilic nature of P-atom. Decrease in toxicity and increase in stability has been noticed with increasing chain length and with chain branching of the compounds.

Further, incorporation of –cl or –CH_3 in the meta position of the aryl ring of parathion -- type compounds reduces the mammalian toxicity appreciably without affecting its insecticidal activity. In higher animal and insects cholinesterase is an essential constituent of the nervous system. Therefore, the capacity of the central P atom of organophosphorus toxicant to phosphorylate the esteratic site of the enzyme cholinesterase contributes to the biological activity of the compound. The phosphorylated enzyme is irreversibly inhibited. This disrupts the normal function of rapid removal and destruction of acetylchloine from the nerve synapse. As a result acetylcholine responsible for transmission of nerve impulses across the synapse accumulates and causes derangement of the nervous mechanisms and death. Symptoms of poisoning due to a organophosphorus compounds are hyperactivity, tremors, convulsions, paralysis and death in insects and muscarinic efforts such as nausea, salivation, lachrymation and myosis, nicotinic effects such as muscular fasciculations and central effects such as giddiness, tremulousness, coma and convulsions in higher animals. The organophosphorus pesticides are classified based on the way in which the phosphorus atom is found in combination into phosphates, phosphonates, phopshonothionates, phosphorothiolates, phosphorothiolothionates (phosphorodithiolates) Phosphorothiolothionates (phosphorodithionate), phosphorothionates, phosphorothiolothionates, phosphoramidate, phosphoramidothiate etc. Some of the common organophosphate insecticides are Phosphamidon, Monocrotophos, Mevinphos, Dimecron, Parathion, Malathion, Fenitrothion,, Fenthionand and Fensulphothion.

CARBAMATES

Carbamates are derivatives of carbamic acid and have an -OCON = group in the molecule. Development of carbamate group has been receiving attention of chemists for the past two decades as result of which some are now on the market. The carbamates that were first marketed are Isolan, a systemic insecticide and dimetilan, a stomach poison, which belong to N,N-dimethyl carbamates. These were followed by the N- methylcarbamates of which carbaryl has become one of the most widely used broad spectrum insecticides. The others include products like propoxur (arprocarb), methiacarb, zectran etc. Insecticides and acricides are also found useful in control of nematodes and snails. The carbamate esters inhibit acetylcholinesterase in insects and mammals. The insecticidal carbamates are derivatives of carbamic acid and dithiocarbamic acid.

ACARICIDES

Acaricides are generally applied against mites. The most common acricide is Chlorobenzilate. It is marketed in trade name Folbex, Akar and Acarben as 25 percent wett able powder.

NEMATICIDES

These chemicals are used for controlling plant parasitic nematodes. Some of the common nematicides are Dibromochloropropane (DBCP). It is marketed as trade names Nemagon and Fumazon. It is highly effective against root knot nematodes when the soil temperature is 21–27°C. It is non-phytotoxic to most of the plants at normal concentrations. It is available as emulsifiable concentrate and granules. This nematicide is very effective for the control of root-knot nematode, cyst nematodes, stylet nematodes, root lesion nematodes and other ectoparasitic nematodes present in the rhizosphere of roots.

RODENTICIDES

Rodenticides are effective against the rat. Zinc phosphide (Zn3Pa) was introduced by the Hooker Chemical Co. in 1943. This rodenticide belongs to inorganic phosphide group and has long been in the use. It is an acute poison for the mammals and birds. It is stable when dry, but decomposes slowly to phosphine in moist air. It is a general poison. It is sold under the trade names of Rumetan, Mous-con, Kilrat. It is generally used in the concentration of 2.5 per cent as a bait for killing field rats, mice and other rodents. The bait is prepared by mixing one part of Zinc phosphide with 40 parts of wheat flour, adding sufficient water to make pellets. It is an acute poison and only one dose is needed to kill the animal. The disadvantage of this compound is that rats come to know of its presence and they develop bait shyness, However, chemical controls have many disadvantages: most have biological activity against many forms of life and therefore can affect non-target organisms; for the same reason, they present various levels of hazard to humans, especially pesticide applicators and other farm workers; most are highly toxic to beneficial insects, such as pollinators and predatory and parasitic natural enemies; both target and non-target insects can develop resistance to insecticides, sometimes very rapidly. Over-reliance on chemicals and diminished use of other control methods have helped push agriculture away from a more natural, balanced state. Therefore, integrated pest management technology has been developed to eliminate the pest population in sericulture

ALTERNATIVES TO INSECTICIDES

As of now many of the important alternatives to insecticide are available. Like the insecticides these too have advantages and disadvantages. Among these, much attention has been drawn towards various aspects of biological and microbial control. According to Batra (1982), the primary goal of IPM is to maintain and enhance the activities of the natural enemies of pest insects, whether they occur naturally or have been added by man to the insect's environment (Batra, 1982). Again, these are not new concepts, e.g. biological control with parasites and predators has been pursued actively in California since the late 19th century stimulated by the initial success in controlling the cottony cushion scale (Icerya purchasi) through the introduction of the Vedalia lady beetles (Rodolia cardinalis). The commonwealth Institute for Biological Control was founded in England in 1927. Bacterial milky spore disease has been used successfully against Japanese beetle (Popillia japonica) larvae since the 1930s. In the area of microbial controls, one of the best-known microbial insecticides originates in the crystalline peptide endotoxin produced by Bacillus thuringiensis (Angus, 1971). Spores of this organism have been marketed by several companies for the control of caterpillars of a number of lepidopterous species since the 1950s. The early expectations were that this selective pathogen would replace many of the synthetic organic insecticides. Unfortunately it proved to be too specific in the field and of limited cost-effectiveness. It is still in use in the USA and abroad, however, on a limited basis. Several other microbial and fungal pathogens have suffered similar limitations and have never enjoyed wide usage in the west. A most extensive effort to develop insect pathogens for practical insect control has been conducted in the Soviet Union, but again with only limited success (Shumakov *et al.*, 1974). Nevertheless, interest in this approach remains considerable. Viral diseases of insects have been known since the 16th century (Cremlyn, 1978) and can be an important natural control factor, in some cases producing epizootics, which decimate insect pest populations. Viral control of certain lepidopterous insects attacking cotton and alfalfa has been realized in recent years with polyhedral inclusion viruses known as baculoviruses. However, viral control of insects at the commercial level is, as yet, regarded as a novelty or specialty at best.

The difficulties in mass culturing the virus, photo stability problems, variable performance and too great a specificity of action have seriously limited this application (Wright, 1976). Several other approaches to insect control also deserve consideration. One of the most important among them is host plant resistance and it is of greater promise for the future (Maxwell and Jennings, 1980). The role which this phenomenon played in saving the French wine industry from the ravages of the grape phylloxera (Daktulosphaira vitifoliae) in the late 19th century is a familiar one. The use of phylloxera-resistant American rootstocks for the European grapes is still the major means of control for this insect. One of the earliest references to host plant resistance as a method for control of Hessian fly (Mayetiola destructor) describes a strain of wheat resistant to the Hessian fly (Havens, 1972, Gallun, *et al.*, 1975). Genetic methods of control, particularly through the release of sterile males, have proved successful in a number of situations, most prominently to control the screwworm fly (Cochliomyia hominivorax) in the southern USA. Other genetic methods such as the use of conditional lethal factors are yet to be perfected, but offer some promise for the future, and are reported by Hoy and McKelvey, (1979).During the past 15 years there has been growing interest in the application of insect pheromones for population monitoring and the disruption of mating in insects. It was conceived as an alternative to the use of chemical insecticides (Henrick, 1977; Konlshi and Ito, 1973). Whereas monitoring practices with pheromones to define the need for timing of insecticide applications or other control measures are being widely accepted and utilized, only limited success has been realized with mating disruption by pheromones at the field level. Although very promising in concept, much further research is needed for the latter to succeed, especially in the area of field delivery and development of formulations with sustained, linear release of the pheromone.

PROSPECTS IN INSECT CONTROL

The control of insects is a goal, which challenges every fibre of man's ingenuity and intelligence, where no permanent victories are likely to be won, and where, currently, our best efforts are not necessarily good enough. However, there is hope for the future. In the area of chemical control agents the comparatively recent discovery and development of the photostable pyrethroids (Elliott, 1978) is leading to profound changes in the use patterns of insecticides. These exceptionally potent, biodegradable compounds may be used in the field at rates as low as 1kg. per ha which is 10 to 100 -fold lower than more conventional insecticides and necessarily leads to a lower burden of chemicals in the environment. Unfortunately, magnificent achievement though they are, the synthetic pyrethroids are not the perfect answer to the insecticide problem since they have high toxicity to many aquatic species, vertebrate and invertebrate, and to many beneficial insects.

The potential for resistance to this group also appears to be high. New natural products, with considerable potency and novel actions against insects, continue to be discovered, such as the plant-derived precocenes, which act to prevent the biosynthesis of juvenile hormones (Bowers, 1981) and the avermectins from soil actinomycetes, which may release the inhibitory neurotransmitter aminobutyric acid in the nervous system (Putter, *et al.*, 1981). The hope for more selective synthetic insecticides is also well-founded. Highly potent inhibitors of chitin synthesis in the insect integument have been developed, some having excellent field effectiveness. As chitin has no essential role in the biochemical economy of vertebrates, these compounds are relatively safe to higher organisms. A second example is based on the discoveries that insect juvenile hormones regulate many developmental functions in insects and that they are unique to arthropods. This has challenged many scientists and research organizations to synthesize and discover more potent synthetic juvenoids for selective insect control. Although success has been limited, due to photo ability and peculiarities of action in important economic pest species, selective uses have been realized with the juvenoid, methoprene (Edwards, and Menn, 1980). Currently this compound is fully registered in the USA and other countries for control of insects of public health and animal importance, those damaging greenhouse plants, and insects infesting post-harvest stored food, feed and fiber. Several other juvenoids are in advanced stages of development, including photostable candidates for possible use in outdoor plant protection. As with other alternate strategies for insect control, it is likely that the full potential of juvenoids in selective and environmentally sound insect control has not been fully realized to date. It is worth pausing here to appreciate the strides that have already been made in the sophistication and efficacy of our control methodologies as illustrated by a specific application of these juvenoids.

HERBAL PEST CONTROL

Many environmental problems such as development of resistance in pests to pesticides, resurgence of target and non-target pests, destruction of beneficial organisms like silkworms, honey bees, pollinators, parasitoids, predators, etc. and pesticides residue in food, fodder and feed, among other things, have attracted attention of entomologists. More than 500 insect species and mites have become resistant to one pesticide or the other. The accumulation of pesticides in different constituents of environment has endangered the survival of many beneficial organisms. Insecticides have been detected in excessive amounts in almost all the food materials including food grains, vegetables, fruits, meat, fish, egg, milk, milk products and even mothers milk. Pesticide poisoning of non-target organisms has led to secondary pest outbreak and pest resurgence. In this context there is considerable global awareness on the need for evolving more and more non-chemical/biological methods of pest control. These techniques will include use of natural enemies, plants and plant products, minerals, ashes, substances of natural origin and traditional pest control practices. As the world commission on Environment and Development; and Interdisciplinary Environment Association have voiced their concern about continuously growing chemification of the environment. The pesticides have become our main stray in sericulture. To counter various ill effects and problems of pesticides farmer should adopt our old traditional farming practices using plant based and other materials of natural origin so as to keep the soil and environment healthy and lively (Randhawa, 1980; Vijayalaxami and Shyam Sundar, 1995).

Use of plants with toxic and repellent action against pests is a common crop protection practice in traditional agricultural systems in India (Banerji *et al.*, 1985; Baskaran and Narayanasamy, 1995a; Elwell and Maas, 1995; Mukerji and Garg, 1988a & b). Long before the introduction of synthetic pesticides, substances derived from plants were employed in pest control. It was estimated that nearly 2400 species of plants in India possessed insecticide properties

(Banerji *et al.*, 1985; Ahmad and Stoll, 1996). Simple methods of preservation by drying and heat-treating the seeds were the most common practices to reduce the pest population. Sometimes plant leaves with repellant activity are mixed with the seed and kept for several months to ward off the pests attack. During Vedic period, plants were used in various forms to keep the pests away from the agriculture fields. A lot of information is available in Asia and Europe during 400 B.C. to 16 A.D. on the medicinal and pesticide properties of a large number of plants. Herbs like *Chrysanthemum marschalli* and *C. roseum* were used as insecticides in the beginning of the 19th century. The importance of chrysanthemums as insecticide was recognized by an Austrian, Anna Rosaner who discovered plenty of dead flies lying around dried chrysanthemums from Dalmatia (USA) and this led to cultivation of the plant to be used as 'Dalmatian insect powder'. This paved way for the discovery of 'Pyrethrum' 'Nicotine' and other plant principles to combat the pests. Oxley in 1948 employed derris root as fish and arrow poison. Systematic cultivation of chrysanthemum in Dalmatia, Japan and East Africa was undertaken. The use of these natural insecticides continued to increase constantly despite constraints by wars and consequently introduction of synthetic insecticides like D.D.T. became increasingly warranted.

Among the plants which are reported to be more commonly used in the pest control, are Neem, Pongamia, Indian privet, Adathoda, Chrysanthemum, Turmeric, Onion, Garlic, Tobacco, Ocimum, Custard apple, Zinger and some other plants given in Table 8.1.

Table 8.1 : Inventory of plants with insecticidal properties found in India

Acacia arabica Lamx.	Gum	General
Acacia concinna	Pods	Repellent
Acacia nilotica L.	Flowers,stem	Contact Antifeedont
Acorus calamus L.	Rhizome	Antifeedont Repellent contact inhibitor
Adhatoda vasica Nees	Leaves	Antifeedont Repellent
Aegle marmelos L.	Leaves	Repellant
Agave americanaL	Leaves	Contact
Ageratum conyzoides L	Leaves	Contact
Allium cepa L.	Bulbs	Repellent
Allium sativum L.	Bulbs	Repellent

Aloe vera L	Leaves	Repellent
Amorphophallus	Rhizome	Repellent
Anabasis aphylla	Leaves	Contact
Anamirta cocculus L.	Leaves	Contact
Annona squamosa L.	Leaves	Contact
Annona reticulata L.	Seed	Contact
Anacardium accidentale L	Bark,	Stem Growth inhibitor
Arachis hypagaea L	Leaves, seeds	Used in poison baits
Areca catechu	Flower	Contact
Argemane mexicana	Flowers	Contact
Aristalochia	Leaves	Repellent
Artemeisia absinthim	Leaves	Antifeedant
Arlocarpus heterophyllus Lam	Leaves/Latex	General
Azadirachta indica A	Seed, fruit, oil	Contact
kernel, bark	repellent	
Bassia lalifolia Roxb	Oil	Synergist
Brassica nigra Rosc.	Seeds	Attractant
Butea monosperma Larnk	Flowers	Contact
Caesalpinia pulcherrima	Flowers	Contact
Callistephus chinensis	Flowers	Contact
Calophylhlm inophyllum	Oil	Contact
Calolropis gigantean	Leaves	Stomach poison
Canna indica	Flowers	Contact action
Canna orientalis Rose	Petals, Flowers	Fumigant
Cannabis sativa L	Leaves	Contact
Capsicumfrulescens L	Fruits	Insecticide
Capparis decidua	Twigs	Repellent
Carica papaya L	Fruits. Latex	Repellent
Carthamus tinctorius	Oil	Contact action
Cassia fistula L	Flowers	Repellent
Callosabruchus chinensis	Leaves	Repellent
Cassia occidentalis	Leaves	Antifeedant
Crocidolomia binotalis	Leaves	Repellent
Casuarina spp	Leaves	Contact
Chrysanthemum	Fruits	Contact
Cinnamomum	Twigs	Attractant
Cinnamomum camphora L	Fruits,	Latex Repellent
Cissus quadrangularis L.	Oil	Contact
Citrus aurantium L	Leaves	Contact
Clerodendron injortunatum	Leaves	Contact action
Clerodendron multiflorum L	Whole plant	Contact
Cocos nucifera	Oil	Contact
Combretum ovalifolium	Leaves	Repellent
Carchorus capsularis	Seeds	Repellent
Crotalariajuncea Buch	Leaves, flowers	Repellent
Cuminum cyminum L	Seeds	Contact
Curcuma domestica	Rhizome	Contact
Cycas revoluta Thumb	Male cone	Contact
Datura strumarium L	Leaves	Contact
Delona regia	Flowers	Contact

Derris elliplica Roxb	Roots	Contact
Callosobnlchus	Flowers	Repellent
Echinochloafrumentaceae	Whole plant	Antifeedant
Eleusine coracana L	Leaves	Contact
Eucalyptus globulus L	Leaves	Contact
Euphorbia sp	Whole plant	Contact
Ficus benghalensls L	Latex	Contact
Ganoderma lucidum	Flowers	Contact
Glyricidia sepium (J) K W	Leaves	Contact
Gossypium indicum L	Seed oil	Contact
Cymbopogon citrates	Leaves	Antifeedant
Helianthus annuus L	Seed oil	Contact
Hibiscus chinensis	Flowers	Contact
Ipomoea carneafistulosa	Leaves	General
Jasminum spp	Flowers	Contact
Jatropha curvus L	Whole plant	Repellent
Justicia adha/oda	Leaves, kernels	Repellent
Kaempfera galanga	Whole plant	Contact
Lantana aculeata L	Leaves	General
Lantana camara L.	Leaves, stem	General
Lobelia nicotionifolia	Leaves	General
LujJa acutangula	Flowers	Contact
Madhuca longifolia var.	Oil	Contact
Mangifera indica L	Leaves (dried)	Fumigant,contact
Maranta arundinacea	Rhizome	Contact poison
Melia azedarach L.	Bark	Attractant
Melilotus indica	Capsule	Repellent
Mentha spicata L	Leaves	Antifeedant
Michelia champaca	Flowers	Repellent
Mundulea sericae Willd.	Leaves	Contact poison
Nerium indicum Mill	Fruits, seeds	Stomach poison
Nicotiana tabacum L.	Leaves,whole	Stomach poison
Nigella sativa	Seed	Contact action
Ocimun basillum	Leaves,Oil	Repellent
Ocimum sanctum L.	Leaves,Oil	Repellent
Opuntia spp.	Leaves	Contact action
Partheniumfruiticosum L.	Leaves	Heliothis tea
Piper betel L.	Leaves	
Piper nigrum L.	Seeds	Stomach poison
Pogostenlon heyneanus	Oil	Contact action
Pogostemon patcholi H	Leaves	Contact
Poinsellia pulcherrima	Flowers	Contact
Pongamia pinnata L.	Oil cake	Antifeedant
Prosopisjulif/ora SW	Leaf	Contact
Ricinus communis L.	Seed oil	Synergist
Ryania speciosa Vahl.	Roots	Contact
Sapindus emarginatus V.	Fruit,leaves	Contact
Sansurrea lappa	Roots	Repellent
Sesamum orientale L.	Seeds	Synergist
Shoenocaulon officinale	Seeds	Contact

Sigesbeckiia orientalis	Leaves	Anti!eedant
Sorghum bicolor L.	Stems	Contact
Slychnos nuxvomica L.	Seeds, leaf	Contact
Tabebuia rosea	Flowers	Contact
Tagetes patula L.	Leaf	Contact
Tanlarindus indica L.	Seed wood ash	Contact
Tecoma indica	Flowers	Contact
Tephrosia candida	Seeds	Contact
Tephrosia purpuria	Leaves	Contact
Tinospora crispo L.	Vines	Contact
Tridax procumbens	Flowers	Contact
Trigonellafoenum-graecum	Seed	Antifeedant
Tripterygium iifordii	Roots	Stomach poison
Vera/rum album	Roots	Stomach poison
Ve/ivera zizanoides	Leaves,roots	Repellent
Vitexnegundo L.	Leaf extract	Contact
Xeronlphis spinosa	Fruits	Antifeedant
Zingiber officinale	Rose. Rhizome	Repellent
Zizyphus mauritinana Lam.	Leaf folder	Contact

Generally, extracts of whole plant or parts of plants are prepared and sprayed; otherwise, they are dried under sunshade, powdered finely and applied as dust. Sometimes mixtures of extracts of more than one plant are made and the extracts are sprayed after allowing certain incubation period so as to enable them to release out the toxicant in the liquid. Among all the plants, Neem (*Azadirachta indica*) is the most promising source of biopesticides. Neem is also called Margosa tree or Indian Lilac. The scientific name, '*Azadirachta indica*' is derived from the Farsi word, 'Azad, and the Hindi word, 'darakhti' which literally means, 'the free tree of India'. The limnoids present in it and its products have made it a harmless to mankind while functioning as insecticide, bactericide, fungicide, pesticide etc. It is likely to provide a solution to many of pest and disease problem in sericulture. Neem owes its toxic attributes to a large number of bitter compounds called meliacins like azadirachtin, nimbin, salanin, meliantroiol, etc., among which azadirachtin is the most efficient. Neem seed kernels are the richest source of meliacin and contain 0.2-0.3% azadirachtin and 0.40% oil though neem leaves, seed, and bark also contain these in smaller quantities. The neem products act as antifeedant, repellent, growth regulator, chemosterilant and toxicant. Vijayalakshmi et al (1995) mentioned that any pest escaping the effect one compound may get affected by the other. Neem has been found effective against more than 200 species of insects like stem borers, locusts, hairy caterpillars, beetles, leafhoppers, plant hoppers, aphids, mealy bugs, whiteflies, mosquitoes and various pests of stored products. Neem cake has the value of fertilizer to plants besides acting against nematode infestation in mulberry. Neem products are highly photodegradable and normally degrade within a week. No problem of development of pest resistance and resurgence has been reported from neem products. Hence they have characteristics suitable for IPM strategy. More than 20 neem-based biopesticides are available in Indian markets (Table 2.1). Thus we have a vast wealth of plants that are rich sources of bioactive compounds and many more might still be lying unexplored.

AZADIRACHTIN

Mainly owing to its various effects on insects, azadirachtin (AZ) is considered the most important active principle in neem seed kernels (NSK). However, the quantity of this compound present in neem seed kernels may vary considerably because of environmental factors and possibly also for genetic reasons. The highest yield of azadirachtin obtained to date was about 10 g/kg of seed kernels (Schmutterer, 1990). Azadirachtin has anti-ovipositional, antifeedant, growth disrupting (growth regulating), fecundity and fitness reducing properties on insects. Azadirachtin is a steroid–like tetranortriterpenoid (limonoid). The first proposal for its structural formula was made in 1972 (Butterworth and Morgan, 1972), but the final elucidation of the complicated molecules was achieved only recently (Broughton *et al.*, 1986, Kraus *et al.*, 1985). According to some authors, Azadirachtin is formed by a group of closely related isomers, called azadirachtin A to azadirachtin G. Azadirachtin A is the most important compound in terms of its quantity in neem seed kernel extracts. Azadirachtin E is regarded as the most effective insect growth regulator (IGR) (Rembolt, *et al.*, 1982). Considerable number of other active compounds were also isolated from neem seed kernels, such as salannin, salannol, salannolacetate, 3-deacetylalannin, azadiractin, 14-epoxyazaradion, gedunin, nimbinen and deacetyl nimbinen (Jones *et al.*, 1989). Most of these compounds showed antifeedant activity in *Epilachna varivestis* biotest (Schwinger *et al.*, 1984). IGR effects were seen in some other ingredients of neem seed kernels, namely 22-23-dihydro-23 b-methoxyazadiractin, 3-tigloylazadirachtol and 1-tiglycol-3-acetyl-11-methoxyazadiractin. Some vilasinin derivatives with strong antifeedant activities were also isolated from neem seed oil. Other compounds with similar properties from the same source are meliantriol, azadiradione, and 14-epoxyazadiradione as well as 6-O-acetylnimbandiol, 3-deacetylsalannin and deacetylazadirachtinol. The IGR effects of the latter on *Heliothis virescens* were as strong as those of Azadirachtin. Salannin was also found in neem oil. Some of the commercial neem formulations are listed in table 8.2.

Table 8.2 : Commercial Neem insecticides

S. No	Trade name	Name of Compound	Manufacturer
1	2	3	4
I.	Bioneem	Azadirachtin 003%	Ajay Biotech Laboratories Pvt LId., Maharashtra
2.	Econeem	Azadirachtin 0.3%	P.J.Margo PlYt. LId, Karnataka
3.	Gillmore Uni-Gel	Azadirachtin 0.3%	Gilmore Inc., USA Neem IGR
4.	Godrej Achook	Azadirachtin 003%	Bahar Agrochem & Feed (P) LId
5.	Margocide CK 20 EC	Azadirachtin 0.15%	Monofix Agro Products LId., Karnataka
6.	Margocide OK 80 EC	Azadirachtin 0.03%	Monofix Agro Products LId., Karnataka
7.	Neemark	Azadirachtin 0.03%	West Coast Herbochem Pvt. LId., Karnataka
8.	Neemazal F	Azadirachtin 5%	Ern Parry (I) LId., Tamil Nadu
9.	NeemazolT/S	Azadirachtin I %	Ern Parry (I) LId., Tamil Nadu
10.	Neembecidine	Azadirachtin 003%	T. Stanes &Co.Ltd.,Tamil Nadu
11.	Neem Gold	Azadirachtin 015%	Southern Petrochemical and industries Corporation LId., Tamil Nadu

1	2	3	4
12.	Nimbasol	Azadirachtin 0.15%	Nimba Foods and Chemicals Pvt.Ltd., New Delhi
13.	Rakshak	Azadirachtin 0.15%	Murkumbi Manufacturing Pvt Ltd.,Karnataka
14	RD-9 Ropellin 93 BC	Azadirachtin 0.03%	ITC Lld., Andhra Pradesh
15.	Sukrina	Azadirachtin 0 15%	Cansler Chemicals Pvt Lld., Tamil Nadu
22.	Neemax	Neem kernel Ecomax	Agrosystems. Maharashtra

Biosol Neem oil, Juerken, Kemissal, Margosal , Neem Plus, Neemguard, Neemicide are the other products commonly available in the market. The other botanical pesticides are pyrethrum, Rotenone, Ryania and Nicotine (Table 8.3).

Table 8.3 : Herbal insecticide Principles

Chemical	Principal Chemical group	Source of plant (organ, content)
Azadirachtin	Alkaloid	Azadirachta indica
Nicotine	Alkaloid	Nicotiana tabacum N. rustica (leaves 5-14%)
Anabasine	Alkaloid	Anabasis aphylla (Leaves 1-26%)
Piperine	Alkaloid	Piper nigrum (seeds)
Veratine	Alkaloids	Schoenocaulon officinale (seeds 2-4%)
veratridine	Alkaloids	Veratrum album (root)
Ryanodine	Alkaloid	Ryania speciosa (Wood 0.16-0.2%)
Wilfordine	Alkaloid	Tripteryglum wilfordii (root) (a mixture of 5 akaloids)
Quassin, Neoquassin	Diterpenoids	Quassia amara, Picrasma Lactones excelsa (Wood)
Picrasmin Sesamin	The crystalline	Sesamum indicum (seed) fraction of sesame oil (0.25%)
Rotenone (ellipton)	Rotenoids Derris	(Deguella elliptica) (root) sumatrol, malaccol deguelin a-toxicarol)
Pyrethrin I	Pyrethrins	Chrysanthemum cinerariaefo/ium
	Pyrethrin II	C. roseum. C. carreum

Apart from this Cinerin I (Flowers 0.7-3 %) and Cinerin II, Jasmolin I and Jasmolin II are also used as insecticides. Garlic acts as a repellent against various pests and is grown as border/ intercrop to prevent pests from going near the main crop. Extract and powder preparations of garlic and onion bulbs are used to check pests in the field and stored grain. Similarly plants like Nochi (*Vitex negundo*), Pongamia (*Pongamia glabra*) , Adathoda (*Adathoda vasica*) and Sweet flag (*Acorus calamus*) are found to be effective against various pests of field crops and in storage. Extracts of *Ipomoea cornea fistulosa*, *Calotropis gigantea* and *Datura strumarium* contain principles toxic to many crop pests. The extract of flowers of champak (*Michelia champaca*) is potent against mosquito larvae. The leaf extracts of *Lantana camara*, tulsi (*Ocimum basilicum, O. sanctum*) and vetiver (*Vetivera zizanoides*) are useful in controlling leaf miners in potato, beans, brinjal, tomato, chillies, etc. Crushed roots of marigold (*Tagetus erecta*) provide good control of root-knot nematode when applied to soil and also mosquitoes. The seed extract of custard apple (*Annona squamosa*) and grape fruit (*Citrus paradisi*) are effective against diamond back moth and colarado potato beetle respectively. Bark extract of *Melia azadarach* acts as potential antifeedant against tobacco caterpillar (*Spodoptera litura*) and gram pod borer (*Heliothis armigera*). Leaf extracts of lemon grass (*Cymbopagon citratus*), argemone (*Argemone mexicana*), cassia (*Cassia occidentalis*). artemesia (*Artemesia absinthium*) and sigesbekia (*Sieges beckiia orientalis*) are strong antifeedants of caterpillar pests like *Crocidolomia binotalis*. Root extract of drumstick (*Moringa oleifera*) inhibits growth of bacteria. The leaf and fruit extracts of bel (*Aegle marmelos*) and leaf extracts of Opuntia, Nochi and Jatropha are potential against various pests. These plants in harmonious integration with other safe methods of pest control like biological control, trap crops and cultural practices etc., can provide eco-friendly and economically viable solutions for pest problems in near future.

TRADITIONAL METHODS

In the past, locusts, birds, beasts, weeds, diseases and unfavorable climate were the causes of plant destruction. The pests were tackled by adopting cultural practices like crop rotation and mixed cropping while scarecrows and sling stones were also used to drive the animals and pests (Randhawa, 1980). In the mid historic period, the pests comprised of locusts, caterpillars, ants, bacteria, nematodes, pigeons, sparrows, parrots, deer and hares. To tackle them, plants like Palas (*Butea frondosa*) and Asoka (*Asoka indica*) were used as insect resistant plants. Diseases transmitted by insects and worms were cured by trimming the affected parts and applying medicines and plastering, fumigation and spraying several healing substances such as Vidanga (*Embelia ribes*) and oil cake of white mustard and animal products like milk and cow dung. From the ancient times, the application of extract from Arka (*Calotropis*) was done to check locust menace. Certain minerals, vegetable and animal products were used to kill the insects and worms. Sprinkling infected leaves with ashes, dust or limewater and a mixture of milk, carcass, vaca, and cow dung in water were also done (Raychaudhury, 1964). Seed treatment with cow dung, milk, juice of *Solanum indicum*, tender coconut water and cow ghee was also prescribed.

INDIGENOUS FARM PRACTICES

Pruning and pollarding of the host plants, removal of dead logs from the filed, cutting and removal of infected parts of the plants, intercropping and trap crop was practiced from long time. Mixed cropping of green gram and groundnut cultivation was most common in ancient times. Trap crops were planted in mulberry filed. Farmers also use a number of traditional pesticides which are either cheap or have no market value a few of which are listed on Table 8.4.

Table 8.4 : Traditional pesticides in practice

Sl. No.	Traditional pesticides	Quantity required per acre (Kg)
1	2	3
1.	Agave flesh extract	25
2.	Ash	15
3.	Brick kiln ash	15
4.	Calotropis leaf extract	1
5.	Cowdung extract	2
6.	Cycas sp. cone	1

1	2	3
7.	Garlic + Kerosene mixture	2
8.	Green chillies extract	2.5
9.	Green chilly and Garlic mixture	21
10.	Ipomoea leaf extract	2
11..	Jatropha plant extract	5
12.	Leaf mixture extract +Asafoetida mixture	1
13.	Lime + Ash mixture	10
14.	Mahuvaoil	31
15.	Neem oil	31
16.	Neem leaf extract	5
17.	Neem seed kernel extract	10
18.	Phenoil + Kerosene mixture	151ltrs
19.	Saw dust + Kerosene mixture	15
20.	Tobacco leaf waste extract	5
21.	Vitex negundo leaf extract	5
22.	Neem leaf powder (2%)	2
23.	Neem leaf powder (4%)	4
24.	Neem seed kernel powder (1%)	1
25.	Neem seed kernel powder (2%)	2
26.	Neem and Datura leaf mixture (2%)	2
27.	Adathoda leaf powder (1%)	1
28.	Adathoda leaf powder (2%)	2
29.	Cannabis leaf powder (3%)	3
30.	Nochi leaf powder (4%)	4
31.	Tobacco leaf waste (2%)	
32.	Ipomoea leaf powder (4%)	4
33.	Turmeric powder (2%)	2
34.	Vasambu rhizome(2%)	2
35.	Eucalyptus & Cinnamomum tree bark (4%)	4
36.	Fly ash (1%)	1
37.	Common ash (2%)	2
38.	Kaolinite clay (Cream) (1%)	1
49.	Kaolinite clay (White)(1%)	1

Based on the performance of traditional pesticides against the pests of host plants in the field, it is advisable to integrate them into the Integrated Pest Management System.

CHAPTER-9

BIOLOGICAL CONTROL

Biological control generally includes the manipulation of one biological organism to control another organism classified as a pest. Biological control is defined as the utilization of natural enemies to reduce the damage caused by the noxious organisms to tolerable level. The natural enemies predators attract practically all-living species, parasites and pathogens–which feed on them in one way or another and in many cases regulate their population densities. Many potential injurious pests are kept at very low levels and never reach economic pest proportions through the effective action of naturally occurring natural enemies, without deliberate intervention by man (Doubt and Smith, 1971. Van Den Bosch, 1978). In the web of nature, the combination of biological control techniques is extensive. Insect pests may be preyed upon or parasitized by other insects. Most insect pests are attacked by bacterial, fungal or viral pathogens. Insects with specialized feeding habits may control specific pests.

The implementation of biological control methods has been categorized into three basic approaches, viz. classical, augmentation, and natural. When a pest is found to not be native to a given area, it may be assumed that the biological organisms that regulated its population dynamics in its native environment are lacking. In such a situation, the classical approach of biological control is employed to determine the pest's native home, locate beneficial organisms that naturally control the pest organism in its native area, and if feasible, import, multiply, release and establish the beneficial organisms in the problem area to facilitate biological regulation of the pest problem. If successful, the importation and establishment of the beneficial organisms will result in a long-term reduction of the pest problem and repeated releases of the beneficial organisms will not be required. Key examples of such accomplishments classical biological control applicable to Ohio include the control of the Cereal leaf beetle on oats and the alfalfa blotch leaf miner on alfalfa by the release and establishment of beneficial parasitic wasps.

The process of importing and releasing beneficial organisms is complex, since many precautions must be taken to prevent the introduction of organisms that may have adverse effects. When beneficial biological organisms are mass reared and released periodically to supplement the natural enemy complex and achieve reduction of a pest problem, the approach is called augmentation. This approach may be applied to pest populations that are either native to the area or of foreign origin. In general, augmentation may be considered when it is economic and feasible to rear, multiple, and release a natural enemy of a pest to the point that reduction of the pest problem is achieved. Successful augmentation efforts have been developed for greenhouse environments where altering the balance between a pest and its natural enemy is feasible. A number of corporations are currently investigating techniques of applying parasitic nematodes to turf for control of soil pests. Application of the augmentation approach to field crops is limited, but a major effort is underway in South Africa to control the Eldana borer on sugarcane via augmentative releases of parasites. Most pest populations are maintained by a number of natural predators, parasites and diseases, which represents natural biological control. If such forces were not in effect, pest populations would overrun us. The balance of crop pest populations and their natural enemies can be significantly influenced by cultural practices and the use of chemicals. Populations of natural enemies can be enhanced by selective use of cultural practices or decimated by indiscriminate use of pesticides. In some cases, pesticides have been developed that effectively control a pest population without having a significant effect on beneficial species. A new pesticide for control of alfalfa weevil is currently approaching registration that effectively kills weevils without harming beneficial parasitic wasps and pollinating bees. An example of natural control research in Ohio includes studies on the impact of predatory ground beetles on early pests of corn such as cutworms and armyworms. One study suggested that the efficacy of a given soil insecticide to reduce cutworm damage may be related to the lack of toxicity of the compound to the ground beetles that prey upon cutworms.

The potential for development of biological controls for a wide range of pest problems is significant. However, development of successful biological control technologies often requires significant investments into research that may or may not readily produce satisfactory results. To date, biological control has not been a marketable product like chemical controls and research efforts into the field have been limited. Furthermore, implementation and evaluation of biological controls are often more complex than that of chemical methods. Recently the utilization of pheromones, sterilized markers, or resistant plants has also been incorporated as tool of biological control All these are of course, important and highly desirable aspect of Integrated Pest management but its methodologies and action are entirely different (Wilson and Huffaker, 1976). In strict sense Biological control should be restricted to the utilization of natural enemies.

SCOPE OF BIOLOGICAL CONTROL

Chemical pesticides play a crucial role in modern sericulture and health care programs. They have been extremely used to control many sericultural pests that transmit a number of diseases. However, the indiscriminate use of chemical pesticides has caused great damage to the ecosystem in several ways, such as accumulation through biomagnifications to alarming toxic levels in the ecosystem; undefined target spectrum and endangers non-pathogenic organisms and higher animals and more importantly development of insect resistant target population, leading to induction of secondary pests. Unilateral reliance on the chemical pesticides is not likely to provide a solution to all our pest problems. Safer, less costly alternatives to chemical control are therefore desirable, as part of an integrated interdisciplinary approach to pest management. Biological control with its firm basis in sound ecological principles and in vast practical experience is by far the most successful and most promising of these alternatives. When successful, the utilization of natural enemies is an inexpensive, non-hazardous means of reducing pest populations and maintaining them permanently below established economic thresholds. Therefore, an alternative approach for chemical pesticides is inevitable. Biological control especially based on use of parasitoids and allelo chemicals products have been emerged as powerful alternatives.

The natural enemies are utilized in three major ways i.e. importation, augmentation and conservation. Although importation has been by far the most successful to date, any of these approach may lead to successful biological control. Success in applied biological control is often dependant on a through understanding of the organisms involved, both injurious and beneficial and their intricate interactions. Basic study of the systematic biology and ecology of the pests and their natural enemies are therefore an integral part of the filed of the biological control (De Bach, 1964). Biological control has been considered as the paramount importance in Integrated Pest Management (IPM) programme and therefore, it has been included in all integrated and multidisciplinary management programme.

NATURAL ENEMIES

Both predators and parasites are animals that feed on other animals, but a predator consumes several host individuals during its development, where as parasites includes its development on single host. Among predators, coccinellid and scarabaeid beetles, lance wings and hemipterans as well as phytoseiid mites are abundantly prevalent in seri ecosystem. Praying mantis, reduviid bugs and ants feed upon the silkworm larvae and causes extensive damage to silk industry. Generally predators are relatively simple, straight forward life styles that have proved most useful in applied biological control. The main group of parasitoids utilized in biological control of sericultural pests is the Hymenoptera (mostly chalcidoidea, Ichneumonoidea and Proctotrupoidea). Most of them are ecto parasite feeding externally upon the host. They live in some protected site – a larva in a leaf mine or a burrow, a pupa in a cocoon an armored scale insect under a wax shield etc. – where they are less likely to be dislodged and loose their host. Many of them also sting and paralyze the host prior to oviposition.

ECOLOGICAL BACKGROUND

Practically all living organisms produce more progeny than would be required to maintain steady population densities. Their populations therefore have inherent property for unlimited exponential growth. Yet, quite obviously, the population of living organisms in nature does not go on increasing indefinitely. To the contrary, they are usually rather stable, maintaining characteristic densities which although fluctuating between certain upper and lower limits, do not change in a given ecosystem overlong periods. The growth curve of a population in a new habitat is not exponential but rather sigmoid or logistic, increasing rapidly of first, then at a slower rate until a state of balance, or dynamic equilibrium, is produced by the interaction of two opposing natural forces; The biotic potential of a species in one hand and environmental resistance on the other. The over all parasitization rate depends upon the individual performance within an area of certain host densities (Hassell, 1986).

Thus a parasitoid responds by increasing the number of the hosts that each individual destroys (functional response), or it responds to increased host density by increasing their own numbers (numerical response) (Soloman, 1949). Numerical response is one of the two essential ingredients in modelling any prey-predator or host parasitoid interactions which are description of the entomophage's density responses (Hassell,1986). This response is usually of more interest than the functional response because it is more often responsible for suppressing the host population (Huffaker *et al.*,1971; Hassell, 1978), although, it has attracted much less modelling effect (Hassell, 1986). A rapid and strong numerical response is of the most important attribute of a successful agent of pest mortality (Coppel and Mertins, 1977). The study of numerical interactions between parasitoid and host population [provides the data for calculating the number of parasitoids needed to regulate the estimated pest population (Knipling and Gilmore,1971).

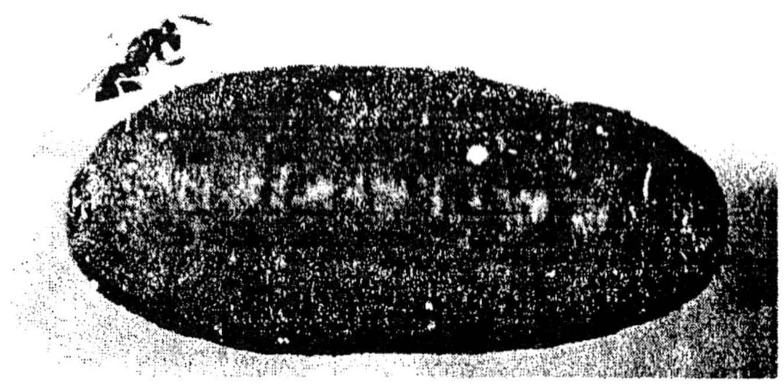

ATTRIBUTES OF EFFECTIVE NATURAL ENEMIES

Till date it is not possible to predict accurately the efficacy of an exotic natural enemy in a new habitat prior to its introduction. This can only be determined. Nevertheless, a clear concept of main attributes to be expected of effective natural enemies would be most helpful in establishing priorities for importation, as well as for research emphasis in augmentation programs (Rosen and Huffaker, 1983). Although the desirable attributes may vary with the type or stage of particular program, some broad generalizations are possible. Thus obviously, an effective natural enemy should be highly density dependent. However, a consequence of density dependence and regulative effects of natural enemies is that the most effective species are often very rare, because they regulate prey populations at low levels, and thereby regulate their own populations at even lower levels. By the same token, a very common natural enemy may be abundant simply because it is effective, in prey population regulation, permitting its prey to achieve high densities on which its own numbers may build up also. (DeBach, 1974). Thus numerical abundance may be a very misleading criterion of natural enemy effectiveness. Apart from this, several desirable attributes have been grouped into four main categories viz. searching capacity, specificity, power of increase, fitness and adaptability (Rosen and Huffaker, 1983).

SEARCHING CAPACITY

Searching capacity manifested by the ability to find out the prey even when it is scarce, is commonly regarded as the most important attributes of an effective natural enemies. Unfortunately it is also the attribute that is most difficult to measure or evaluate. The concept has been commonly expressed in terms of natural enemy's "area of discovery" (Nicholson and Belly, 1935).

SPECIFICITY

Although some out standing successes of biological control have been achieved through utilization of natural enemies of related species, the empirical record of biological control projects indicates that most success have been achieved with fairly specific natural enemies (Dout and DeBach, 1964). Having co-evolved with its prey, a specific natural enemy is more likely to be better attuned to its physiology, behaviour, habitat preferences, pattern of its dispersion and phenology than a more generalized feeder. It is therefore more likely to be highly density dependent in relation to its particular prey (Huffaker, 1970). On the other hand general feeders have the decided advantage of being able to survive even during period of scarcity of given host. Host specificity is quite rare among natural enemies. Very few species are known to be strictly monophagous, whereas most are oligophagous and many are truly polyphagous. However, the widely held notion that predators, in contrast to parasites, are not host specific is definitely incorrect. Many coccinellidae predators are narrowly specific and the same appears to be true for certain predatory mites. The host searching ability of the parasitoid is influenced by various factors viz. habitat selection, host findings, host acceptance and host suitability (Doutt, 1964). Habitat selection is first and sometimes most important step in sequence. Before actually searching for its host, a natural enemy may seek a certain environment, a particular type of vegetation or host plant etc. In locating its preferred habitat it may respond to various chemical and physical cues, which may be related to either the host plant or the host itself or to their interaction (Hagen *et al.*, 1976; Vinson, 1981). Sometimes, this is as far as the selection process goes.

Many ichneumon parasites are known to be habitat specific, attacking a wide range of host occurring in their preferred microhabitat (Townes, 1972). Chalcidoid parasites, usually exhibit marked host specificity. Generally host searching ability and host acceptance are physiological selection and is mediated by responses to physical and chemical cues. Host suitability is another vital aspect especially for those who indiscrimately attack many host species. The success and failure of parasitization is largely determined by the host suitability in such parasitoids. However, certain predators may feed on the host that is poisonous to them or that will adversely affect their development or fecundity. Others may perceive the composition of an unsuitable prey and stop feeding on it. The host plant may influence host suitability. Oviposting parasitoids may inject venom that modifies the host and render it suitable for parasite development (Vinson, 1975). In any case the ability to recognize and avoid unsuitable host is highly desirable attribute of effective natural enemies. In parasitoids this would also involve the recognition of parasitized hosts and avoidance of super parasitism.

POWER OF INCREASE

A natural enemy that has received host regulation in a stable environment that has reached the hypothetical steady state would only require a power of increase sufficient for replacement of the parent population in each generation. The actual power of increase of a natural enemy in the field may be affected by its fecundity and rate of development, as well as by such other factors as its searching capacity and adaptability to the condition of the particular habitat.

FITNESS AND ADAPTABILITY

To be effective in a new habitat, an introduced natural enemy should preferably be pre-adapted to it. However, the possibility of possible gradual post-colonization adaptation cannot be discounted. A well-adapted natural enemy should not require any essential requisites that are not present in the new habitat, should be able to tolerate the prevailing climatic conditions, and should be effectively synchronized with the biology and phenology of its host in the habitat. Specific features required for synchronization may vary with the habitat. Diapauses for instance may be mandatory in one area, whereas in another area it may prove to be detrimental. In general precise synchronization is more likely to occur in a host specific natural enemy than in a general feeder. Ideally, a natural enemy should be adapted to all the habitats and niches occupied by the target pest. It should frequent all the host plant and tolerate all the climatic regimes that its host does, and should equally effective in all of them. Unfortunately, this is usually not the case. It is quite common to find a widespread, polyphagous pest attacked by different species of natural enemies in the different parts of its range. In fact, the more host specific an enemy is, the more narrowly adopted it is likely to be. Just as the natural enemy should be well adapted to the various natural aspect of the ecosystem, it should also be adapted to cope with man made hazards, such as pesticide treatments. Resistance to various pesticides is known to have developed in the field in the number of predatory mites (Croft and Brown, 1975; Hoy, 1979) and further research is likely to discover additional cases among various natural enemies. Such resistant species strains should be regarded as highly desirable biological control agents (Rosen and Huffaker, 1983).

OTHERS ATTRIBUTES

Various other attributes have been ascribed to effective natural enemies, but they all may be regarded as manifestations of the four main attributes discussed here. Competitive ability has often been mentioned as desirable. One should distinguish here between extrinsic or exploitable competition, involving efficiency in the exploitation of pest population by a population of natural enemies, and intrinsic or interference competition which, in parasitoids, involve competition between the larval forms or in a given individual host. Superiority in exploitative competition is largely determined by a natural enemy searching efficiency, whereas superiority in interference competition is a form of adaptiveness that may be less desirable than the capacity to discriminate parasitized hosts and avoid such competition altogether. The old controversy about the relative value of the predators and parasitoids in the biological control is largely useless. Indeed, more parasitoids appear to have the attributes of effective natural enemies than do predators. However, a host specific, adaptable predator with high searching capacity and high rate of increase is certainly likely to be more effective than a parasitoid that is poor in this respects. The main attributes of the effective natural enemies - searching capacity, specificity, power of increase and adaptability–are of course related. Thus, the factors effecting the searching behaviour of natural enemy may also determine its prey specificity, whereas searching efficiency of a population of natural enemies is also a function of the numbers searching, hence their power of increase and their fitness (Huffaker. *et al.,* 1970). To be an effective natural enemy does not have to be superior in all these attributes. Even in outstanding biological control successes the controlling natural enemies often exhibited deficiencies in some of them (Messenger, 1976).

UTILIZATION

Biological control may be achieved through the adaptation of effective exotic natural enemies to a pest's habitat or through enhancing the efficiency of existing natural enemies. Importation, including the search for, transfer, quarantine handling, colonization and establishment of exotic natural enemies, has been by far the most important and most promising approach to date, and has accounted for the great majority of outstanding success in applied biological control. It is also the least expensive method of natural enemy utilization. However, in certain instances an already established natural enemy, whether indigenous or introduced, may show considerable promise but fall short of fulfilling its potential due to inadequacies of its own attributes or the environment. The efficacy of such natural enemies may be sometimes enhanced through various method of augmentation, or direct manipulation of their populations, such as periodic colonization, genetic improvement, or the use of semio chemicals that affect their performance. Another possible approach is through conservation, or manipulation of the environment, either by adding of asking requisites or by mitigation of various detrimental factors.

IMPORTATION

In spite of its successful record and great promise as an effective method of pest control, the importation of exotic natural enemies has been largely neglected in most countries (DeBach, 1972). Very few efforts have been made to search for and utilized the natural enemies of many important pests. Even well-known natural enemies of proven high efficiency still await transfer into many areas where their hosts are serious pests. There is enormous untapped potential of natural enemies to be utilized in importation projects. Our knowledge of the natural enemies fauna of large part of the world is woefully incomplete. In the parasitic Hymenoptera, it has been estimated that between 70 and 90 per cent of all extant species are still undiscovered, and that no biological information whatsoever is available for 97 % of all species (Kerrich, 1960; Townes, 1972). The same is probably true for other groups of natural enemies. All these unknown species may be regarded as potential weapons to be added to the arsenal of applied biological control. Most of the major pests of modern sericulture are organisms of foreign origin. Extensive intercontinental travel and commence over the last several centuries have resulted in the advertent transfer of numerous potential pests into new areas, often unaccompanied by the parasites and predators that had co-evolved with them in their native lands. Free form the remaining influence of their natural enemies, such avoiding species may increase rapidly and assume much more serious pest proportions than they had in their original habitats.

Importation of natural enemies that they have left behind them is likely to elevate the problem. Such introduced pests are therefore, prime targets for applied biological control projects. However, it should be emphasized that native pests may also be amenable to biological control by importation of natural enemies. Instead quite a few important pests have been brought under complete biological control in their native habitats through the importation of the natural enemies of the related exotic pests (DeBach, 1974). Sound systematic is an essential pre-requisite to the success of any importation project. When live natural enemies are being sought or are actually transferred from one region to another, the correct identification of both the pest and natural enemy species and the recognition of cryptic species or intra–specific entities may of course, be of utmost importance. Many serious failure of biological control projects have resulted from inadequate biosystematics (Rosen, 1985). When the land of origin of an exotic pest is known and accessible, it is obviously likely a starting point in the search of natural enemies. The search for natural enemies should be extended throughout the range of the distribution of the pest, as well as to the related species.

The most important and certainly the most controversial, issue in natural enemy importation involves the number of species to be introduced against any given pest. Most biological control experts recommend a policy of multiple importations. As many primary natural enemies as are available should be imported and released, in the hope that the best one will become established and control the pest. This empirical approach has been based both on the practical experience of the applied biological control projects and on ecological theory. However, several authors have criticized it as unscientific and have raised the specter of inter-specific competition among natural enemies reducing their over all impact. Rather, than import natural enemies indiscriminately, they recommend that only the most suitable species be pre-selected and introduced into a given habitat. In most successful importation projects, it is a single and very few species of natural enemies that have been responsible for ultimate biological control. However, it is virtually impossible to predict which of the available candidates from importation will prove to be the most effective ones in a new habitat. Even very similar eco systems are never exactly alike, and any subtle difference in any of the numerous interacting environmental factors operating in them may affect the performance of natural enemy. At present state of our knowledge, neither field studies in the native habitat nor elaborate test in the laboratory may serve as a reliable basis for such prediction (Rosen and Huffaker, 1983). From practical standpoint, then, pre-selection of the best: species simply not feasible. The crucial question however, is whether competition among natural enemies may indeed be detrimental to the final out come of an importation project. The investigation of earlier workers indicate that addition of an effective natural enemy may reduce the impact of other species, but in all documented cases the over all effect of natural enemy complex has always been increased (Doutt and De Bach, 1974). The policy of multiple importations is further supported by the ecological principles of competitive displacement and co-existence (DeBach and Rosen and DeBach, 1979, DeBach, 1964a). When two or more species of natural enemies are ecological homologous – i.e., they occupy the same ecological niche, attack the same host in same manner and have the same ecological requirements – competition between them in a given habitat will result in displacement of the less adapted species by the one capable of producing more progeny under the particular condition of that habitat. The winner of such inter-specific competition is likely to be the species that is most capable of exploiting available resources, including the pest population. In other words, the most effective agents of biological control agent are likely to displace the others. Different species may end up winners in different habitat conversely; when the competing species are not homologous – e.g., when they attack different alternative hosts, or have different climatic tolerances etc. – they will be able to co-exit in the same habitat and complement each other's performance in spite of competition among them. Thus, in either case, whether through competitive displacement or through complementary action, the multiple importation of natural enemies is likely to result in more effective biological control. This has been demonstrated time and again by successful importation projects carried out against many serious pests (DeBach *et al.*, 1971).

On the other hand the historical record does not suppose sequence theory, contending that only by importing the entire complex of natural enemies, attacking all stages of pests in native habitat, can biological control be achieved. Generally, single effective natural enemy is usually responsible for biological control in any given habitat, sometimes aided by one or two additional species. Multiple importations are required simply in order to make sure that these "best" species are discovered. Sometimes same species is attacked by different species of natural enemies but there is a large variation in their population. Various adaptive strains may exist in different habitats, and there may be considerable genetic variation in its various attributes. A broad genetic spectrum would certainly increase the chances of an imported natural enemy becoming established and effecting biological control in a new habitat. Theoretically, peripheral populations may be potentially more adaptable, and are therefore more desirable source of imported material than those from centre of distribution, but such detail information about population structure is usually lacking (Messenger, 1976). In any case it would seem advisable to introduce as many as large sample of natural enemies as possible, from many different habitats and localities as possible. Other than basic systematic and biological information, no elaborate ecological studies are required prior to the importation of natural enemies, and such study should certainly not delay actual importation. Verification of primary status of natural enemy is of course, of crucial importance, but this may be done in quarantine after importation. Information about voltanism, diapauses, specificity etc., can be useful in planning, release and establishment strategies, but this too may be obtained as at later stage. The search for natural enemies should be preferably carried out in undisturbed environments. Back yard gardens and uncultivated tracts are sometimes more likely to yield as greater variety of species than commercial plantations. Unless pesticides resistance material is especially sought, chemically treated area should be avoided.

To prevent the inadvertent introduction of any undesirable organism, including hyper parasitic species, all imported material should be handled under strict quarantine (Rosen, 1985). Under exceptional circumstances field collected material of well-known exotic species may be directly released in the field, provided that they have been individually examined and identified by an expert. Ideally any newly established species should be propagated for at least one generation in quarantine, to confirm its primary status and to conduct the necessary taxonomic and biological studies. In any event, only reasonably well-identified primary natural enemies may be released from the quarantine facility; Careful record and vouchers specimens of all importations should be kept for future reference.

AUGMENTATION

Augmentation of established natural enemies may be accomplished of mass reared or field collected parasites and predators through periodic colonization or genetic colonization of their attributes or through their chemical manipulation of their behaviours. Direct manipulation of parasitoids are usually more complicated, laborious and expensive than the importation of exotic species, and should therefore be embarked upon only if in–depth ecological studies and a thorough cost benefit analysis have unequivocally demonstrated their feasibility (DeBach, 1971., Rabb *et al.*, 1976). More specifically the nature of damage and economic threshold of the pest should be determined. The potential efficacy of natural enemies should be demonstrated and the factors preventing it from fulfilling its potential should be clearly understand, the proposed manipulative technique should be demonstrably effective, and the cost of augmentation should be more than the balanced by the increased control attained.

PERIODIC COLONIZATION

Periodic colonization of natural enemy may involve either inoculation or inundative releases. Inoculation release consists of relatively small numbers of natural enemies, intended to propagate in the target habitat so that their progeny would affect control for several subsequent generations. Inundative releases, on the other hand consists of very large numbers and are aimed at achieving immediate control by the related natural enemies, themselves, not by their progeny. Either insectaries reared or field collected natural enemies may be used for both types of releases. Inoculative release may be appropriate when a natural enemy population is severely decimated periodically by adverse weather condition, cultivation practices or pesticidal treatment. The use of *Cryptolemous montrouzerii* Mulsant, a coccinellid predator of mealy bugs, on mulberry may serve as a example (Rosen and DeBach, 1979; DeBach and Hagen, 1964). This Australian lady beetle was introduced into California as early as 1892, but has become permanently established only in a narrow costal zone, and was not very effective even there.

However, inoculative releases of 10-20 beetles per tree at the beginning of the summer are usually sufficient in control most mealy bug infestations occurring on Citrus. Another approach is to make small periodic releases throughout the most of the year, to augment an existing natural enemy population that is rendered ineffective by prevailing weather conditions. *Aphytis* sp. for instance are important natural enemies of the California red scale *Aonididella aurantii* (Maskell), on citrus in California and elsewhere, but are incapable of effecting complete biological control in certain climate zones (Rosen and DeBach, 1979). Inoculative releases of about 4,00,000 wasps per acre (1 million per hectare) per year in 10 monthly increments provide effective control in some such areas in California (De Bauch, 1964a & b) whenever the augmentation of an established natural enemy population by periodic inoculative releases is considered, attention should be paid to the selective numbers. Even a seemingly residual population of natural enemies may be rather large. In a heavy infestation of scale insect on citrus, a low incidence of parasitism (1-2%) may in fact represent several hundred thousands or even millions of live parasites per hectare. The efficacy of such releases should therefore be clearly demonstrated by large-scale field experiments. When the initial pest infestations at beginning of the season are patchy and irregular, it is sometimes advisable to inoculate the pest population along with, or even before, the release of natural enemies, in order to create a more uniform light infestation on which the natural enemies may became firmly established. This "pest-in-pest" method, which proved experimentally successful with inoculative releases of primary mites against the clamers mite. Inundative releases of natural enemies may be regarded as application of "biotic insect ides". By their very nature such large-scale releases are considerably more expensive than their importation or inoculative releases. It is most effective against univoltine pests or against pests that cause economic damage only during limited period of year (De Bach and Hagen, 1964). However, a natural enemy utilized in inundative releases need not possess all the attributes desired of a species imported for permanent establishment or even of a species used in inoculative releases. Thus in as much as only the immediate performance of the release natural enemies is of direct concern, their ability to reproduce, disperse or over winter in the field may be relatively less important on the other hand, rearing cost is of outstanding importance in an induntative release. Indundative releases have been attempted with quite a few natural enemy species *Trichogramma* is the most common. The species of *Trichogramma* are minute, polyphagous endoparasites of various insect eggs. Their most searching abilities are limited and are usually not very well synchronized with the phenology of their hosts. They may be easily reared in large numbers in the eggs of the Angoumois grass moth *Sitotroga cereallela* .

GENETIC IMPROVEMENT

When a potentially effective natural enemy is deficient in some crucial adaptive attribute, genetic improvement of that attribute may be attempted. Selective breeding and hybridization may be carried out in the laboratory to speed up the process and even to obtain results that might never occur in nature (De Bach 1964c). Many adaptive races are known to exist and there usually is considerable genetic variation in natural populations. Some parasitoids have been genetically improved with remarkable success. Hybridization was also used to improve natural enemies but without outstanding field success. The recent development and most successful utilization of pesticide resistant strains of phytoseiid predator *Metaseiculus occidentalis* have been a major break through. Field population of this were found to have acquired resistant to organophosphorus pesticides and sulphur and op-resistant strains from Washington and Utah were introduced into apple orchard in several countries, where such resistance had not occurred (Croft, 1976).

Strain resistant to carbaryl and permethrin were then developed by selective breeding and were released in pesticide treated deciduous fruit tree orchards and vineyards. These artificially selected strains have now been successfully established and have demonstrated their efficacy as biological control agents in the face of commercial pesticide application (Rosen, 1985). In any programme of genetic improvement, the attributes to be improved should be clearly defined. Adequate genetic variability either natural or induced should be present in the stock to be selected and effective rearing methods should be available for the target species. Release strategy for improved strain may depend on nature of genetic mechanism involved. When the improved trait is governed by single dominant gene (as in the case with carbaryl resistance in *M. occidentalis*), the new strain may simply be released into an established or natural population. The potential for genetic improvement of natural enemies is almost unlimited. Selection for pesticide resistance is only the first and most obvious step in this direction. With further research emphasis, this may well became one of the major approaches to applied biological control.

CHEMICAL MANIPULATION

It is well-established fact that host searching and acceptance behaviours of natural enemies are mediated by various kairomonal cues emanating from the host itself, from the host plant, from other organism associated with host or from various interaction between them. Recent studies have indicated that chemical manipulation has great impact on the modifying the behaviour of the parasitoids in the natural condition.

ROLE OF BIOLOGICAL CONTROL IN IPM

Integrated Pest Management is an art of the possible. It is an attempt to achieve a realistic solution to a very serious dilemma. The continuing population explosion has confronted mankind with two major problems: imminent starvation on the one hand, and the threat of the worldwide environmental pollution on the other. It is imperative that what ever is done to alleviate one problem does not aggravate the other. Thus, in order to feed the ever-increasing population of the world and avoid starvation, effective pest management programmes should be developed but such programme should tend to decrease, rather than increase, environmental pollution. Biological control is of course, highly desirable. It is an ideal form of pest management–permanent, inexpensive, non-hazardous. Unfortunately, however, biological control cannot be expected to provide an immediate solution to all pest problems at present time. No effective natural enemies, are currently available for certain, key pests, until such are discovered, chemical pesticides or some other means of control will continue to be required. IPM provides a reasonable compromise, taking into account both the desirability of the biological control and necessity of some form of chemical control.

Effective IPM may be achieved through development of a various program of applied biological control, in combination with judicious use of relatively selective pesticides, only when absolutely necessary and in the least disruptive modes of application. Other selective tactics - e.g. cultural controls, autocidal methods, or utilization of semiochemicals should be incorporated in the programme wherever applicable. IPM represents a holistic approach, recognizing the unity of the eco system and harmonizing all available measures in an attempt to optimize pest control and crop production. Utilization of natural enemy should regard as the backbone of any IPM program. The value of naturally occurring natural enemies cannot be overemphasized, and their conservation should be first goal of IPM. However, merely, introducing modifications in the existing chemical control programmes to conserve natural enemies does not in itself constitute IPM. Other forms of applied biological control, such as the various methods of augmentation, should also be included whenever feasible. In particular, any viable programme of IPM should include the active importation and establishment of exotic natural enemies as one its major components. In summary, classical and augmentative biological controls are not available for very many pest problems - especially severe pest problems that demand immediate attention. However, natural biological control is in effect in most situations, and it is important that every effort be implemented to enhance such biological activity wherever it exists.

CHAPTER-10

SAMPLING SURVEILLANCE AND FORECASTING

The pest population of silkworm and its host plants has increased and frequently there is out break of new pest. It causes extensive damage to silk host plants, which resulted in the deterioration of quality and quantity of leaves of silk host plants and ultimately fluctuation in cocoon production (Singh and Thangavelu, 1994). In this context, pest and predator forecasting may provide an opportunity to minimize the losses, particularly to reduce expenditure involved in pest management. Among the various strategies of Integrated Pest Management, Pest sampling, surveillance and forecasting is most important to know accurate information about the distribution, population density and dynamics of pest population in the target field. Pest forecasting is useful in making tactical pest management decisions such as what control measure is to be taken up and when (Knipling, 1973; Norton, 1980). Pest monitoring through field surveys and surveillance helps in tackling build up of any pest/disease and employment of pheromone traps and lures ensured minimum and need based application of pesticides instead of prophylactic and calendar based pesticide spray schedule thus reducing the load of pesticides formed the basis of IPM. The present methodology adopted for the survey and surveillance, developed in early 1970s, need to be modified. Computer-aided pest forewarning systems should be developed (Aisagbonhi, 1989). Various types of surveys are conducted under the surveillance programme.

The most common among them are qualitative and quantitative survey. Generally in sericulture, qualitative surveys precede quantitative ones, which are commonly employed in insect pest management. Such surveys aim at establishing numerical abundance of an insect pest population in space and time for prediction of future population trends and for damage assessments. In quantitative surveys, counting of insects or damage caused by them is carried out by estimating their population density using sampling techniques. A sampling technique is a method used to collect information of a single sample, where as sampling programme is the procedure for employing sampling techniques in space and time. Sampling of pest population has been used for a variety of purposes. It is, therefore, essential to have a clear perception of the objectives of a sampling programme, which include pest related crop damage assessment, determination of efficiency of control methods, analysis of diversity and dynamics of pest eco system, study of biology and ecology of pest, construction of life table, optimization of sampling programme, defining economic injury level, developing pest forecasting, and IPM decision making. IPM provides a format to know the time for implementation of pest control strategies and if required, which one, how and when to apply. This is done by surveying or monitoring pest density (d) or plant damage (x), and establishing relationship between yield (y) and pest density (d) or the damage (x).

$$Y = f(d) \text{ or } Y = f(x)$$

Where f denotes a function of pest density (d) or plant damage (x). From economic point of view such as value of crop and the cost of pest management, the pest density at which it is economical for a farmer to control pest is found. This is economic threshold level (ETL). The pest level (d) inflicting plant damage (x) is compared with the ETL, and if (d) is greater than ETL, action to reduce pest attack is taken.

SAMPLING TECHNIQUE FOR IPM

Integrated Pest Management calls for integrating all available tools in cost effective, environmentally friendly and sustainable manner. The philosophy and principle under lying the concept of IPM is to reduce mankind dependence on chemical pesticides by identifying, integrating and employing other methods which are cost effective, eco friendly and lasting in efficacy (Sharma, 1991). There are many sampling techniques as there are pests (Ananthakrishnan, 1981; Andrewartha, 1961; Cochran, 1977; Ruesink and Kogan, 1982). Good account of sampling methods (Atwal and Bains, 1974; Kogan and Herzog, 1980, Kuno,1991; Mathew, 1984; Perry, 1989; Wilson, *et al.*, 1989) and a few survey manuals are also available (Puntener, 1981). Sampling methods for individual pest species have been developed on gall insect, stem borer, aphids, leaf roller and various major defoliators etc. There is no universal method that is efficient for all categories on all the food plants of silkworms, which are growing under diverse ecological conditions. Sampling methods should be quick, simple, inexpensive, and measure the actual pest population or damage as accurately as possible. Sampling technique may be divided into direct counts on the plants or in the environment and indirect counts of some effects caused by pests, such as plant injury or damage. Both can be assessed as an index or on a rating scale for saving time, labour and cost.

DIRECT SAMPLING METHODS

A direct count of insect population is the most widely applied method of sampling on silkworm host plants. Larger insects are easily counted in their habitat. There are several ways of direct counting methods but it is counted on standard unit, usually area of the ground (e.g. number of leaf folder larvae per square meter). If counted on non-standard unit, viz. plantation row, number of leaves, shoot, stem, internodes, the unit is converted to a square meter unit. However, various insects are counted on the part of the plants. Nymphal population of gall insect *Trioza fletcheri minor* on the leaves of *T. arjuna* and *T. tomentosa* are counted per unit of the leaf. Stem borer population on tasar silk host plants are counted by cutting the damaged plants. Sampling in the environment, is conducted to relate catches by sweep net, suction trap, light trap or pheromone trap to actual pest population densities on the ground of correcting for difference in the trap of differences in the surroundings. Soil inhabiting insects are collected by digging the soil with soil core sampler to a standard depth collects soil-inhabiting insects. Pests are then separated from the soil by dry methods; such as the hot water jacketed Barlese's funnel or the

Tullgren funnel heated by electric light bulb or oil heater; pests move to collecting tube at the bottom of the funnel (Chiarappa, 1974). In the wet extraction methods, as in the Salt- Hollick method, samples are soaked, shaken, with a detergent and insects are floated off in salt solutions, such as magnesium sulphate and separated by centrifugation (Cochran, 1971). Some insects may be driven off the soil by repellents such as formalin or with an insecticide. In case of termites, placing paper on the soil attracts the insects. When soil samples are taken along the length of row crop, pest densities are expressed on the basis of area.

USE OF TRAPS

Various types of trap have been devised for collecting and killing different types of insects. The cricket trap, the house-fly trap, electric traps, air suction trap and light traps are most common (Wood,1984; Karel *et al.,* 1983) Light traps sampling is used frequently for sampling insect pests in the silkworm-rearing field. Three different colours of light, i.e., blue, yellow and red along with natural light has been tested to find out the degree of photo attraction of seven pests associated with non mulberry silkworm host plants. All the seven species showed high degree of attraction towards natural light and least for blue. Marked difference has also been observed in the behaviour of predators and parasitoids for yellow and red coloured lights. Bhattacharya *et al.*, (1995) reported similar observation of some insects associated with lac. Traps are normally set for a period of time and then catches are picked up. They are used continuously and the number of insects collected depends upon the density over a period of time. Traps can be either attractive or passive in mode of collection of insects. Attractive traps like visual traps and bait traps rely on a physical or chemical stimulus to lure insects into them. Passive traps like malaise, pitfall, sticky traps, water pan and suction traps on the other hand collect insect incidentally (Barakad Burkholder,1976; Singh *et al.*, 2000). Light traps are the most convenient traps to collect the coleopteran and lepidopteron-defoliating insects in sericulture (Abenes, 1990).

A light trap basically consists of a light source above a funnel and a container below to collect the catch, and is equipped with a protective roof. The light source is either electric bulb, or a fluorescent tube. Generally short wave lengths of light spectrum attract noctuids, the so-called black light emitting ultraviolet lights are now widely used in most traps. Once inside the trap, insects may be collected for counting or killed in collecting container. Insecticides such as Dichlorvos (or Dichlorvos impregnated vinyl strips, e.g., vapona) are sometimes placed in the container for killing the insects, thus reducing pest load. A commonly used light trap is the "Robinson trap". This trap uses a mercury discharge ultraviolet electric bulb. Rothamsted Light trap commonly known as Chinsurah light trap is used thorough out in India under Pest Surveillance Programme. The size of the insect catch is governed by three main factors, viz. density of the adult population in the environment, the responsiveness of the insect towards the trap, and the efficiency of the light trap used (Weathershan,1989). Another common method for collection of insect is the use of attractants in the rearing field. It is precise, specific and ecologically sound for pest management in sericulture. Mostly, these are the natural products but certain synthetic attractants are also known. Analogues of certain natural products have also been successfully synthesized. Attractants traps are used to catch the uzi lfy, ichneumon fly, wasps, mantis and several other parasites and predators of silkworm. These traps are based on natural or synthetic chemicals, which stimulate insects to get attracted in the rearing field. Chemical attraction by crop plants or the chemical constituents of it, or by attractant chemical emitted by an insect is being used in the assessment of insect pests. Chemicals that attract several species are called kairomones and those that attract one species are called pheromones. Sugar and Propionitrile and Sinigrin are good attractants for uzi fly, *Exorista sorbillans* and *Blepharipa zebina*. Ethyl propionate and phenols are used for elterids and scarabaeid beetles.

PHEROMONE TRAPS

Nowadays Pheromone traps are also used in assessing pest populations. These traps exploit the chemical olfactory stimulants (pheromones or chemical sex attractants) emitted by insects to attract the opposite sex (Borden,1985). Such sex pheromones are often excreted by females (Gypsy moth to attract males) or in some insects emitted by the males (e.g. ball worm) to attract females

(Karlson and Butinandt,1959). Pheromones are now available commercially for a wide range of lepidopterous pests, certain beetles and scale insects (Hallett *et al.*, 1999, Rajapakse, *et al.*, 1999; Shorey *et al*, 1967). In India, different types of traps have been evaluated and some of them have also been now commercialized by the industry. Pheromone lures dispensed in different types of dispensers placed in traps are used either for monitoring or mass trapping specific insect pests of a crop. Even though different kinds of traps have been designed, tried out and evaluated for over a period of three decades in India in different crops, dry traps including funnel type with sleeves have been widely accepted by the farming community on account of their efficiency and high capture capacity. Funnel trap is the best for obtaining maximum captures and this is further improved by fixing a baffle inside the design of the hood on the canopy of the trap. The slippery surface of the funnel further enhances the capture inside the polyethylene sleeves.

Funnel traps are effective for trapping adults of *Helicoverpa* and *Spodoptera* sps. (Pawar *et al.*, 1988).Insects have the tendencies to be attracted towards various colour (Suresh *et al.*; 1989). Therefore, in sericulture various colour traps are used to catch various pests. This is some sort of behaviour modifying strategy to monitor various insect pests existing in the silkworm rearing field especially in the area where non mulberry sericulture is practiced. Some small foliar pests are attracted to yellow. Yellow traps are effective in attracting various species of aphid, Jassids and bugs. The best colour shape and size of trap is determined through trials. Colour is sometime combined with water traps. A malaise trap consists of a netting tent with one open side. A small container with preservative is placed in the upper corner or highest point of the tent. The insect flies or crawl into the trap through the open side of the trap and move up the netting and are collected in the container. Since the insect enter the trap accidentally, these traps are effective for highly active insects such as adult flies and hymenopterans and are unsuitable for small insects such as aphids (Wall, 1989). Water traps are shallow, plastic or metal trays, 5-8 cm deep and 0.1 m^2 in area filled with water, detergent and an oil film. These traps are placed in or near the crop. Sometimes a particular colour, pheromone, or some other attractants incorporate. Water traps are more effective, if raised above ground level, but the height depends on species being sampled and the time of the experiment. The traps are must be protected from weather. The trap refilled regularly. The shape and size of the trap is standardized as per the requirement in the field area. Aphids, plant hoppers are easily trapped in such trap.

STICKY TRAPS

Sticky traps made up of paper, plastic, metal sheet, etc., are commercially available where in insects are captured when they alight and come in contact with the glue on the sticky trap surfaces (Taneja and Jayaswal, 1983). Sticky traps are made by spreading a suitable adhesive or sticky substance (e.g. tree banding grease or car grease) on an plastic card or paper sheet (David and Birch,1989). The shade of sticky traps may be flat, cylindrical, triangular or spheres, depending on the habits and size of the insect species and the form of attractant. Cylindrical and spherical traps are efficient for passive collection, while triangular or flat structures are used for pheromone trapping. After the captured insects are killed, these traps become messy as often trapping surfaces get coated with unwanted debris. These traps have proved to be very effective in trapping many species of pests, especially aphids, hoppers, flies, hymenopterans, beetles and mites. Fly paper is the most commonly used sticky trap. Sticky red spheres have been used for estimating adult maggot population in the field. Cylindrical sticky traps have been used for trapping mustard aphids, plant hoppers, etc., in India. The protection of sticky traps from rain and dust is essential for their efficient functioning. Most of the time, it is combined with the attraction of colour or pheromones. Depending upon the sticky material used, the catch is washed off with kerosene or a suitable solvent and the insects are identified and counted. Generally pitfall traps are used for pests on soil surface, such as ground beetles, collembolans and spiders. They consist of smooth sided containers such as glass or plastic jars sunk into groud with their top level on the ground surface. The catch pests falls accidentally into the trap. When baited, they attract individuals from considerable distances. They need protection from predators, which will otherwise devour the catch, and rain, which floods the traps.

The use of sweep net is a common technique for sampling pest populations, especially for highly mobile insects such as plant bugs from mulberry gardens. It gives a good catch per unit time at a very low cost of equipment and labour without damaging the crops. Efficiency of a sweep net varies with species, height of plantation, weather (especially wind speed, air temperature and solar radiation), time of day, and style of sweeping. The mouth of the net must be closed down as sweeping is completed. The catch is expressed as average number of pests per sweep. Sweeping can give repeatable results if the diameter of the net opening and the number, extent and frequency of sweeps are constant. Hence, standardizing the sweep, particularly if more than one individual is doing the sweeping is desirable (Ruesink and Haynes, 1973, Jacobson, 1972). Sampling with sweeping net is conducted using a 40 cm diameter sweep net giving a pendulum swing, one stroke for a step while walking with a casual pace. Various crops require different sweeping techniques. For example, in mulberry the net is swept across the top of the plants about 20 cm deep. Sweeping is not suitable for pests normally found near the base of the plants.

SUCTION TRAP

Suction trapping is applied only at that place where attraction to light or chemicals is not useful. It is conducted at different levels above the plantation. It provides a format to know the quantity of the pests. Suction traps are of two types, fixed and portable. Flying insects are sampled with field type of suction traps. They usually consist of a metal sieve cone into which air is sucked by an electric fan. In some other suction pumps, cones are enclosed for protection, the air is drawn through, and the opening may be at different heights to sample different kinds of populations. A control mechanism is often used to segregate the catch into the time periods. Another popular method for insect trapping method is also known as graphic method. It is based on the principle that as removing individuals reduce the population from it, the catch per unit time decreases. The population size can be estimated from the rate at which the catch decreases. The rate at which trap catches falls off are be directly related to the size of the total population (unknown) and the number removed (known) (Huffaker, 1980). Direct sampling methods are influenced by many factors viz. the number of insect present, the type and stage of silk host plants; the biology, feeding behaviour, sex ratio, abiotic factors, sampling time and lack of uniformity among individuals in their sampling techniques.

INDIRECT SAMPLING METHODS

Population density of pest complex existing on silkworm and its host plants are assessed easily and quickly on the basis of the effects of damaged caused on the plants. It is called indirect sampling methods or population indices (Santhwood,1978). This kind of insect sampling is mostly used for indexing insect populations. The differences between incidence (the number of plants damaged) and intensity or severity (the degree or extent of damage) are generally worked out to measure the effects of insect pests attack. The incidence of pest is a discrete measure, whereas intensity is continuous and finite. Percentages with constant base of 100 are often used. There are a number of ways of measuring the effects of damage caused by the insects on the whole plants, stem, leaves and roots. The number or percentage of plants attacked, missing or showing signs of damage is recorded as an indication of the population of the pest. It is simple and quick method for stem borers of non-mulberry silkworm host plants, cutworms in mulberry, beetle larvae and other soil pests. Sometimes, the number or percentage of wilted plants or dead central shoots (dead hearts), for example, in mulberry plants, indicate the intensity of attack by lepidopterous stem borers, while shoot flies of defoliating beetles. The number of galls indicates the intensity of attack by gall midges; the number of exit holes or a length of tunnels has also been used. Leaves having holes, spots, mines, rolls, leaf scrapping or lesions etc. are measured for caterpillars, semiloopers, case worms, leaf folders, leaf miners, leaf beetles, bugs, orthoptrans, mites, etc. The actual leaf area damaged can be measured by using a dot matrix grid, photography or by a planimeter, electronic scanning, area integration device such as "Lincar" or by comparison with the undamaged leaf using length x breath x constant formula (Andrewartha, 1961). Soil inhabiting insects were sampled by calculating the effects of root pests. The damage caused in the root system, in the percentage nodes below the soil surface having injury, percentage roots pruned below the soil surface and the length of root damaged. Damage to tuberous root is measured by counting lesions of areas of damage on the surface.

RELATIONSHIP BETWEEN DIRECT AND INDIRECT METHODS

If an indirect method of pest management is used, its relation to a direct method such as pest density per unit area of the ground is examined so that quick and simple indirect assessments can be made to measure the actual populations of a pest. The relation between the percentage of stem borers attack and the borer population per square meter is established by counting the borer population on the plant (Kuno,1991).

MEASUREMENT OF POPULATION

There are many techniques that take advantage of some characteristic behaviour of the target species to maximize the chances of collecting a representative sample of population. The most common techniques for sampling insect in the field crops are visual counts; sweep nets, suction trapping and ground cloth. Generally, the population counts obtained by a relative sampling method is calibrated against result of absolute population counts (Rousink and Kogan, 1982). Absolute estimation of the population refers to the density per unit area in the habitat, viz., and the number of individual per square meter. This method is very useful in insect population dynamics studies (e.g., for constructing life tables, establishing oviposition, and mortality rates etc.), however, because of the difficulty in measuring it and the cost involved, it is not widely used in IPM programmes (Rausink and Kogan, 1980). In relative methods, the population is measured in unknown units, thus allowing comparison only in space and time.

The methods employed are either the catch per unit effort or various forms of the trapping, for example, the number of brown planthopper (BPH) per sweep or insects per some fixed number of sweeps with a sweep net at a particular place and time. Relative methods have many advantages over the absolute methods, being less expensive, they require simple equipment, provide a lot of data, and are appropriate for IPM programmes (Heong, 1981). In population sampling, the dispersion of insects in a population is as important as its variance and standard error. Dispersion is the pattern of spread of insects in an area. It gives information about population dynamics of the pest and may influence the way a population is to be sampled in an area (Taylor, 1984). Thus the knowledge of distribution pattern of an insect in a field is important in developing a sampling programme. The dispersion of individuals of a population is generally uniform, random or clumped. Most common ways insects are dispersed in a environment are "random" and "clumped" (Sutherst,1991). In random "dispersion" an individual insect has as good a chance of being present in an area as in another, and individuals do not affect each other's presence in an area. Random dispersion of insects occurs most often in a relatively uniform environment. However, the most common type of insect population dispersion is "clumped" (contagious). Clumping dispersion in a population indicates that if one individual of the species is found, chances are good that there are others in the same vicinity (Torii,1967; Shepard *et al*, 1986). Clumping may be caused by either environmental (e.g., uneven habitat) or behavioural (e.g., mating and feeding) factors or both. Random and clumped dispersions can be statistically described by mathematical models. The most common models used in entomology are the Poisson model to describe a random dispersion and the negative binomial model to describe a clumped dispersion or positive binomial model to describe uniform dispersion. It is essential to select a sample unit, and preliminary sampling is carried out to determine if the insect dispersion (distribution pattern) is poisson (random), negative binomial (clumped), positive binomial or other form of distribution (Waters, 1955; Ruesink and Kogan, 1980).

Insect sampling for IPM programmes is generally taken up at the stage of insect development that causes damage to the crop involved and the likely management strategy for that insect. This provides information on insect populations for taking up management measures to suppress it immediately. However, in some cases, a stage prior to the damaging stage is selected so that prediction of pest population can be made and this gives enough time to take up the management strategy. For example, stem borer eggs are counted to provide information on whether stem borer larvae infestation in mulberry plant is likely to be increased or declined in the next several days. Further, it may be important to know whether the eggs are white or dark in colour because dark eggs hatch in next few hours, where as white eggs will require two to three days to hatch. Similarly, sampling non feeding adult stage helps in predicting subsequent likely damaging larval population. However, such predictions based on number of non-damaging stage of a pest are not always reliable because beneficial insects, climatic conditions, and other factors can suppress the development of an insect pest population, changing a potentially high infestation into a harmless one at the damaging stage.

PEST SURVEILLANCE AND FORECASTING

Pest surveillance programme play an important role in IPM because they help in pest forecasting and decision making to counteract pest infestation and avoid economic loss in crop yields. In fact, pest forewarning and forecasting are the outcomes of an effective surveillance programme. Pest forewarning is an early warning system based on surveillance information from immediate past. Pest forecasting, on the other hand, requires not only pest surveillance data from field over the years but also information concerning biotic and a biotic factors that affect host and pest interactions and ultimately damage the crop. Thus such surveillance programme has assisted in establishing seasonal abundance of pests rather than data required for pest forecasting programme. Pest surveillance and forecasting programme is conducted on a particular pest problem on a crop or on a major pests of a crop as a whole in an individual field or in a region (Pruthi, 1970). A regional programme is more useful where "infrequent wide spread outbreaks" of pests occur due to mobility of migrant pests and favourable weather conditions. Regional Sericultural Research Stations through coordination and cooperation from Regional Extension centres monitor pest population level and issue forewarning to farmers when an outbreak is expected. For pests that reach outbreak levels regularly during each season and for outbreak years of "infrequent pests", a field level surveillance programme is more useful. Typically, a surveillance programme for pest forecasting consists of data collection, data processing and storage, and information delivery

(Nordlund, 1981). Most developed countries have a well-established and effective pest surveillance and forecasting programmes for the last 50 years or more (Yoshimeki, 1978). Pest forecasting is useful in making tactical pest management decisions such as what control measure is to be taken up and when (Norton, 1980). For regulatory measures, forecasting information can be used as an early warning indicating when intensive monitoring should begin. Simple analysis of data is required for providing such information. The decision to make control recommendations has in the past been based, for example on the economic threshold levels (Zadoks, 1983; Karel *et al.*, 1993). Efficient pest management information delivery systems are required for providing pest information to extension workers and farmers (Beeden, 1972). In India, the data collection, processing and storage and information delivery system for pest surveillance programme have so far been carried out manually.

PEST SURVEILLANCE IN INDIA

Pest surveillance programme in India are generally carried out by the government agencies, especially the Directorate of Plant Protection, Quarantine and Storage (DPPQS), Ministry of Agriculture, Govt. of India through its 26 integrated pest management centers spread in 22 States/Union Territories of the country. A Pest Surveillance programme of DPPQS is to assist State Governments in pest surveillance on major crops. Various crop based Institutes and Directorates of the Indian Council of Agricultural Research (ICAR) also carry out pest surveillance programme of DPPQS is to assist State Governments in pest surveillance on major crops. In sixties and seventies, DPPQS organized rapid roving surveys on various crop in collaboration of State Departments for detecting green Leafhopper population, as well as natural enemies of insect pests in some part of India for advising suitable control measures. Following the success of the programme the surveys were extended to different States. Similar programme have also been taken up for important pests and diseases of various economically important crops by various ICAR institutes in the country from time to time. Central Silk Board has started pest surveillance programme at its main Research institutes in collaboration with its extension units. Several advances have been made in the recent past on specific pest situations modeling, pest simulation modeling by studying its behaviour under controlled or experimental conditions, and pest model design system through non-linear development functions in the case of certain major pests in other crops across the globe. Similar studies for at least major pests of silk host plants like defoliators, leave sucking insect and gall insect, weevil and leaf caterpillars will be highly rewarding to mange their populations well below the economic threshold levels. Regular sampling is done for making IPM decisions by plant protection specialists and farmers.

Since IPM approach is knowledge and skill oriented programme, training of grass root level extension workers is the prerequisite for effective transmission of the message to the farming community. Pest monitoring through field surveys and surveillance helps in tackling build up of any pest/disease and employment of pheromone taps and lures ensured minimum and need base application of pesticides instead of prophylactic and calendar based pesticide spray schedule thus reducing the load of pesticides formed the basis of IPM. IPM also called for integration and utilization of biological methods of control using biocontrol agents such as parasites, predators, pathogens and ecofriendly neem based products (Upadhyay *et al.*, 1998). The farmers involved in these activities are trained for identifying the harmful and beneficial insects, sampling of crop field and recording of the field data. The data sheets are designed to provide required information on pests, their natural enemies, growth stages, related factors of crop phonology, location, weather conditions etc. All these information are entered in the computer and based on proper programming the occurrence of pest population is forecasted and accordingly control measures are applied. It helps to increase the productivity of silk. Pest monitoring through field surveys and surveillance helps in forecasting the population build up of pest. It reduces the load of pesticides application and forms the basis of Integrated Pest Management in sericulture. Common sampling techniques for quantifying pest populations and damage caused by them are reviewed emphasizing the need for quick and simple sampling methods. Various direct and indirect sampling methods for establishing pest populations are discussed and methods have been discussed to use indirect sampling method under IPM programme in sericulture. The use of pheromone lures and traps forms one of the important ingredients of integrated pest management, which calls for integration of all available methods in a cost effective and environmental friendly manner offering consistent efficacy. Silkworms feed on the variety of silk host plants and spin cocoons. Each silk host plant is attacked in the field by number of insect pest species. Several pests are common to mulberry, tasar, oak tasar, muga and eri host plant but pest status and seasonal abundance differs from each crop. The key pests are serious perennially occurring persistent species which cause considerable yield loss every year on large areas and require control measure. Regular occurrence of minor pest is noticed but sudden increase in its population is not known. The occasional pests are sporadic but potential causing sufficient damage. Silk losses due to attack of all the pests have not been calculated. However, information on pest biology and ecology, and control practices being practiced is available but the period of outbreak of major pests and predators on silkworms and its host plant needs to be reinvestigated. Pest and predators forecasting based on surveillance information may provide an opportunity to minimize the losses, particularly to reduce expenditure involved in pest management.

CHAPTER-11

INTEGRATED PEST MANAGEMENT

Pest management is the science of preventing, suppressing, or eradicating biological organisms that are causing a problem. Today the term pest management is often used in place of such terms as pest control, plant protection, or other equivalent expressions. The term pest is often used to refer to invertebrate (arthropods, slugs, etc.) as a group in contrast to plant diseases and weeds. However, the term pest should apply to any unwanted biological organisms - especially when the problem is associated with agriculture, sericulture or environmental issues. Pest management practices may be classified according to the *approach* or the *method* used to deal with a pest problem. In terms of approach, pest management practices may be designed to prevent a problem, suppress a problem, or eradicate a problem. In regard to method, pest management practices may be classified in a number of categories of which the most common are chemical, cultural and mechanical, biological, and legal. The term Integrated Pest Management (IPM), implies integration of approaches and methods into a pest management system, which takes into consideration the ecology of the environment and all relevant interactions that pest management practices may have upon the environment in which one or more pest problems may exist. When IPM principles are applied to a given pest problem, it is generally assumed that environmental impact and economic risks have been minimized. Since IPM considers all applicable methods, it is also assumed that emphasis on chemical methods may be reduced when effective non-chemical alternative methods are available. As a result, implementation of IPM principles and practices is advocated in various federal and state regulations affecting pesticides.

Integrated Pest Management (IPM) is the art of the possible to achieve the realistic solution of pest population without any harmful effect of human beings. Silk host plants and silkworm larvae are attacked by large number of pests and predators. The continuing incidence of pest population has confronted sericulturist with two major problems: decrease in the silk production, and the threat of environmental pollution on the other (Singh, 1989; Singh *et al.*, 1990; Mandal and Singh, 1990). It is imperative that whatever is done to alleviate one problem does not aggravate the other (Perkins,1982. Singh *et al.*, 1989). Thus, in order to increase the silk production and to avoid environmental pollution, effective pest management program is urgently needed. IPM provides a reasonable compromise, taking into account both the desirability of biological control and necessity of some form of chemical control, in combination with judicious use of relatively selective pesticides, only when absolutely necessary and in the least disruptive modes of application (Singh and Thangavelu, 1991). Other selective tactics *viz.* cultural controls, autocidal methods or utilization of semiochemiclas – is incorporated in the program wherever applicable (Teng and Heong, 1988). IPM represents a holistic approach, recognizing the unity of the ecosystem and harmonizing all available measures in an attempt to optimize pest control and crop production (Wilson and Huffaker,1976; Apple and Smith, 1976; Rosen, 1985). Utilization of natural enemies is regarded as the backbone of any IPM program (DeBach and Hagen, 1964; Van Lenteren, 1986).

The value of naturally occurring natural enemies cannot be overemphasized, and their conservation is the first goal of IPM (Debach and Huffaker, 1971). Modern sericulture is moving towards sustainability, a key development where increased emphasis is placed on the integration of many different techniques to achieve stable, long-term production. The sustain view that sericultural crops could be made to produce more by usage of synthetic chemical is outdated. Disintegration of this idea has allowed the concept of IPM (Integrated Pest Management) to play a vital role in modern sericulture. Mulberry and commercially exploited non-mulberry silkworms are attacked by large number of insects, which causes serious depredation in silk productivity. There is much variation in the pest population and extent of damage in each variety.

Mulberry silkworm (*Bombyx mori*) is completely domesticated and feeds on mulberry plant (*Morus alba*) (Rangaswamy *et al.*, 1976; Ullal and Narshimhanna, 1978). The larvae of tasar silkworm (*Antheraea mylitta*) feeds mostly on arjun (*Terminalia arjuna*), asan (*Terminalia tomentosa*) and sal (*Shorea robusta*) where as oak tasar (*Antheraea proylei*) prefers to feed on various species of *Quercus* which are abundantly available in north eastern and western region of India (Jolly *et al.*, 1979). Muga silkworm (*Antheraea assama*) is very specific to region of Assam and feeds mainly on som (*Machilus bombycina*), soalu (*Litsaea polyantha*) (Thangavelu *et al.*, 1988). Eri silkworm (*Philosamai ricini*) is a domesticated variety and multivoltine in nature. It feeds on castor (*Ricinus communis*), kesseru (*Heteropanax fragrans*), simalu (*Manihot utilissima*) and payam (*Evodia fraxinifolia*) (Sarkar, 1988). Low yields of silk have been attributed to variety of factors, of which loss caused by insect pests has been considerable. Over 150 insect pests have been reported to attack food plants. Shoot fly, gall midges, hairy caterpillars and mealy bug occur in serious proportions to cause reduction in silk productivity. Most of the time, several parasites and predators also attack silkworm. Uzi fly, Ichneumon fly and other predators cause heavy loss to silk industry. Dermestid beetle infests cocoons. One of the main reasons for low production of silk is the pest problems associated with silkworms and its host plants (Singh *et al.*, 2000).

IPM METHODOLOGIES

- Survey should be conducted in mulberry and non-mulberry sericultural areas and data should be generated on the incidence of pest population on silkworms and its food plants.

- The data from survey/ surveillance programme should be utilized regularly to decide the type and level of IPM intervention needed by the farmers.

- Biological control agents should be screened and population dynamics of host and parasitoid should be determined. Host searching ability, sex ratio and reproductive ability of the parasitoid should be calculated. Culture technique of the host and parasitoid should be developed.

- To release the biological control agents based on the pest load, weather conditions, cultivation practices and cropping systems suitable technology should be developed.
- The efficiency of IPM may be assessed by conducting survey before and after the treatment of each IPM tactic.
- Residual toxicity of the pesticides in the various crops may be assessed and bioremediation of pesticide pollution problems should be studied.
- IPM technology package may be refined and fine-tuned to meet the local and seasonal requirement of the area. Soundness of the technology may be tested and compared with farmers' practice from socio-economic and environment points of view.
- Sustainable sericultural IPM practices should be developed and greater productivity improved silk quality and enhanced income and profitability may be achieved.

PEST INCIDENCE

The major insect orders known to be the pests of mulberry and non-mulberry silkworm host plants are Lepidoptera, Hemiptera, Coleoptera, Thysanoptera, Orthoptera and Isoptera (Sengupta *et al.*, 1990). Mealybug, *Maconellicoccus hirsutus* (Green) (Hemiptera, Pseudococcidae) causes 'Tukra' disease in mulberry. The leaf roller, *Diaphania pulverulentalis* (Lepidoptera: Pyralidae), has been noticed as a severe pest of mulberry, *Morus alba* (Geetha Bai *et al.*, 1997; Marimadaiah and Geethabai, 2000). The Bihar hairy caterpillar, *Spilosoma obliqua* Walker (Lepidoptera, Arctiidae) is a polyphagous pest and sporadic in nature. The tiny caterpillars of first two instars feed gregariously, but as they grow older, they disperse widely in search of food. The cutworm, *Spodoptera litura* is a polyphagous pest infesting several crops including mulberry. Unilateral control measures against *M. hirsutus*, *S. obliqua* and *D. pulverulantalis* mainly include the application of chemical insecticides. The major insect pests belonging to the order coleoptera reported causing damage to mulberry are *Sthenias grisator*, *Apriona* spp., *Myllocerus* sp. etc. Coleopteran white grub, *Holotrichia serrata* was reported to damage mulberry root extensively. The larvae feed on the roots and under ground portion of stalk while the adults feed on the foliage of trees and shrubs. The affected shoot dries up, resulting heavy losses to mulberry. The wingless grasshopper, *Neorthacris acuticeps nilgiriensis* Uvarov (Orthoptera, Acrididae) is a serious pest in rain fed mulberry plantations. About 15 species of mites belonging to families of Tetranychidae and Eriophydae have been reported to be key pests of mulberry (Pruthi and Mani, 1945; Narayanaswamy *et al.*, 1996). Tetranychus equitorium and Aceria mori cause considerable damage to mulberry in India (Mohanasundaram and Sivagami, 1983). Several polyphagous pests, which damage mulberry, tasar, oak tasar, muga and eri food plants have been reported (Singh *et al.*, 2000).

All parts of the tree and all stages of tree growth are subjected to insect attack. Different insect species infest different parts, but all stages of growth are subjected to all types of insect attack. Defoliators, leaf minors and gall forming insects are considered as group cause most damage to older trees. *T. arjuna* and *T. tomentosa* leaves are generally attacked by gall insect (*Trioza fletcheri minor*) (Singh *et al.*, 1994) and stem borer *(Psiloptera fastuosa, Aeolesthes holosericea* and *Batocera* sp.) but several defoliating insects have also been reported to cause extensive damage to leaves during rearing period. (Singh *et al.*, 1991; Singh *et al.*, 1992). Tiwary *et al.*, (2000) reported cerambycid borer infestation on sal plantation in India. Oak tasar silkworm *Antheraea proylei* and its food plants are attacked by large number of pests. The major categories of insect destroyers of oaks have been grouped as sap sucking, defoliating, meristem feeding, acorn feeding and gall forming insects (Singh *et al.*, 2000). Som (*Machilus bombycina*) and Soalu (*Litsarea polyantha*) are the primary food plant of muga silkworm and abundantly available in Assam. Pest attacks from nursery to mature plants. The biology and control measures of some of the major pests associated with som and soalu plants have already been reported (Singh and Sen, 2001). The castor white fly (*Trialeurodes ricini)*, castor semi looper *(Achaea janata Linn.*) and Castor Hairy caterpillar *Euproctis lunata are* serious pest of castor plants. Apart from castor it has been reported on *Terminalia arjuna, T. totmentosa* and Muga food plants. Castor seed and capsule borer *(Dichocrocis punctiferalis)* damages the seed extensively. Thrips *(Thrips tabaci)* is world wide known insect and causing serious damage to various plants in their growing stage. It has been reported as a major pest of kesseru , Sepium and Ailanthus. Jassid *(Ambrosia bagatelle)* commonly known as leaf hopper is widely distributed in India and is the most destructive pest of castor, som and soalu plants. Castor Leaf Hopper *(Empoasca flavescens)* causes damage to the leaves (Singh *et al.*, 2000).

SILKWORM NATURAL ENEMIES

Silkworm larvae are attacked by several predators and parasites (Singh and Thangavelu, 1994; Jolly *et al.*, 1979 and Dhar *et al.*, 1989). Predator consumes several host individuals during its development, where as parasites includes its development on single host. The most important among them are stink bug (*Canthecona furcellata*), mantis (*Hierodulla bipapilla*), wasps (*Vespa orientalis*) and ants (*Oecophylla smargdina*). Predators are mostly confined in the silkworm-rearing field and kill the early instars silkworm larvae. Some predators have biting or chewing mouthparts to devour their host viz. preying mantis, where as others such as stink bug or reduviid bug use piercing and sucking mouth parts to feed upon the body fluid of silkworm larvae. Many predators are agile ferocious hunters, actively seeking their host on the branches of silkworm host plant viz. wasps and ants and certain hunters have specially adapted seizing organs; such as barbed forelegs of mantids. Many species are predacious in both nymphal and adult stages, although not necessary on the same kind of host. Among parasites uzi fly (*Exorista bombycis* and *Blepharipa zebina*), Ichneumon fly (*Xanthopimpla pedator*) and *Apanteles* are the most important parasites of silkworm larvae (Singh and Thangavelu, 1991; Singh *et al.*, 1993).

PROBLEMS FOR PLANT PROTECTION

Plant protection in sericulture has all the characteristics of different phases of plant protection, *viz.* subsistence, exploitation, crises, disaster and IPM phases. Early to 1950, it can be considered as subsistence phase when plants were grown without modern plant protection chemicals. The exploitation phase started with the use of DDT and BHC sometimes after 1950 and the recommendation at that time were prophylactic schedule of insecticides treatment on the host plants. The crises phase started with the outbreak of tussock moth, Bihar hairy caterpillar, leaf roller, weevils, gall insects, stem borers and uzi fly. The serious outbreak of uzi fly in West Bengal and Karnataka were symptoms of disaster phase. In many south-east Asian country it is found that insecticide use preceded out breaks of secondary pest like leaf roller. More over insecticide poisoning a farm worker become a serious issue and the chemical used against mulberry plants induced insecticide resistance in number of disease vector that breed in flooded field (Teng and Heong, 1988). The main issue here is a lack of comprehensive system for monitoring pest out breaks, area coverage, extent of loss and the case study on reason for out breaks. A structural organization with technical input and a corresponding system for documentation and subsequent retrieval and dissemination of data will go a long way in pinpointing the developments on plant protection and crop within state/country, the reason for fair out breaks and to advise the policy makers on decision making (Smith *et al.*, 1976).

Silkworms are very much sensitive to insecticides; therefore, extensive and intensive use of insecticides on silkworm host plant is harmful. Insecticides use resulted in decreases in pest population in sericulture areas but its repeated use has created resistance to several insects. However, resistance to insecticides, particularly in non-mulberry pests and poisoning of farm workers, farm animals and environmental pollution has not been clearly documented. The pollution in the drinking water was not properly assessed. The assessment of residue level in leaves and silkworm sometimes increased above economic threshold level and resulted in high mortality of silkworm larvae. Even though the danger due to insecticides has been pointed out, they have definitely contributed to the increased yields in cocoon production and in general, insecticides have contributed their part in green revolution. The issue here is can Sericulturist do away the pesticides? The farmers in sericulture are small and marginal farmers and they are more concern in saving their crop rather than worry about pollution and environmental issues. Even the loss of crop in a single season will immerse them in debts. They will continue to use pesticides no matter how dangerous they are to his health and to their livestocks. Hence the pest management strategy that has to be adopted should be sustainable and different and to make non-chemical methods collectively more effective. The chemicals should be used sparingly and as a last resort.

PESTICIDE USE PATTERN

Many industrialized countries have enforced stringent pesticide regulations and developed alternative pest management approaches as a result of which pesticide use in these countries has shown a declining trend. Consequently, the magnitude of contamination of food materials has also slowed down. However, many developing countries continue to use persistent pesticides in agriculture and public health programmes, and the contamination of different components of environment continues to be excessive and pervasive. The production of Indian pesticides industry has almost remained stable at 85,000 MT during 2008-09. In value terms, the size of the Indian pesticide industry was estimated at Rs.98 bn for 2008, including exports of Rs.48 bn. Per hectare consumption of pesticide is low in India at 381 grams when compared to the world average of 500 grams. Low consumption can be attributed to fragmented land holdings, lower level of irrigation, dependence on monsoons, low awareness among farmers about the benefits of usage of pesticides etc. India, being a tropical country, the consumption pattern is also more skewed towards insecticides which accounted for 62% of the total pesticide consumption in 2008. Rice is the highest pesticides consuming crop. Of the total pesticides consumption, 25.9% is consumed by rice. Andhra Pradesh is the highest pesticides consuming state (23%) followed by Punjab & Maharashtra. With the advent of the Integrated Pest Management (IPM) technique, the use of biopesticides and Genetically Modified (GM) seeds has increased. Globally, GM seeds are used mainly for commercial crops like cotton, maize, soyabean and canola. In India, Bt cotton is widely used and the acreage stood at 7.61 mn ha for 2008, a growth of 23% over the previous year. Use of GM seeds may diminish the use of insecticides but the use of herbicides may improve. Though the value of IPM has been well recognized, very little is being adopted at field level. The Ministry of Agriculture, Government of India is concerned very much with the slow progress in IPM as there is raising demand for chemical pesticides. The pesticide industry has estimated the demand for the current year at over 86,460 tones. The trend in the use of chemical pesticides in India in the last one-decade shows greater use of insecticides, which are mostly health hazardous and eco-destabilizing. Hence, a wholesome technology to suit the small and marginal farmers in various agricultural regions has to be developed/improved and adopted. Chemical fertilizers and pesticides are the major chemical inputs used in farming.

REASON FOR IPM APPLICATION

The pest problem has increased due to poor maintenance of food plants, use of higher doses of fertilizers and unhygienic condition of silkworms rearing. In order to reduce pest population and resulting plant damage, several methods are known but due to easiness in application farmers prefer to apply synthetic insecticides. However, repeated and frequent application of modern synthetic insecticides has created problems of pests resurgence and out breaks, insecticide residues, development of insecticide resistant strains, phyto-toxicity and hazards to non-target species including natural enemies and other beneficial organism, alternation in pest species population dynamics, environmental degradation and dispersion of natural balances associated with over reliance on chemical control (Bartlett, 1964; Croft and Brown, 1975; Flint and Van Den Bosch, 1981). In the last decade, much of the attempt has been made to achieve excellence in genetic potential of crops in sericulture system. Large quantity of chemical fertilizers and pesticides were used to increase leaf production and pest reduction but ultimately it led to soil degradation, water contamination and loss of biological diversity (Langeweg, 1989; DeBach, 1964a, b)). Keeping in view the sensitivity many established pesticides were discarded on account of development of phenomenon of resistance in the target insect species and adverse environmental problems or health hazards which has become unacceptable, squander valuable but finite resources have generally lead to unsustainable sericulture.

The findings of the earlier workers indicate that Integrated pest management (IPM) is only one solution which has got all positive attributes of an innovation and highly suitable in sericulture and it is the remedy for menace in sericulture. IPM calls for integrating all available tools in cost effective, environmental friendly and sustainable manner (DeBach, 1974; Baker and Cook, 1974; Clausen, 1978). The philosophy and principal underlining the concept of IPM is to reduce mankind dependence on chemical pesticides by identifying, integrating and employing other methods which are cost effective, eco-friendly and lasting efficacy (Sharma and Batra, 1989). Since IPM approach is knowledge and skill oriented programme, training of the grass root level extension functionaries is the prerequisite for effective transmission of the message to the farming community.

Pest monitoring through field surveys and surveillance help in tackling pest population and employment of cultural, mechanical and ecological practices ensures less utilization of pesticides (Norton and Mumford, 1983; Perkins, 1982). The application of minimum and need based pesticides instead of prophylactic spray schedule form the basis of IPM. In true sense, IPM is called for integration and utilization of biological methods of control using biocontrol agents such as parasites, predators, pathogens and eco-friendly neem based products (Huffaker, 1980; Huffaker and Smith, 1980; Singh and Maheshwari, 2002 and Upadhyay *et al.*, 1997). The IPM approach encompasses adoption of all available methods, techniques, skills and strategies of pest management in a harmonious manner based on seri-ecosystem analysis and field observations. The pest surveillance and monitoring and biological control are the major components of this programme. (Hazarika *et al.*, 1994). IPM strategy involves integration of components such as genetic, chemical, botanical pesticides and socio economic factors. IPM should not mean putting together on paper a set of independent recommendations given by different scientists. IPM will be effective only if the component techniques are developed by agronomists, breeders, entomologists, plant pathologists and social scientist together (Upadhyay *et al.*, 1997; Goodwell, 1984). The IPM technology is now being extended to several crops like rice, cotton, pulses, oil seeds, sugar cane, ground nuts certain vegetables and fruit crops. The physical targets for IPM covering all activities have also been fixed.

CONCEPT OF IPM

The concept of IPM, which combines all possible manners in a compatible and harmonious manner, has gained prominence (Smith *et al.*, 1976; Van Lenteren and Woets, 1988). Natural enemies play an important role in suppressing pest population in the crop whenever suitable condition prevails for their survival, development, conservation and multiplication in any seri-ecosystem. Thus biological control is considered as an essential component of IPM as it is economical, effective and eco-friendly. Some biological control agents, when used alone or in combination with less persistent insecticides and botanicals have proved better than insecticides. With the increasing importance of sustainable sericulture, the concept of Integrated Pest Management (IPM) for sustainable development has emerged. In the recent past, plant protection scientists, as well as the farmers have identified pest management methods, which are ecologically non-disrupting and stable. Concurrently, mulberry varieties with at least moderate resistance/tolerance to pests and diseases have been developed and cultivated. Applying the principles of 'organic farming', several non-chemical methods have become popular among the farmers. Simple cultural practices like intercropping, trap cropping and crop rotation have been found to provide adequate protection from pest damage with no additional cost and without harmful effects on the environment. The farmers who glamoured for chemical pesticides in the sixties and seventies are now disillusioned with these poisonous eco-destabilizing substances; and are now on the look out for sensible and bio-rational methods of IPM.

IPM AS TOOL

Integrated Pest Management (IPM) is an ecologically based, environmentally conscious method that combines, or integrates, biological and non biological techniques to suppress weeds, insects, and diseases (Nordlund, 1997). Successful implementation of IPM in pest management was introduced in the 1960 (Smith and Reynolds, 1965). Integration of multiple pest suppression techniques has the highest probability of sustaining long term crop protection (Kenmore, 1991; Sharma *et al*., 1996). Much of the technologies and data analysis procedures have been developed about those strategies and tactics most appropriate for use in implementing specific IPM systems. These include economic thresholds, sampling technology, modeling, natural control, geographic distribution, effects of pest migration and movement, host resistance, and pesticides. IPM's basic framework is acknowledged to the natural controls. These include natural enemies, weather, climate, and food resources and semiochemicals. Natural enemies play an important role in regulating populations of all pest classes (Vinson, 1981).

APPROACHES TO PEST MANAGEMENT

Prevention is the best approach for pest management. When a pest problem is anticipated and action is taken to prevent a significant problem from occurring, the approach is termed prevention. The action taken may include any applicable method proven to prevent or reduce the probability of a significant pest problem from occurring. The prevention approach may include either chemical or non-chemical methods. When a farmer applies a granular soil insecticide at planting time, it is assumed that the treatment will prevent a significant loss in yield due to the presence of soil insect or nematode pests. After prevention is the suppression stage. After a pest problem has been detected, any action taken to suppress the pest population is termed a suppression approach. In practice, few treatments totally eliminate a pest problem, but the pest population is reduced to a point at which it is no longer perceived as a problem.

Thus, any action ranging from treatment of pest infestation on silk host plants or silkworms may be regarded as suppression. Post emergence application of herbicides to reduce emerging weed populations is regarded as suppression. The use of chemical methods is generally associated with suppression practices, but non-chemical methods may also be employed to suppress a pest problem. When a pest problem must be totally eliminated from a designated area, the approach is termed eradication. If a new pest such as uzi fly is detected in a silk growing growing area, regulatory agencies may implement widespread actions to totally eliminate the pest problem before it becomes established to a point that it can no longer be eradicated. When a serious insect pest problem is detected in a commodity of foreign origin, fumigation tactics may be employed to totally eliminate the presence of the unwanted pest from stock identified as infested. In general, the eradication approach does not apply to elimination of an established pest population from a large area. In sericulture, Integrated Pest Management is a comprehensive approach to pest control that uses combined means to reduce the status of pests to tolerable levels while maintaining a quality environment. The main purpose of the IPM programme is to reduce crop losses, increase farmer income, reduce pesticide use and protect environment, reduce pesticide residues, improve ability to monitor pests, and increase involvement of farm workers in IPM decision making (Van Lenteren, 1983). It helps in maintenance of quality environment and conservation of natural ecosystem and non-agricultural environment (air, water, soil, wild life and plant life). Pest management strategy has been divided in two groups *viz.* reductionist (pesticide treadmill) and holistic (IPM treadmill) (Tait and Lane, 1987; Tait, 1987).

REDUCTIONIST APPROACH

The reductionist approach contains routine and rational pest management system. Routine pest management implies the use of pesticides as prophylactic measure, regardless of pest incidence and will probably require higher levels of pesticide use than any other strategy in a given set of circumstances. Rational pest management requires that each pesticides application be justified on scientific, technological and or economic grounds (Tait, 1980). This will generally involves lower levels of pesticides use than a routine system under similar cropping condition.

HOLISTIC APPROACH

Holistic (IPM treadmill) approach is the most suitable procedure. Integrated pest management systems consider the interactions amongst the whole range of organisms with beneficial, neutral and pest status, the long-term aim being to increase the level of pest suppression which is achieved by natural, as opposed to chemical means (Singh *et al.*, 2001). Most IPM system will involve the use of pesticides but at a lower level than routine and rational pest management.

PEST MANAGEMENT METHODS

Pest management practices are grouped under various method categories including (1) biological, (2) chemical, (3) cultural and mechanical, and (4) legal. Additional categories may be defined, but we will attempt to group most practices under the four mentioned.

BIOLOGICAL CONTROL

Biological control generally includes the manipulation of one biological organism to control another organism classified as a pest. In the web of nature, the combination of biological control techniques is extensive. Insect pests may be preyed upon or parasitized by other insects. Most insect pests are attacked by bacterial, fungal or viral pathogens. Insects with specialized feeding habits may control specific weeds. The implementation of biological control methods has been categorized into three basic approaches viz. classical, augmentation, and natural. When a pest is found to not be native to a given area, it may be assumed that the biological organisms that regulated its population dynamics in its native environment are lacking. In such a situation, the classical approach of biological control is employed to determined the pest's native home, locate beneficial organisms that naturally control the pest organism in its native area, and if feasible, import, multiply, release and establish the beneficial organisms in the problem area to facilitate biological regulation of the pest problem. If successful, the importation and establishment of the beneficial organisms will result in a long-term reduction of the pest problem and repeated releases of the beneficial organisms will not be required.

The process of importing and releasing beneficial organisms is complex, since many precautions must be taken to prevent the introduction of organisms that may have adverse effects. When beneficial biological organisms are mass reared and released periodically to supplement the natural enemy complex and achieve reduction of a pest problem, the approach is called augmentation. This approach may be applied to pest populations that are either native to the area or of foreign origin. In general, augmentation may be considered when it is economic and feasible to rear, multiple, and release a natural enemy of a pest to the point that reduction of the pest problem is achieved. Successful augmentation efforts have been developed for greenhouse environments where altering the balance between a pest and its natural enemy is feasible. Application of the augmentation approach in sericulture is limited, but a major effort is underway to control the uzi fly.

Most pest populations are maintained by a number of natural predators, parasites and diseases, which represents natural biological control. If such forces were not in effect, pest populations would overrun us. The balance of crop pest populations and their natural enemies can be significantly influenced by cultural practices and the use of chemicals. Populations of natural enemies can be enhanced by selective use of cultural practices or decimated by indiscriminate use of pesticides. In some cases, pesticides have been developed that effectively control a pest population without having a significant effect on beneficial species. The potential for development of biological controls for a wide range of pest problems is significant. However, development of successful biological control technologies often requires significant investments into research that may or may not readily produce satisfactory results. To date, biological control has not been a marketable product like chemical controls and research efforts into the field have been limited. Furthermore, implementation and evaluation of biological controls are often more complex than that of chemical methods. In summary, classical and augmentative biological controls are not available for very many pest problems - especially severe pest problems that demand immediate attention. However, natural biological control is in effect in most situations, and it is important that every effort be implemented to enhance such biological activity wherever it exists.

Biological control utilizes natural enemies such as parasites, predators, pathogens or competitors, deriving its energy directly from the pests themselves. It is acknowledged to the best type of pest control (Lloyd, 1986). It is environmentally safe alternative to the chemical means of control. It is a very effective method of control offering a long-term protection. The biological control strategies were born in a citrus grove in 1889, in the city of Los Angels, California (David, 1985). The release of 129 imported Australian Vidalia beetles resulted in dramatic reduction of the cottony cushion scale which has threatened California citrus industry. The technique of releasing an important organism that establishes itself and spread to permanently control a pest is today known as classical biological control concept (Rosen and DeBach, 1981). Successful classical biocontrol means that no further costs are required to keep the pest under control. The process of locating the place of origin of the non native pest and then finding and introducing natural enemies from its place of origin presents obvious ecological and logistical challenges. Therefore, exhaustive testing of the introduced pest , predator or parasite is essential before being release to be sure it will not harm non-target organisms. Generally, even after meeting these challenges, there are certain other factors viz. climatic differences, pesticide use, disturbances of habitat by other agricultural operations, and or the removal of non-crop vegetation that might otherwise offer food and shelter to the natural enemies are the important factors for survival of the parasitoid population in its natural habitat (Narendran, 2001; Van Lenteren and Woets, 1988). Large number of parasitoids has been screened against all the major pests and parasites of silkworms and their food plants (Singh and Maheshwari, 2002).

The coccinellid predator *Cryptolemous montrouzierie,* is a voracious natural predator of the mealy bugs which are hard to be controlled by conventional insecticides. It is known to feed on about 1000 eggs or 300-500 mealy bugs nymphs. Psyllids are the major pest of arjun and and plant (Raman *et al.*, 1997; Singh and Thangavelu, 1994). The common parasitoids screened against them are *Aprostocetus niger* and *Tetranichus indicus* (Singh *et al.*, 1995). The eggs of stem borer, *Spiloptera fastosa* is attacked by *Szelenyiola batocerae* (Singh and Kulshreshta, 1990). Silkworms are attacked by various predators and parasites (Singh and Thangavelu, 1991a, b). The most common predator attacking tasar silkworm is *Canthecona furcellatta* (Singh and Sinha, 1989). Several scelionid egg parasitoids have been screened but *Psix striaticeps* and *Trissolcus* sp. have been recorded to be highly potential and effective in minimizing the stink bug population in the rearing field (Thangavelu and Singh, 1992). Several uzi fly parasitoids *viz. Nesolynx thymus*, *Trichopria* sp., *Dirhinus anthracia*, *D. himalayanus*, *Spilomicrus karnatakensis*, *Trichomalopsis apanteloctena*, and *Pediobius sp.* have been screened and field trial has been conducted to minimize uzi fly population in sericulture (Kumar *et al.*, 1992a, b, c; Ram Kishore *et al.*, 1992 and Jyothi *et al.*, 1992a, b). The natural enemy complex is diverse and there is evidence to believe that pest population of plants is maintained at low levels mainly because of their regulating activities. The pupal parasitoid *Pediobius foveolatus* Crawford is an important parasitic hymenopteran that is practically used for control of uzi fly. Among chalcididae, *Brachymaria lasus* has been found to be a major natural enemies of the pupal stages of *Blapharipa zebina*, a serious parasite of tasar and muga silkworm (Singh *et al.*, 1995). Among pteromalid, *Trichomalopsis apanteloctena* has been reported as the highly potential parasitoid (Singh and Thangavelu, 1995; Singh *et al.*, 1995). All most all of them are ectoparasiotoid - feeding externally upon the host. Ectoparasitoids most frequently occur on hosts that live in some protected site – a pupa in a cocoon – where they are less likely to be dislodged and loose their host. Many of them also sting and paralyze the host prior to oviposition.

CHEMICAL CONTROL

Chemical control is generally applied to control the pest population. Pesticides are the most readily recognized method of pest management. The institution of the pesticide industry is a relatively recent development of the last half of the 20th century. The evolution of pesticide products changes significantly with each decade. Advancements in the development of biological pesticides may alter the field significantly. The range of risks and benefits attributed to pesticides will remain a key issue of society. The use of pesticides will remain as a dominant method to be incorporated in future pest management programs. Sericulturist faced problems due to extensive and intensive use of insecticides on silkworm food plants. Insecticide use resulted in increases in pest population in mulberry and non-mulberry food plants. Resistance to synthetic insecticides against major pests *viz.* tussock moth, gall midges, leaf roller and stem borer has forced the entomologist to switch over to plant originated insecticides in sericulture. Among the plants, which are reported to be more commonly used in pest control, are Neem, Pungamia, Indian privet, Adathoda, Chrysanthemum, Turmeric, Onion, Garlic, Tobacco, Ocimum, Custard apple, Zinger and some other plants. Usually extract of whole plant or parts of the plants are prepared and sprayed, otherwise, they are dried under sunshade, powdered finely and applied as dust.

Sometimes mixtures of extracts of more than a plant are made and the extracts are sprayed after allowing certain incubation period to enable them to release out the toxicant in the liquid. Among the plants, neem is the most promising source of biopesticides. Neem owes its toxic attributes to all large number of bitter compounds called meliacins like azadirachtin, nimbin, salanin, meliantriol, etc. among which azadirachtin is the most efficient. Neem seed kernels are the richest source of meliacin and contain 0.2% - 0.3% azadirachtin and 0.4% oil though neem leaves, seeds and bark also contain these in smaller quantities. The neem product acts as insect antifeedant, repellent, growth regulator, chemosterilant and toxicant. Any pest escaping one effect may be killed by other (Vijayalakshmi, *et al.*, 1995). Neem has been found effective against more than 200 species of insects like stem borers, hairy caterpillars, podborers, beetles, leafhoppers, planthoppers, aphids, mealybugs and whiteflies (Singh and Sinha, 1993). Application of azadirachtin to maggots of the uzifly, *Blepharipa zebina* completely disrupted subsequent development to pre pupae, pupae and adults. In a dose and stage dependent manner azadirachtin caused a delay and inhibition of puparium formation, loss of weight, pupal death, prevention of emergence of adults and malformation to adult structures (Singh and Thangavelu, 1996; 1998). Neem cake has manurial value to plants besides acting against nematode infestation in mulberry. Neem products are highly photodegradable ensuring their nonexistence in environment however, its action can be extended up to 3-7 days in the field. There is no problem of resistance and resurgence. Hence, they have characteristic suitable for IPM strategy.

CULTURAL AND MECHANICAL CONTROL

Prior to the advent of chemical pesticides, humanity relied primarily on cultural and mechanical methods of pest management. With the development of pesticides, the relative impact of various cultural and mechanical practices on pest populations was often overlooked. As public interest in environmental issues expands, the impact of cultural and mechanical pest management practices is receiving greater attention. Cultural practices are integral part of mulberry, tasar, oak tasar and muga cultivation which include pruning, plucking, cleaning and stirring of soil around the bush during the winter (Singh *et al.*, 2000). These practices directly influence the pest build up. Pruning and pollarding also contribute towards converting the host plant into a bush (Srivastava *et al.*, 1999). Works on impact of these practices on the activities on natural enemies are rare. However, Singh *et al.*, (1992a, b) observed that these practices are much more useful for redemption of pest complex in sericulture. The use of light trap for the control and for monitoring of schafers, weevils and tussock moths has been well documented (Singh and Thangavelu, 1991; 1993 and 1994). During winter period, when the rearing of tasar silkworm is suspended, the gall infected leaves are plucked and burnt in the field. It is found to be the most convenient way to minimize the pest population.

Further, some regular farming practices, viz. weeding, inter-cultivation; pruning and pollarding ensure further reduction in gall insect population (Singh *et al.*, 1992; 1992a; Raman *et al.*, 1997). Similar cultural and mechanical practices have been adopted for vapourer tussock moth *Notolophus antique* (Singh *et al.*, 1991; Singh and Thangavelu, 1993). In order to minimize the pest population in the field, mechanical collection of egg mass and early stage caterpillar are most effective cultural control programme. Further, an attempt has made to use the light trap as one of the devices in integrated pest management in sericulture. For this purpose pest-O-flash an illumination device is used in the field to attract, weevils, stem borers, and defoliating insects. The most prominent among them are *Anomala blanchardi, Myllocerus viridinus*, and *Crinorrhinus nebulosus*. Removal of dead barks from the food plants is also effective in minimizing the schaffers beetle and stem borer's population. Crop rotation, tillage practices, barriers, hedgerows, traps, and other forms of environmental modifications all influence the incidence of pest problems. Crop host resistance to pests - which may be considered a cultural practice - remains a key factor in many pest management programs. Sanitation is the basis for pest management in sericulture. Sanitation is also applied to crop production when removal of sources of pest populations is a critical factor. Examples of cultural and mechanical practices in pest management are abundant. Every time a grower rotates a crop or tills the land, weed, insect, nematode, slug and disease pest problems are affected.

LEGAL

Regulatory actions are often employed to prevent immigration of foreign pests or to prevent the dispersal of established pests. Such actions are termed legal methods of pest management.

FEED BACK IN IPM

Holistic and reductionist pest management system can both be defined in terms of positive feed back. In reductionist system, pest problems are main driving forces on crop protection decision making. As they increase, or are perceive to increase, pesticide use increases, leading to decline in natural controls which in turns lead to an increase in the pest problems followed by further increase in pesticide use. In holistic system, pest control by natural factor is the controlling influence on decision making. As this increases, pest problems decline, hence pesticides use declines, leading to a further increase in pest control by natural factors. The switch from system which driven by a pest problems to one which is driven by natural controls is unlikely to be smooth one. It must involve a discontinuity of some kind to jolt the thinking of the decision-maker (Tait, 1980; Van Den Bosch, 1982).

STRATEGIES AND TACTICS

The application of pest management concepts begin with the development of a strategy. A pest management strategy is the over all plan to eliminate or alleviate a pest problem. The particular strategy developed depends on the particular life system of the pest and the crop involved. Recently some important strategies *viz.* planting of cover crops, providing nectar- producing plants and sources of alternate hosts in and around fields, and inter planting different crops to provide habitat diversity are all management techniques that lead to the build up of natural enemy populations and result in enhanced biological control of pests.

MARKET POTENTIAL

IPM market should be developed to avoid excessive use of pesticides. Sticky traps, pest-O-flash and pheromones lure chemicals and Tricho cards should be popularized and provided as tool for controlling various pests and parasites in sericulture. The potential for promotion of pheromone technology is an important aspect. It may be popularize at grass root level and mass awareness should be created among farmers for further transfer of technology from laboratory to land.

FUTURE OUTLOOK

The pesticide application as prophylactic measures is very common among farmers. In most of the states, farmers very religiously follow the package of practices evolved on the application of pesticides on a prophylactic basis. These package of practices need to be reviewed immediately to incorporate components of IPM technology for the benefit of farmers. Research progress on the use of IPM strategies is being made. There is a greater potential for kairomones in contributing significantly to pest management, however, the uses of kairomones should not be considered in isolation from other control measures and component of IPM.

Integrated Pest Management (IPM) and Integrated Crop Management (ICM)

Integrated Pest Management (IPM) is a systems approach to pest management where decision-making assumes top priority. Its success depends on how prudent the decisions are in determining when and how a single or group of control measures are to be taken ensuring that they can work complimentary to each other and avoiding harm from one another. IPM integrated pest management approaches, methods and disciplines to minimize environmental impact, minimize risks, and optimize benefits. It is a systems approach to pest management that utilizes decision-making procedures based on either quantitative or qualitative observations of the pest problem and the related host or habitat. A key concept in IPM programs is the application of decisions making processes to determine when a chemical pesticide or other action is needed or not. Such decisions depend on evaluation of the pest problem often in a quantitative manner. In the evaluation of agricultural crop pests, the point at which the economic benefit of pesticide use exceeds the cost of treatment is commonly referred to as the economic threshold. Academic definitions of the threshold concept may vary from discipline to discipline. Another term commonly accepted is action threshold, which is commonly applied to set of conditions where action is warranted and may be based more on practical experience and judgement than on refined mathematical models relating biological and economic parameters.

Since IPM decision making depends on field observations, the role of the pest scout, pest management advisor, or field biologist emerges in such programmes. Although do-it-yourself field observations may be widely practiced, most IPM programs require a person in the field to collect relevant information on the pest populations in question and related parameters concerning the crop or host habitat. Integrated Crop Management (ICM) is an expansion of the IPM concept. Such a programme expands the principles of IPM to include other agricultural decision-making tasks such as fertilizer and soil water management. An ICM program would include an IPM component to deal with pest management decisions plus address remaining issues applicable to the total crop production process. The overall concept of general pest management programme relates to prevention, suppression and eradication of pest from the field. Preventive action may include application of any method to prevent or reduce the probable occurrence of a pest attack and this may be either chemical or physical or any appropriate method. Application of soil insecticides while planting is to prevent possible damage from pests to the plants under establishment.

Cleaning and disinfection of grainages and grainage appliances is to eliminate sources of infestation and prevent infestation from within the environment. Control includes any action taken to suppress or eliminate the pest from the environment. In practice, it is a rare situation where a pest is eliminated from the environment, but often a pest is reduced to a level at which it can no longer cause damage above economic threshold level. Application of chemicals is often associated with suppression, but non-chemical methods are also employed to considerable extant. Heat treatment or sun drying is the most common practice employed to reduce insect infestations. When a new pest is noticed, it becomes the primary responsibility of every one to make all out efforts to eliminate it. When a serious pest of foreign origin is detected in a commodity, fumigation is resorted to eliminate it from the infested stock. Similar is the case with any pest of public health importance and if detected in a hospital or food establishment, all out efforts are required to eradicate the pest.

BIBLIOGRAPHY

Abenes, H. L. P. and Z. R. Khan (1990) Attractiveness of light colour to selected predators of rice pests. *Int. Rice Res. Newl.* 15, 24 – 25.

Ahamed, A., Chandrakala, C. A., Maribashetty, M. V., Gururaj, V.G. and Raghuraman, R. (2000) Influence of feeding mealybug infested mulberry leaves on food and water utilization and cocoon yield in a bivoltine race of silkworm *Bombyx mori*. Journal of *Experimental Zoology,* 3, 49-54.

Ahmad, S. and Stoll, G. (1996) Biopesticides, In: Biotechnology Building on Farmers Knowledge. eds. Loske Bunders, Berts Haves Kort and Win Hiemsha, MacMillan Education Ltd., London, pp. 52-79.

Ahmed.S.(Ed.) (1993) Neem *(Azadiractin indica)* for pest control and rural development in Asia and the Pacific. Special Session on Neem from the 17th Pacific Science Congress.1991. In Press.

Aisagbonhi, C. I. (1989) A survey of the destructive effect of *Macrotermes bellicosus* Smeathman (Isoptera: Termitinae - Macrotermitinae) on coconut seednuts at NIFOR, Benin, Nigeria. *Tropical Pest Mgt.* 35, 380-381.

Alam, S., Islam, M.N., Alam, M.Z. and Islam, M.S. (1998) Effectiveness of three insecticides for the control of the spiralling whitefly, Aleurodicus dispersus Russell (Homoptera: Aleyrodidae) of guava. *Bangladesh Journal of Entomology,* 8, 53-58.

Allard, R. W. (1960) *Principles of Plant Breeding*. John Wiley &Sons, New York and London.

Alphen, J., J. M. Van. and L. E. M. Vet (1986) An evolutionary approach to host finding and selection; in *Insect parasitoids*. Waage, J. K. and D. J. Greathead (eds.), pp. 23-61, Academic press, London.

Ananthakrishnan, T. N. (1981) Recent advances in entomology in India. S. Viswanathan publishers, Madras, India.

Andrewartha, H. A. (1961) Introduction to the study of animal populations. University of Chicago Press, Chicago, USA.

Andrewwartha, H.G. and L.C. Birch, (1954) The Distribution and Abundance of Animals. University of Chicago Press, Chicago, IL., USA

Angus, T. A. (1971). *Bacillus thuringiensis* as a microbial insecticide. In Naturally Occurring Insecticides. Edited by M. Jacobson and D. G. Crosby. Pages 463-497. Marcel Dekker, New York.

Anonymous (1941) The Destructive Insect Pests Act, 1914. Gazette of India

Anonymous (1976) FAO Agricultural Services Bulletin, Manual of Sericulture, FAO of the United Nations, Rome.

Anonymous (1982). History and present status of pesticide usage in Japan. *Japan Pestic. Information* 40, 53-65.

Anonymous (1989) Plants, fruits and Seeds (Regulation of Import into India) Order1989. Gazette of India No. 706, Part II- Section 3-Sub-Section (ii).

Anonymous (1996) Improved Tropical Sericultural Technology suitable for hot humid regions of Tamil Nadu – Etiology and Management of Tukra disease associated with mealybug, *Maconellicoccus hirsutus* in Mulberry plantation. Department of Entomology, Annamalai University, Tamil Nadu.

Arthur, A. P. (1962) Influence of host tree on abundance of *Itoplectis conquisitor* (say.) (Hymenoptera : Ichneumonidae), a polyphagous parasite of the European pine shoot moth *Rhyacionia buoliana* (Schiff.) (Lepidoptera: Olethrentidae). *Can. Ent.* 94, 337-347.

Ascher K.R.S.(1993) Non conventional insecticidal effects of pesticides available from the neem tree. *Azadiracta indica* Archs Insect Biochem Physiol. 22,433-449.

Atwal, A. S. and S. S. Bains (1974) Applied animal ecology. Kalyani Publishers, Ludhiana, India.

Babu, C. R. and C. M. S. Chauhan (1994) Enhancement of tropical tasar silk production insect herbivory interaction. Department of Science and Technology, project report by Department of Botany, University of Delhi, Delhi, India.

Bandopadhyay, U. K., Santhakumar, M. V., Das, K. K. and Saratchandra, B. (2002) Determination of economic threshold level of whitefly *Dialeuropora decempuncta* (Quaintance and Baker) in mulberry, *Morus alba*. *Int. Indust. J. Entomol.* 4(2) : 133 – 136.

Bandyopadhyay, U.K. and Santhakumar, M.V. (2000) Efficacy of some insecticide alone and to combination with neem-oil against the whitefly, *Dialeuropora decempuncta* infesting mulberry. *Journal of Entomological Research,* 24, 325–329.

Bandyopadhyay, U.K., Das, N.K. and Saratchandra, B. (1999) Studies on the use of carbaryl, monocrotophos and quinalphos for control of whitefly *Dialeuropora decempuncta* (Quaintance and Baker) (Hymnoptera : Aleurodidae) on mulberry Morus alba L. *Journal of Advanced Zoology* 20; 100-102.

Bandyopadhyay, U.K., Riana, S.K., Raina, Chakraborty, N., Santhakumar, M.V., Sen, S.K. and Saratchandra, B. (1999) New record of a homopteran pest on mulberry (*Morus alba*). *Sericologia*, 39, 319–321.

Bandyopadhyay, U.K., Santakumar, M.V., Das, K.K. and Saratchandra, B. (2001) Yield loss in mulberry loss in mulberry due to sucking pest whitefly, *Dialeuorpora decempuncta* Quintance and Baker (Homoptera: Aleyroididae). *International Journal of Industrial Entomology*, 2, 75-78.

Banerji, R., Misra, G. and Nigam, S.K. (1985) Role of indigenous plant material in pest control, *Pesticides* 3: 17-23.

Barak, A. and W. E. Burkholder (1976) Trapping studies with dermestid sex pheromones. *Environ. Entomol.* 5, 111-114.

Baskaran, V. and Narayanasamy, P. (1995a) Traditional Pest Control, Caterpillar Publications, Mariyappa Nagar, Tamilnadu, India p.91.

Baskaran, V. and Narayanasamy, P. (1995b) Traditional practices contain store grain pests, In : Proc.National Seminar on Organic Farming, Agricultural College and Research Institute, Madurai, p. 131.

Batra, S. W. T. (1982). Biological control in agroecosystems. Science 215, 134-139.

Beeden, P. (1972) The pegboard an aid to cotton pest scouting. *PANS* 18, 43-45.

Beevi, N. D., Janarthanan, R. and Natarajan, K. (1992) Efficacy of some insecticides against eggs of *Maconellicoccus hirsutus* Green on mulberry *Morus alba* L., *J. Insect Sc.* 5, 114.

Beevi, S.P., Lyla, K.R. and Vidya, P. (1999) Report of *Encarsia* (Hymenoptera: Aphelinidae) on spiralling whitefly *Aleurodicus dispersus* Russell (Homoptera: Aleyrodidae).Insect Environment 2,35-36.

Beine, B. P. (1984) Avoidable obstacle to colonization in classical biological control of insects. *Can. J. Zool.* 63, 743-747..

Benson, J. F. (1973) Intra specific competition in the population dynamics of *Bracon hebetor* Say (Hymenoptera : Braconidae). *J. Anim. Ecol.* 42, 105-124.

Berlinger, M.J. (1986) Host Plant Resistance to *Bemisia tabaci.* Agriculture, *Ecosystems and Environment*, 17, 69-82.

Bhattacharya, A., Y. D. Mishra, A. H. Naqvi and A. K. Sen (1995) Attraction of some insects associated with lac towards various coloured lights. *J. Insect Sci.* 8, 205 - 206.

Bidmon,H.J., Kauser,G., Mobus,P., Koolman,J. (1987) Effect of azadiratin on blowfly larvae and pupae. *See Ref. 80. pp.253-71.*

Borden, J. H. (1985) Aggregation pheromones; in *Comprehensive insect physiology, biochemistry and pharmacology.* Kerkut, G. A. and L. I. Gilbert (eds.), pp. 257 – 285, Pergamon press, Oxford.

Bosque, C and J. E. Rabinovich (1979) Population dynamics of *Telenomus fariai* (Hymenoptera : Scelinoidae) a parasite of chagas disease vectors. VII. Oviposition behaviour and host discrimination. *Can. Entomol.* 111, 171-180.

Botha, J., Marc, P. and Hardic, D. (2004) Silver leaf whitefly *Bemisia tabaci* (Biotype B) with reference to related whitefly in western Australia. Farmnote No. 35, Department of Agriculture Western Australia, Perth.

BottrelL, D. G. and Adkisson, P. L. (1977). Cotton insect pest management. *Ann. Rev. Ent.* 22, 451-481.

Bowers, W. S. (1981). How anti-juvenile hormones work. *Amer. Zool.* 21,737-742.

Bragg, D. (1974) Ecological and behavioural studies of *Phaeogenes cynarae*, ecology, host specificity, search and oviposition and avoidance of super-parasitism. *Ann. Ent. Soc. Amer.* 67, 931-936.

Braman, S. K. and K. V. Yeargan (1989) Reproductive strategy of *Trissolcus euschisti* (Hymenoptera : Scelionidae) under conditions of partially used host resources. *Ann. Entomol. Soc. Amer.* 82, 172-176.

Braunholtz, J. T. (1998). Crop protection: the role of the chemical industry in an uncertain future. *Phil. Trans. Roy. Soc. London.* 295, 19-34.

Brodbeck, B. and Strong, D. (1987) Amino acid nutrition of herbivorous insects and stress to host plants. In : Barbosa, P. and Schulltz, J. C. (eds) Insect Out breaks. Academic Press, San Diego and London, pp. 347 – 363.

Burgess, H. D. (ed.) (1981) *Microbial Control of Pests and Plant Diseases*, 1970 – 1980. Academic Press, London.

Butterworth,J.H., Morgan.E.D., Percy.G.R., 1972. The structure of azadirahtin, the functional groups. *J.Chem.Soc.* Perkin Trans.1:2445-50.

Caltagirone, L. E. and Huffaker, C. B. (1980) Benefits and risks of using predators and parasites for controlling pests. *Ecol. Bull.* 31,103-109.

Camors, F. B. Jr. and T. L. Payne. (1972) Response of *Heydenia unica* (Hymenoptera : :Pteromalidae) to *Dendroctonus frontalis* (Coleoptera : Scolylidae) pheromones and a host tree terpene. *Ann. Ent. Soc. Amer.* 65, 31-33.

Carson, R. (1962). Silent Spring. Houghton-Miftlin, Boston.

Cate, J. R., and Hinkle, M. K. (1993) *Integrated Pest Management: The Path of a Paradigm*. Washington, D. C. : National Audubon Society.

Chakraborty, N., Santha Kumar, M.V. and Das, N. K. (1996) Field efficacy of exotic predator, *Cryptolaemus montrouzieri* (Mulsant) in controlling the mealybug, *Maconellicoccus hirsutus. Integrated Pest Management and Sustainable Agriculture* (ed. S. C. Goel), pp. 132-138. The Uttar Pradesh Zoological Society, Muzaffarnagar, India.

Cherian, M. C., (1917) Notes on three hemipterous from South India. *Ind. Jour. Ent.* 3, 115-119.

Chiarappa, L. (1974) Crop loss assessment methods. Food and Agricultural Organization (FAO) and Common wealth Agricultural Bureau International, Wallingford, Oxford.

Chien, C. C., L.Y. Chou, and S. C. Chiu (1984) Biology and natural enemies of *Hydylepta indicata* in Taiwan. *J. Agric. Res. China.* 33, 181-189.

Chien, C.C., Chou, L.Y. and Chang, S.C. (2000) Introduction, propagation and liberation of two parasitoids for the control of spiraling whitefly (Homoptera: Aleyrodidae) in Taiwan. *Chinese Journal of Entomology,* 20, 163-178.

Chu, Y. I (1975) Rearing Density of *Eocanthecona furcellata* with special consideration to its mass production (Aspinae : Pentatomidae). *Rostria* 24, 135-140.

Chua, T. H. (1979) A comparative study of the searching efficiencies of a parasitoid a hyperparasite. *Res. Popul. Ecol.* 20, 179-187.

Clausen, C. P (1978) Introduced parasites and predators of arthropod pests and weeds : A world review. Agriculture hand book. USDA., Wanshington.

Clunies-Ross, T. and Hildyard, N. (1992) *The Politics of Industrial Agriculture:* A report by the Ecologist. Earth scan, London.

Coaker, T. H. (1987) Cultural Methods: the crop. In : Burn, A. J., Coaker, T. H., and Jepson, P.C. (eds) *Integrated Pest Management.* Academic Press, London, pp. 69 –88.

Cochran, W. G. (1977) Sampling techniques. John Wiley & Sons, New York.

Cohen, S. and Berlinger, M.J. (1986) Transmission and cultural control of whitefly-borne viruses. Agriculture, Ecosystems and Environment, 17, 89-97.

Coppel, H. C. and J. M. Mertins (1977) Biological Insect Pest Suppression. Springer verlag, Berlin.

Coulson, J. R., Klaasen, W., Cook, R. J ., King, E. G., Chiang, H. C., Hagen, K. S. and Yendol, W. G. (1982) Notes on the biological control of pests in China, 1979. In : *Biological Control of Pests in China.* United States Department of Agriculture, Washington, D. C., USA.

Cremlyn, R. (1978). Pesticides. Preparation and Mode of Action. John Wiley & Sons, New York.

Cremlyn, R. (1979) Pesticides: Preparation and Mode of action. John Wiley and Sons, New York.

Croft, B. A and A. W. A. Brown (1975) Response of arthropod natural enemies to insecticides. *Ann. Rev. Ent.* 20, 285-335.

Croft, B. A. and A. W. A. Brown (1975) Response of arthropods natural enemies to insecticides. *Ann. Rev. Ent.* 20, 285-235.

Crowdy, S. H. (1977) 'Translocation' in systematic fungicides (Ed. Marsh,R. W.) 2nd Edn., Logman, London, p. 92.

Datta, R. K. and P. K. Mukherjee (1978) Life histroy of *Trcicholyga bombycis* (Diptera : Tachnidae) a parasite of *Bombyx mori* L. (Lepidoptera : Bombycidae). *Ann. Ent. Soc. Am.* 71, 767-770.

David, C. T. and M. C. Birch, (1989) Pheromones and insect behaviour; in *Insect pheromones in plant protection.* Justum, A. R. and R. F.S Gordon (eds.), pp. 17 - 35, John Wiley & Sons Ltd.

David, B.V. and Regu, K. (1995) Aleurodicus dispersus Russell (Aleyrodidae: Homoptera), a whitefly pest new to India. Pestology, 19, 5-7.

Davis, C. J. (1967) Progress in the biological control of the southern green stink bug *Nazena viridula* variety *smaregdula* (Fabricious) in Hawaii (Hymenoptera : Pentatomidae). *Mushi* 39, 9-16.

DeBach, P. (1974) Biological Control by Natural enemies. Cambridge University Press, Cambridge, U.K.

Debach, P. (1964a) Success trends and future possibilities. In : Biological Control of Insect Pests and Weeds. Edited by P. DeBach, Pages 673 – 713, Chapman and Hall, London.

DeBach, P. (1964b) The scope of biological control; In *Biological Control of Insects Pests and Weeds*. DeBach, P. (ed.), pp. 3-20, Chapman & Hall, London.

DeBach, P. (1964c) Success trends and future possibilities; in *Biological Control of Insects Pests and Weeds*. DeBach, P. (ed.), pp. 673-713, Chapman & Hall, London.

DeBach, P. (1966) The competitive displacement and coexistence principles. Ann. Rev. Ent. 11, 183-212.

DeBach, P. (1971) The theoretical basis of importation of natural enemies. *Proc. XII Internat. Ent.* 2, 140-142.

DeBach, P. (1972) The use of imported natural enemies in pest management ecology. Proceeding Tall Timber Conference on Ecological Animal Control by Habitat Management 3(1971), 211-233.

DeBach, P. (1974) Biological Control of Natural Enemies. Cambridge University Press, London and New York.

DeBach, P. and C. B. Huffaker (1971) Experimental techniques for evaluation of the effectiveness of natural enemies; in *Biological Control*. Huffaker, C. B. (ed.), pp. 113-140, Plenum Press, New York and London.

DeBach, P. and K. S. Hagen (1964) Manipulation of entomophagous species; in *Biological control of Insect Pests and Weeds*. DeBach, P.(ed.), pp. 429-458, Chapman and Hall, London.

DeBach, P., and. Rosen, D. (1991) *Biological Control by Natural Enemies.* Cambridge, U. K. : Cambridge University Press.

Dent , D. R. (1992) Scientific programme management in collaborative research. In Haskell, P. T. (ed.) *Research Collaboration in European IPM Systems*. British Crop Protection Council, Farnham, UK, pp. 69-76.

Dent , D. R. (1995) Integrated Pest Management. Chapman & Hall, London.

Devaiah, M. C., R. Govindan and K. C. Narayanaswamy (1992) Life cycle of uzi fly *Exorista bombycis*; in *Recent Advances in Uzi Fly Research*. ChannaBasavanna, G. P., G. Veeranna and S. B. Dandin (eds.), pp. 1-12, Karnataka State Sericulture Development Institute, Bangalore, India.

Dhingra, R K. (2002) Corriander as a mix crop in mulberry nursery. *Indian Silk* 41,46.

Douressamy, S., Chandramohan, N., Sivaprakasam, N., Subramanian, A. and Sundra Babu, P.C. (1997) Management of spiralling whitefly. Indian Silk, 36, 15-16.

Doutt, R. L. and DeBach, P. (1964) Some biological control concepts and questions. In Biological Control of Insect Pests and Weeds. Edited by P. DeBach. Pages 118 – 142. Chapman and Hall, London.

Doutt, R. L. and Smith, R. F.(1971) The pesticides syndrome – diagnosis and suggested prophylaxis. In Biological Control. Edited by C.B. Huffaker. Pages 3 – 15, Plenum Press, New York and London.

Doutt, R. L. (1964) Biological characteristics of entomophagous aduls; in *Biological control of insect pests and weeds*. DeBach, P. (ed.), pp. 145-167. Reinhold, New York.

Doutt, R. L. and smith, R. L. (1971). The pesticide syndrome diagnosis and suggested prophylaxis. In Biological Control. Edited by C.B. Huffaker. Pages 3-15. Plenum Press, New York and London.

Drummond, R. O., George, J. E. and Kunz, S. E. (1998) *Control of Arthropod Pests of Livestock : A Review of Technology*. CRC Press Inc., Boca Raton, Florida.

Edwards, J. P and menn, J. J. (1980). The use ofjuvenoids in insect pest management. In Chemie der Pftanzenschutz und Schadlingsbekamp fungsmittel. Edited by R. Wegier. Vol. 6, pages 185-214. Springer, Berlin.

Elliott, M. (1977). Synthetic pyrethroids. In Synthetic Pyrethroids. Edited by M. Elliott. ACS Symposium Series No. 42. Pages 1-28.American Chemical Society, Washington, DC.

Ellis, B. R. (1993) Advances in Pest Control. *Alpine Garden Society Quarterly Bulletin*. 61, 98-102.

Elwell, H. and Maas, A. (1995) Natural Pest and Disease Control. Natural Farming Net work, Zimbabwe,pp. 128.

Elzen, G. W. (1983) Isolation, identification and bioassay of plant chemicals mediating searching behaviour by an insect parasitoid. Ph. D. dissertation. Texas A & M University, college station, Texas.

Elzen, G. W., H. J. Williams, S. B. Vinson (1984b) Role of diet in host selection of *Heliothis virescens* by parasitoid *Campoletis sonorensis* (Hymenoptera : Ichneumonidae). *J. Chem. Ecol.* 10, 1535-1541.

Elzen, G. W., H. J. Williams, S. B. Vinson (1984a) isolation and identification of cotton synomones mediating searching behaviour by parasitoid *Campoletis sonorensis*. *J. Chem. Ecol.* 10, 1251-1254.

Elzen, G. W., H. J. Williams, S. B. Vinson (1983) Response by the parasitoid *Campoletis sonorensis* (Hymenoptera : Ichneumonidae) to chemicals (Synomones) in plants implications for host location. *Environ. Ent.* 12, 1872-1876.

Etebari, K., Matindoost, L. and Singh, R.N. (2004) Decision tools for mulberry thrips Pseudodendrothrips mori (Niwa, 1908) management in sericultural regions: An overview. Entomologia Sinica, 11, 1–13.

Eto, M. (1974). Organophosphorus Pesticides: Organic Chemistry and Biological Activity. CRC Press, Cleveland, Ohio.

Fest, C. and schmidt, K. J. (1973). The Chemistry of Organophosphorus Pesticides. Springer, Berlin.

Fisher, R. A. (1930) The Genetical theory of natural selelction. Oxford. University Press, London.

Flint, M. L. and R. Van den Bosch (1981) Introduction to integrated pest management. Plenum press, New York, USA.

Gajendra Babu, B. and David, P.M.M. (1999) New host plant records and host range of the spiralling whitefly, Aleurodicus dispersus Russell (Hemiptera: Aleyrodidae). Madras Agricultural Journal, 86, 305-313.

Gallun, R. L., starks,K. J. and guthrie, W. D. (1975). Plant resistance to insects attacking cereals. Ann. Rev. Ent. 20, 337-357.

Gautam, R. D. (1998) Use of exotic Coccinellids for the management of hibiscus mealybug *Maconellicoccus hirsutus* (Green) in the Caribbean Region. *Proceedings of the First Seminar on the Hibiscus Mealybug*, Centeno, Trinidad and Tobago, April 12, 1998. Ministry of Agriculture, Land and Marine, Port of Spain (Trinidad and Tobago).

Geetha, B., Loganathan, M., Swamiappan, M. (1998) Record of spiraling whitefly Aleurodicus dispersus Russell in Tamil Nadu. Insect Environment, 4, 55.

Georghiou, G. P. (1982). The Occurrence of Resistance to Pesticides in Arthropods. An Index of Cases Reported through 1980. FAO Plant Production and Protection Series. FAD, United Nations, Rome.

Georghiou, G. P. and saito, T. (1982). Pest Resistance to Pesticides: Challenges and Prospects. Plenum Press, New York.

Gerling, D. (1986) Natural enemies of B. tabaci, biological characteristics and potential as biological control agents: a review. Agriculture, Ecosystems and Environment, 17, 99-110.

Ghose, S. K. (1972) Biology of the mealy bug, *Maconellicoccus hirsutus* (Green) (Pseudococcidae : Hemiptera). Indian Agric., 16(4), 323 – 332.

Glass, E. H. (1976). Pest management: principles and philosophy. In Integrated Pest Management. Edited by J. L. Apple and R. F. Smith. Pages 43-44. Plenum Press, New York.

Goolsby, J., Legaspi, J.C., and Legaspi, B.C. (1996) Quarantine evaluation of exotic parasitoids of the sweetpotato whitefly, Bemisia tabaci (Gennadius). Southwestern Entomologist, 21, 13-21.

Goolsby, J.A. and Ciomperlik, M.A. (1999) Development of parasitoid inoculated seedling translants for augmentative biological control of silver leaf whitefly (Homoptera: Aleyrodidae). Florida Entomologist, 82, 532-545.

Goolsby, J.A., Ciomperlik, M.A., Legaspi, B.C. Legaspi, J.C. and Wendel, L.E. (1998) Laboratory and field evaluation of exotic parasitoids of Bemisia tabaci (Biotype 'B') in the Lower Rio Grande Valley of Texas. Biological control, 12, 127-135.

Gope, B. (1981) A promising predator of Bihar hairy caterpillar and bunch caterpillar. *Two leaves and a bud* 28, 47-48.

Goring, C. A. I. (1977). The costs of commercializing pesticides. In Pesticide Management and Insecticide Resistance. Edited by D. L. Watson and A. W. A. Brown. Pages 1-33. Academic Press, New York. ter? Ent. News 56,85-88.

Gowda, D.K.S., Manjunath, D., Kumar Pradip, and Datta, R. K. (1996) *Spalgis epius* Westwood (Lepidoptera: Lycaenidae) - a potential predator of mulberry mealy bug, *Maconellicoccus hirsutus. Insect Environ.* 2, 87.

Gowda, R. R. and Devaiah, M. C. (1985) Beetle pests infesting sotored silk cocoons. Mysore J. Agric. Sci., 29(1): 20 –21.

Greathead, D.J. and A. H. Greathead (1992) Biological control of insect pests by insect parasitoids and predators: the BIOCAT database. *Biocontrol News and Information* 13, 61-68.

Griffiths, D. (1977) Model for avoidance of superparasitism. *J. Anim. Ecol.* 46, 543-554.

Hagen K.S. (1962) Biology and ecology of predaceous coccinellidae. Annual Review of Entomology, 7, 289-319.

Hagen, J. S., Sawall, E. F. Jr. and Tassan, R. L. (1971) Use of food sprays to increase effectiveness of entomophagous insects. *Proceedings of Tall Timber Conference on Ecological Animal Control by Habitat Management.* Tallahassee, Florida, USA, pp. 59-81.

Hagen, K. S., Bombosch, S. and McMurthy, J.A. (1976) The biology and impact of predators. In Theory and Practices of Biological Control. Edited By C.B. Huffaker and P.S. Messenger. Pages 93 – 142. Academic press, New York, San Francisco and London.

Hall, W.J. (1921) The hibiscus mealybug in Egypt in 1925 with notes on the introduction of *Cryptolaemus montrouzieri*. Ministry of Agriculture, Egypt. Technical Science Series, Entomology Section Bulletin 70:1-15.

Hall , W. J. (1921) The hibiscus mealybug. Ministry of Agriculture, Egypt, Technical Science Series, Entomology Section Bulletin 17:1-28.

Hall, R. W., Ehler, L. E. and Bisabri-Ershadi, B. (1980) Rate of success in classical biological control of arthropods. *Bull. Entomol. Soc. Am.* 26, 111-114.

Hallett, R. H., A. C. Oehlschlager and J. H. Borden (1999) Pheromone trapping protocols for the Asian palm weevil, *Rhynchophorus ferrugineus* (Coleoptera : Curculionidae). *Internatl. J. Pest Mgt.* 45, 231-237.

Hamilton, W. D. (1967) Extraordinary sex ratio. *Science* 156, 477-488.

Hamon AB. (2000). Pictures of pink hibiscus mealybug. *FDACS-DPI. http:/ /www.doacs.state.fl.us/pi/enpp/ento/pink.htm* (7 October 2001).

Harpaz, I. (1973). Early entomology in the Middle East. In History of Entomology. Edited by R. F. Smith, T. E. Mittler and C. N. Smith. Pages 21-36. Annual Reviews, Palo Alto, California.

Hassell, M. P. (1971) Mutual interference between searching insect parasites. *J. Anim. Ecol.* 40, 473-486.

Hassell, M. P. (1978) The Dynamics of Arthropod Predator-Prey Systems. Princeton University Press, London.

Hassell, M. P. (1986) Parasitoids and population regulation; in *Insect Parasitoids.* Waage ; J. K. and D. J. Greathead (eds.), pp. 201-224, Academic press, London.

Hassell, M. P. and R. M. May (1974) Aggregation in predators and insect parasites and its effect on insect stability. *J. Anim. Ecol.* 43, 567-594.

Havens, J. N. (1972). Observations on the Hessian fly. Trans. Soc. Agron. New York I, 89-107.

Hedin, P. A. (1982). New concepts and trends in pesticide chemistry. *J. Agric. Food. Chem.* 30, 201-215. McKelvey, Jr. Pages 17-34. John Wiley & Sons, New York.

Heinz K.M. and Parrella, M. P. (1991) A shortcut with sticky traps. Grower Talks. 55 ; 40 – 42,45.

Henrick, C. A. (1977). The synthesis of insect sex pheromones.Tetrahedron 33,1845-1889.

Heong, K. L. (1981) The uses and management of pest surveillance data. *Malays. Agric. J.* 52, 65-89.

Heong, K. L. (1988) Computerized surveillance systems in pest management; in *Movement of pests and control strategies*. PLANTI Proceedings No. **3**, pp. 375-386, PLANTI, Serdang, Malaysia.

Heong, K.L., A. Manza, J. Catindig, S. Villareal and T. Jacobsen, (2007) Changes in pesticide use and arthropod biodiversity in the IRRI research farm. Outlooks Pest Manage., 18: 229-233.Direct Link

Hepard, M., E. R. Ferrer, P. E. Kenmore. and J. P. Sunmangil (1986) Sequential sampling: Plant hoppers in rice. *Crop Prot.* 5, 319-322.

Hoelmer, K.A., Osborne, L.S. and Yokomi, R.K. (1994) Interactions of the whitefly predator Delphastus pusillus (Coleoptera: Coccinellidae) with parasitized sweet potato whitefly (Homoptera: Aleyrodidae). Environmental Entomology, 23, 137–139.

Hoffmann, M.P. and Frodsham, A.C. (1993) *Natural Enemies of vegetable insect Pest*. Cooperative Extension, Cornell University, Ithaca, NY.

Hokkanen, H. M. T. (1985) Success in classical biological control – Critical review. *Pl. Sc.* 3, 35-72.

Horstadius, S. (1974) Lennaeus, Animals and Man. *Biological Journal oj Linnaean Society.* 6, 269-275.

Hoy, M. A. (1979) The potential for grenetic improvement of predators for pest management programs. In Genetics in Relation to Insect Management. Edited by M. A. Hoy and J.J. Mc Kelvey, Jr. Pages 106 – 115. The Rockefeller Foundation.

Hoy, M. A. and McKelvey, J. J. JR. (1979). Genetics in Relation to Insect Management. Rockefeller Foundation, New York.

Hoy, M.A. (1992) Biological control of arthropods: genetic engineering and environmental risks. Biological Control, 2, 166–170.

Huffaker, C. B. (1977) Augmentation of natural enemies in the People,s Republic of China. In *Biological Control by Augmentation of Natural Enemies*. Edoted by R.L. Ridgway and S.B. Vinson. Pages 329 – 339, Plenum press, New York and London.

Huffaker, C B. (1985) Biological control in Integrated Pest Management; An entomological perspective. In: Hoy, M. A. and Herzog, D. C (eds) *Biological Control in Agricultural IPM Systems*. Academic Press, Orlando and London, pp. 13 – 24.

Huffaker, C B. and Kennett, C. E. (1969) Some aspect of assessing efficacy of natural enemies. *Canadian Entomologist* 101, 425 – 447.

Huffaker, C. B. (1971) Biological control. Plenum Press, New York.

Huffaker, C. B. (1980) New technology of pest control. John Wiley, New York.

Huffaker, C. B. and Kennett, C. E. (1969) Some aspects of assessing efficiency of natural enemies. *Can. Ent.* 101, 425-447.

Huffaker, C. B., M. Van De Vrie and J. A. Mc Murthy (1970) Ecology of tetranichid mites and their natural enemies: a review. II. Tetranichid populations and their possible control by predators: an evaluation. *Hilgardia* 40, 391-458.

Huffaker, C.B., P. S. Messenger and P. DeBach (1971) The natural enemy component; in *Natural control and theory of biological control*. Huffaker, C.B. (ed.), pp. 16-67, Plenum Press, New York.

Hughes, R. D., L. T. Woolcock and M. A. Hughes (1992) Laboratory evaluation of parasitic hymenoptera used in attempts to biologically control aphid pests of crops in Australia. *Ent. Exp. Appl.* 63, 177-185.

Husain, M. A. and Lal, K. B. (1940) The bionomics of *Empoasca devastans* Distant on some varieties of cotton in the Punjab. *Ind. J. Entom.* 2, 123-136.

Isman, M. B. (2006) Botanical insecticides, deterrents, and repellents in modern agriculture and increasingly regulated world. *Annu. Rev. Entomol.* 51,45-66.

Jacobson, M. (1972). Insect sex pheromones, Academic press, New York.

Jacobson, M. (1975) *Insecticides from plants: A review of the literature,* U.S.D.A., Washington, D.C. pp. 138.

Jacobson,M. (1986) The neem tree: natural resistance par excellence. ACS Symp.Ser.296:220-32.

Jallali, S.K. and Singh, S.P. (1989) Biotic potential of three coccinellid predators on various diaspine hosts. Journal of Biological Control, 3, 20–23.

Jayaraj, S., Ananthakrishnan, T. N. and Veeresh, G. K. (1994) Biological Pest Control in India: Progress and Perspectives, Rajiv Gandhi Institute for Contemporary Studies Pub. No.2, pp. 101.

Johnson, N. F. and L. Masner (1985) Revision of genus *Psix kozlov* and Le (Hymenoptera : Scelionidae). *Systematic Entomology* 10, 33-58.

Jolly, M. S. (1967) Pests in tasar research. Bull. Central Silk Board 2, 1-16.

Jones,P.S., Ley,S.V., Morgan, E.D., Santafianos, D. (1989) The chemistry of the neem tree. See Ref.26. pp. 19-45.

Joronin, K. E.(1981) Ecological aspects of the behaviour of *Telenomia* (Hymenoptera : Scelionidae); in *Insect behaviour as a basis for developing control measures against pests of field crops and forests*. Prístavku, V. P. (ed.), pp. 36-41, Amerid, New Delhi, India.

Joshi.B.G., Ramaprasad.G, Sitaramaiah,S. (1982) Effect of neem seed kernel suspension on Telenomus remus, an egg parasite of Spodoptera litura. Phytoparasitica 10:61-63.

Kajita, H. Samudra, I.M., Naito, A. (1991) Discovery of the spiraling whitefly Aleurodicus dispersus Russell (Homoptera: Aleyrodidae) from Indonesia, with notes on its host plants and natural enemies. Applied Entomology and Zoology, 26, 397-400.

Karel, A. K., D. K. Garg and V. K. Baranwal (1993) Development of database management system and simulation modeling for rice IPM. Progress report, National Centre for Integrated Pest Management, Faridabad, India.

Karel, A. K., M. E. Quentin and D. L. Mathews (1983) Alternatives to the conventional use of chemical pesticides; in *Resource efficient farming methods*. PP. 104-110, Rodale Press, Emmaus, PA, USA.

Karlson, P. and A. Butinandt (1959) Pheromones (Eclohormones) in insects. *A. Rev. Ent.* 4, 9-58.

Kaur, R., Mir, M. R M., Khan, A. and Nazir, S. (2002) Intercropping of mulberry with saffron. *Indian Silk* 41, 5-6.

Kerrich, G. J. (1960) The state of our knowledge of the systematic of the hymenoptera parasitica with particular reference to the British fauna. *Trans. Soc. Brit. Ent.*, 14, 1-18.

Kerrich, G.J. (1960) The state of our knowledge of the systematic of the hymenoptera parasitica with particular reference to the British fauna. Transactions of the Society for British Entomology, 14, 1-18.

Khan, M.A. and Niight, M. (1989) The first record of Dermestes undulates Brahm (Coleptera : Dermestidae) feeding on the cocoons of silkworm, Bombyx mori L. Indian J. Seric. 28(2) : 277 -278.

King, P. E. (1961) A possible method of sex ratio determination in the parasitic hymenopteran *Nasonia vitripennis. Nature* 189, 330-331.

Kirk, A.A., Lacey, L.A. and Goolsby, J.A. (2001) Classical biological control of Bemisia and successful integration of management strategies in the United States. Virus-Insect-Plant Interactions (eds. K.F. Harris, O.P. Smith & J.E. Duffus), pp. 309-329. Academic Press, San Diego.

Klein, A. R. and Dunkel, F.V. (2003) New pest managegement frontiers: linking plant medicine to traditional knowledge. *American Entomologist.* 49, 7-73.

Knipling, E. F. (1979) The basic principles of insect population suppression and management. USDA, Washington, USA.

Knipling, E. F. and J. E. Gilmore (1971) Population density relationship between hymenopteran parasites and their aphid hosts-theoretical study. *USDA Tech. Bull.* 1428, 1-34.

Kogan, M and McGrath, D. (1993) Integrated Pest Management: present dilemmas and future challenges. In : *Anais 140 Congreso Brasileiro de Entomologia.* SEB, Piracicaba, SP, Brasil, Sociedade Entomologica do Brasil.

Kogan, M. and D. G. Herzog (1980) Sampling methods in soyabean entomology. Springer Verlag, New York.

Konlshi, M. and Ito, Y. (1973). Early entomology in East Asia. In History of Entomolagy. Edited by R. F. Smith, T. E. Mittler and C. N. Smith. Pages 1-20. Annual Reviews, Palo Alto, California.

Krishnaswamy, S., M. S. Jolly and R. K. Datta, (1964). A study on the fly pest infestation of the larvae and cocoons of *Bombyx mori. Indian J. Seric.* 3, 7-12.

Kuhr, R. J. and dorough, W. (1976). Carbamate Insecticides: Chemistry. Biochemistry and Toxicology. CRC Press, Cleveland, Ohio.

Kumar P., Ram Kishore, Noamani, M. K. R., Sengupta, K. (1992) Effect of feeding tukra affected mulberry leaves on silkworm rearing performance. *Indain J. Seric.* 31(1) : 27 –29.

Kumar, A., P. Kumar, B. D. Singh, K. Sengupta (1991) Parasites of uzi fly *Exorista sorbillans* Wiedemann (Diptera : Tachinidae). VI. Record of a new *Dirhinus* sp. on uzi fly and notes on its biology. *Sericologia* 31, 251-252.

Kumar, P., A. Kumar, M. K. R. Noamani and K. Sengupta (1989) Parasite of uzi fly *Exorista sorbillans* Wiedemann (Diptera : Tachinidae) a new record. *Curr. Sci.* 58, 821-822.

Kumar, P., M. S. Jolly, and V. V. Reddy (1986) Efficacy of nylon net enclosure in containing uzi fly (*Tricholyga bombycis* Beck.) infestation to silkworm, *Bombyx mori* L. *Indian J. Seric.* 25, 74-77.

Kumar, P., M. S. Jolly, P. P. Sharma, V. V. Reddy and M. V. Samson (1986) Ovicidal effect of organic acids against the uzi fly *Tricholyga bombycis* (Diptera : Tachinidae). *Sericologia* 26, 175-188.

Kumar, P., O. K. Ramadevi, B. D. Singh and M. S. Jolly (1989). A new record of pupal endoparasite, *Exoristobia philippinensis* Ashmead (Hymenoptera : Encyrtidae) of the uzi fly, a serious parasite of silkworm. *Curr. Sci.* 58, pp. 212.

Kumar, P., O.K. Ramdevi, K. Banerjee, M. K. R. Noamani and M. S. Jolly (1988) Record of a new pupal parasite *Spilomicrus karnatakensis* Sharma of *Tricholyga bombycis* Beck. *Sericologia* 28, 415 - 416.

Kumar, P., Ram Kishor and K. Sengupta(1990) Parasitoids of uzi fly *Exorista sorbillans* Wiedemann (Diptera : Tachinidae). XI Degree of seasonal parasitization of the puparia of uzi fly *Trichopria* sp. (Hymenoptera : Diapriidae). *Indian J. Seric.* 29, 188-193.

Kumar, R and K. G. Mukerji (1996) Integrated disease management-future perspectives; in. *Advances in Botany.* Mukerji, K. G., B. Mathur, B. P. Chamola and P. Chitralekha (eds.), pp. 335-347, APH Publishing Corp. New Delhi, India.

Kuno, E. (1991) Sampling and analysis of insect populations. *Ann. Rev. Entomol.* 36, 285-304.

Ladden, J. L. (1970) Physiological aspects of host tree favourability for the wood wasp, *Sinex noctilo. Proc. Ecol. Soc. Aust.* 3, 147-149.

Lambkin, T. (1998) Spiralling whitefly threat to Australia. Quarantine Bulletin, 8, 1-5 (Department of Primary Industry, Brisbane, Australia).

Lecomte, C. and E. Thibout (1984) Etude Olfactometrique delaction de diverses substances allelochemiquws vegetales dans la recherché de I hote par *Diadromus pulchellus* (Hymenoptera : Inchneumonidae) *Ent. Exp. Appl* 35, 295-303.

Levin, B. R. (1979). Problems and promise in genetic engineering in its potential applications to insect management. In Genetics in Relation to Insect Management. Edited by M. A. Hoy and J. J. McKelvey, Jr.Pages 170-175. Rockefeller Foundation, New York.

Lewis, W. J. and D. A. Nordlund (1980) Employment of parasitoids and predators for fall armyworm control. *Fla. Entomol.* 63, 433 – 438.

Lewis, W. J., R.L. Jones, D. A. Nordlund and A. N. Saprks (1982) Kairomones and their use for the management of entomophagous insects. I. Evaluation for increasing rate of parasitization by *Trichogramma* sp. in field. *J. Chem. Ecol.* 1, 343-347.

Li, L.Y. (1994) Worldwide use of Trichogramma for biological control on different crops: a survey. Biological control with egg parasitoids (eds. E. Wajnberg & S.A. Hassan), pp. 37-54, CAB International, Wallingford.

Lisansky, S. G. (1990) Green growers guide. The world wide directory of agro- biochemicals, CPL Press, New York, U.K.

Lokkers, C. (1986) The distribution of weevers ant, *Oecophylla smaragdina* Fabricous (Hymenoptera : Formicidae) in Northern Australia. *Aust. J. Zool.*, 34: 683 – 687.

Madden, J. L. (1970) Physiological aspects of host tree favourability for the wood wasp, *Sirex noctilo. Proc. Ecol. Soc. Aust* .3, 147-149.

Mani, M. (1988) Distribution, bio-ecology and management of grape mealybug, *Maconellicoccus hirsutus* (Green), with special reference to its natural enemies. *Haryana Agric. Univ.* 14, 269-270.

Mani, M. (1988) Bioecology and management of grapevine mealy bug. Indian Inst. Hort. Res. Tech. Bull., 5 : 32.

Mani, M. (1989) A review of the pink mealybug — *Maconellicoccus hirsutus* (Green). *Insect Sci. and its Appl.* 10, 157-167.

Mani, M. (1993) Studies on mealybugs and their natural enemies in ber orchards. *J. Biol. Cont.* 7, 75-80.

Mani, M. and Krishnamoorthy, A. (1997) Discovery of Australian ladybird beetle (Cryptolaemus montrouzieri) on spiralling whitefly (Aleurodicus dispersus) in India. Insect Environment, 3, 5-6.

Mani, M. and Krishnamoorthy, A. (1999) Natural enemies and host plants of spiralling whitefly Aleurodicus dispersus Russell (Homoptera: Aleyrodidae) in Bangalore, Karnataka. Entomon, 24, 75-80.

Mani, M. and Krishnamoorthy, A. (2000) Population dynamics of spiraling whitefly, Aleurodicus dispersus Russell (Aleyrodidae, Homoptera) and its natural enemies on guava in India. Entomon, 25, 29-34.

Mani, M., Dinesh, M.S., Krishnamoorthy, A. (2000) Presence of Encarsia spp. on spiralling whitefly Aleurodicus dispersus (Russell) in peninsular India. Insect Environment, 6, 100.

Manjunath, D., Kumar Pradeep, Ram Kishore, Noamani, M. K., Satya Prasad, K. Narayanswamy, K.C., Datta, R. K. (1993) Towards understanding tukra and its management. Indian Silk 32(3) : 6 –9.

Manjunath, T. M. (1986) Role of commercial insectaries in promoting biological control; in. *Biological control in tropics.* Hussain, M.Y. and A. G. Ibrahim (eds.), pp. 219-294, University Press,. Malaysia.

Martin, J.H. (1990) The whitefly species Aleurodicus dispersus and its rapid extension of range across the Pacific and South East Asia. MAPPS Newsletter, 14, 36-37.

Mathews, G. A. (1984) Pest management, Longman press, New York & London.

Mathys, G. and Baker, E. A. (1980) An appraisal of the effectiveness of quarantines. Annu. Rev. Phytopathol., 18: 85 –89.

Matteson, P. C., Gallagher, K. D. and Kenmore P. E. (1994) Extension of Integrated Pest Management for plant hoppers in Asian irrigated rice: empowering the user. In : Denno, R. F. and Perfect, T.J. (eds) *Plant hoppers their Ecology and Management.* Chapman & Hall, New York and London, pp. 656 – 685.

Maxwell, F. G. and Jennings, P. R. (1980) Breeding Plants Resistant to Insects. John Wiley & Sons, New York.

Maxwell-Lefroy, Harold (1909) Indian Insect Life. Today and Tomorrow Publishers

Mcewen, F. L. (1978). Food production -the challenge for pesticides. BioScience 28,773-777.

McEwen, P. K., Jervis, M. A. and Kidd, N. A. C. (1994) Use of sprayed L- tryptophan solution to concentrate numbers of the green lacewing *Chrysoperla carnea* in olive tree canopy. *Entomologia Experimentalis et Applicata* 70, 97 – 99.

Meesenger, P.S., Wilson, F. and Whitten, M.J. (1976) Variation, fitness and adaptability of natural enemies. In Terms and Practices of Biological Control. Edited by C.B. Huffaker and P.S Messenger. Pages 209 – 231. Academic press, New York, San Francisco and London.

Metcalf, R. L. and Metcalf, E. R (1992) *Plant Kairomones in Insect Ecology and Control.* Chapman and Hall, New York, USA.

Metcalf, R. L. and Metcalf, R A. (1993) *Destructive and Useful insects,* 5th edn. McGraw-Hill, New York, USA.

Metcalf, R. L. (1982) Insecticides in pest management. *Introduction to Insect Pest Management,* Pp. 217-277 (eds R. L. Metcalf and W. H. Luckmann). New York: Wiley & Sons.

Metcalf, R. L. and luckmann, W. H. (1975). Introduction to Insect Pest Management. John Wiley & Sons, New York.

Mishra, C. S. (1919) Tukra disease of mulberry. Proc. 3rd Ent. Mtg. Pusa pp. 610 618.

Mishra.P.K. (1983) Studies on the biological efficacy of extracts of different parts of neem Azadiracta indica A Juss. M.Sc.thesis, Post Graduate school, Indian Agricultural Research Institute, New Delhi, India.

Monteith, L. G. (1955) Host preference of *Drinobohemica* Mesn. (Diptera : Tachinidae) with a particular references to olfactory response. *Can. Ent.* 87, 509-530.

Monteith, L. G. (1964) Influence of the health of the food plants of the host on host finding by tachinid parasites. *Can. Ent.* 96, 1477-1482.

Morse, S. and Buhler, W. (1997) IPM in developing countries: the danger of an ideal. *Integrated Pest Management Reviews.* 2, 175-186.

Mound, L.A. and Halsey, S.H. (1978) White fly of the World. John Wiley and Sons, New York. p. 340.

Muhamad, R. and G.T. Chung, (1993) The relationship between population fluctuations of Helopeltis theivora Waterhouse, availability of cocoa pods and rainfall pattern. *Pertanika J. Trop. Agric. Sci.*, 16: 81-86.

Mukerji, K.G. and Garg, K.L. (1988a) Biocontrol of Plant Diseases, (eds.) Vol. I, CRC Press Inc., Florida,USA.

Mukerji, K.G. and Garg, K.L. (1988b) Biocontrol of Plant Diseases, (eds.) Vol. II, CRC Press Inc., Florida,USA.

Muller, T. F. (1983) The effect of plants as the host relations of specialist parasitoid of *Heliothis* larvae. *Ent. Exp. App.* 134, 78-84.

Muralikrishna, M. (1999) Bio-ecology, host range and management of spiralling whitefly, Aleurodicus dispersus Russell (Homoptera: Aleyrodidae). MSc. (Ag.) Thesis, University of Agricultural Sciences, Bangalore, India, p. 67

Napompeth, B. (1987) Biological control and integrated pest control in the tropics - an overview. *Acad. Naz. Sci.* 50, 415-428.

Narayanasamy, P. (1994a) Development and use of mycoinsecticide from indigenous fungal pathogen against the brown planthopper (Nilaparvata lugens (Stal) problem in rice, Final Project Report, Ministry of Environment & Forestry, Govt of India, pp.95.

Narayanasamy, P. (1994b) Studies on the utility of lignite fly ash as an insecticide and an adjuvant in insecticide formulations, Final Project Report Tamil Nadu State Council for Science and Technology, Govt. of Tamilnadu, Chennai, pp. 144.

Narayanasamy, P. (1995) Mycoinsecticide: A novel biopesticide in Indian scenario, Biotech. Develop. Rev. pp. 15-20.

Narayanasamy, P. (1997) Development and use of fly ash pesticides, In : Fly Ash in Agriculture, Proc.Natl. Sem. on use of Lignite Fly Ash in Agriculture, ed.

Narayanaswamy, K.C., Ramegowda, T., Raghuraman, R. and Manjunath, M.S. (1999) Biochemical changes in spiralling whitefly (Aleurodicus dispersus Russell) infested mulberry leaf and their influence on some economic parameters of silkworm (Bombyx mori L.). Entomon, 24, 215-220.

Navasero, R. C. and E. R. Oatman. (1989) Life history, immature morphology and adult behaviour of *Telenomus solitus* (Hymenoptera : Scelionidae). *Entomophage* 34, 165-177.

Nayar, K.K., T.N. Ananthakrishnan and B.V David (1976) General and Applied Entomology Tata McGraw Hill Publishing Company Limited New Delhi

Nettles, W. C. ,Jr. (1979) *Eucelatoria* sp. Females factors influencing response to cotton and okra plants. *Environ. Ent.* 8, 619 - 623.

Neuenschwander, P. (1994) Spiralling whitefly Aleurodicus dispersus, a recent invader and new cassava pest in Africa. African Crop Science Journal, 2, 419-421.

Nicholson, A. J. and Bailey, V.A. (1935) The balance of Animal populations. Part i. Proc. Zoological Sco. London 1935, 551-598.

Nishida, T. (1956) An experimental study of the ovipositional behaviour of *Opius fletcheri* Silvestri (Hymenoptera : Braconidae), a parasite of the melon fly. *Proc .Hawaii. Ent. Soc.* 16, 126-134.

Nordlund, D. A. (1984) Biological control with entomophagous insects. *J. Ga. Entomol. Soc. Second Suppl.* 19, 14 –27.

Nordlund, D. A. and C. E. Sauls (1981) Kairomones and their use for the management of entomophagous insects. XI Effect of host plants on Kairomonal activity of frass of *Heliothis zea* larvae for the parasitoid *Microplitis croceipes. J. Chem. Ecol* .7, 1057-1061.

Nordlund, D. A. (1981) Semi chemicals: A review of the terminology; in *Semio chemicals: Their role in pest control.* Donald, D.A., R. L. Jones, and W. J. Lewis (eds.), pp. 13-28, Wiley–Interscience Publication, New York.

Nordlund, D.A. (1984) Biological control with entomophagous insects. *Journal of the Georgia Entomological Society*, 19, 14–27.

Nordlund, D.A. and W.J. Lewis (1976) Terminology of chemical releasing stimuli in intraspecific and interspecific interactions. *J. Chem. Ecol.* 2, 211-220.

Norton, G. A. (1980) The role of forecasting in crop protection decision-making : An economic view point. *EPAO Bulletin.* 10, 269-274.

Norton, G. A. and Mumford, J. D. (1983) Decision making in pest control. In : Coaker, T.H. (eds) *Advances in Applied Biology*, Vol. VIII. Academic Press, London, pp. 87-119.

Norton, G. A., J. Holt and J. D. Mumford (1993) Introduction to pest models; in *Decision Tools for Pest Management*. G. A. Norton, and J. D.Mumford (eds.), pp 89-99, CAB International, Wallingford, U.K.

Norton, G.A. and Mumford, J. D. (eds) (1993) *Decision Tools for Pest Management*. CAB International, Wallingford, UK.

Okuda, M. S. and K. V. Yeargon (1988a) Habitat partitioning by *Telenomus podisi* and *Trissolcus euschisti* (Hymenoptera : Scelionidae) between herbaceous and woody host plants. *Environmental Entomol.* 17, 795-798.

Okuda, M. S., and K. V. Yeargon. (1988b) Intra and Interspecific host discrimination in *Telenomus podisi* and *Trissolcus euschisti* (Hymenoptera : Scelionidae). *Ann. Entomol. Soc. Amer.* 81, 1017-1020.

Palaniswami, M.S., Pillai, K.S. Nair, R.R. and Mohandas, C. (1995) A new cassava pest in India. Cassava Newsletter, 19, 6-7.

Panda, N. and Khush, G. S. (1995) *Host Plant Resistance to Insects*. CAB International, Wallingford, UK, IRRI, Manila, Philippines.

Pant, C. P. (1960) Some aspects of bionomics of *Erias* sp. at Kanpur. *Agra Univ. J. Res.* 9, 31- 40.

Parnell, F. R. (1935) Origin and development of U4 cotton. Empire cotton growing. *Crop Review.* 12, 177-182.

Parrella, M.P., Paine, T.D., Bethke, J.A., Robb, K.L. and Hall, J. (1991) Biological control of sweet potato whitefly (Homoptera: Aleyrodidae) on commercially grown poinsettia stock production. Environmental Entomology, 20, 713–719.

Parthasarathy, R. and Narayanasamy, P. (1997) Record of Aspergillus terreus Thorn, on rice grasshopper Hieroglyphus banian (R) in India, Int. Rice Res. News 22(3): 33-34.

Patil, G. M. and R. Govindan (1984) Biology of uzi fly *Exorista sorbillans* (Wiedemann) (Diptera : Tachinidae) on eri silkworm *Samia cynthia ricini* Boisduval. *Indian J. Seric.* 23, 32-37.

Paulson, G.S. and Kumashiro, B.R. (1985) Hawaiian Aleyrodidae. Proceedings of the Hawaiian Entomological Society, 25, 103-129.

Pawar, C. S., S. Sithanantham, V. S. Bhatnagar, C. P. Srivastava, and W. Reed (1988) The development of sex pheromone trapping of

Heliothis armigera at ICRISAT, India. *Trop. Pest Manage.* 34, 39 – 43.

Perkins, J. H. (1980). The quest for innovation in Agricultural Entomology, 1945-1978. In Pest Control: Cultural and Environmental Aspects. Edited byD. Pimentel andJ. H. Perkins. American Association for the Advancement of Science Selected Symposium 43. Pages 23-80. Westview, Boulder, Colorado.

Perkins, J. H. (1982) *Insect , Experts and Insecticide Crisis: the Quest for New Pest Management Strategies.* Plenum, New York, USA.

Perry, J. N. (1989) Population variation in entomology. *Sampling Entomologist* 108, 184-98.

Pimentel, D. (1963) Introducing parasites and predators to control native pests. *Canadian Entomologists.* 95, 785-792.

Pimentel, D. (1978) World Food, Pest Losses, and the Environment. American Association for the Advancement of Science Selected Symposium No. 13. Westview, Boulder, Colorado.

Polaszek, A., Abd-Rabou, S. and Huang, J. (1999) The Egyptian species of Encarsia (Hymenoptera: Aphelinidae) – a preliminary review. Zoologische Mededeelingen, Leiden, 73, 131-163.

Prathapan, K.D. (1996) Outbreak of the spiralling whitefly Aleurodicus dispersus Russell (Aleyrodidae: Homoptera) in Kerala. Insect Environment, 2, 36-38.

Price J. D. (1970). Applied biology as an evolutionary process. Ann. Appl. Bioi. 66,179-191.

Price, P. W., C. E. Bouton, P. Gross, B. A. McPherson, J. N. Thompson and A. E. Weis.(1980) Interactions among three tropic levels: influence of plants on interactions between insect herbivores and natural enemies. *Annu. Rev. Ecol. Syst.* 11, 41- 65.

Price, P. W. (1981) Semiochemicals in evolutionary time; in *Semiochemicals : their roles in pest control.* Nordluand, D.A., R. L. Jones, and W. J. Lewis (eds.), pp. 251-279, Wiley, New York.

Price.J. F., Schuster.D.J., and P.M.McClain (1990) Azadiractin from neem tree Azadiracta indica A.Juss) seeds for management of sweet potato whitefly (Bemista tabaci(Gennadius) on ornaments Proc.Fla.State Horti.Soc. 103,186-188.

Pruthi, H. S. (1970) Text book of agricultural entomology. Indian Council of Agricultural Research, New Delhi, India.

Puntener, W. (1981) Manual for field trials in Plant protection. Documenta, Ciba Geigy, Basle.

Putter, I., macconnelL, J. G., reiser, F. A., haidri, A. A., ristich, S. S. and dybas, R. A. (1981). Avermectins: novel insecticides, acaricides and nematicides from a soil microorganism. Experientia 37,963-964.

Rabb, R. L., R. E. Stinner and R. Van Den Bosch (1976) Conservation and augmentation of natural enemies; in *Theory and practices of biological control*. Huffaker, C. B. and P. S. Messenger (eds.), pp. 233-254, Academic Press, New York, San Francisco and London.

Rabb, R.L. , Stinner, R.E., and Van Den Bosch, R. (1976) Conservation and augmentation of natural enemies. In Theory and Practice of Biological Control edited by C.B. Huffaker and P. S . Messenger. Pages 233 – 254, Academic press, New York, San Francisco and London

Rabb, R.L., Stinner, R.E. and van den Bosch, R (1976) Conservation and augmentation of natural enemies. Theory and Practices of Biological Control (eds. C.B. Huffaker & P.S. Messenger), pp. 233-254. Academic Press, London.

Rai, P.S. (1978) *Canthecona furcellata* (Wolff) (Pentatomidae : Heteroptera): A predator of leaf feeding caterpillars of rice. *Curr. Sci.* 47, 556-557.

Rajapakse, C. N. K., N. E. Gunawardena, and K. F. G. Perera (1999) Pheromone baited trap for the management of red palm weevil, *Rhynchophorus ferrugineus* F. (Coleoptera: Curculionidae) population in coconut plantation. *COCOS* 13, 54-65.

Ramamurthy, T.V. (1995) Vedic agriculture (natural farming/ecological farming/organic farming processes)for the guidance of modern farming, Keynote papers and Extended Abstract of 2nd congress on Traditional Science & Technologies of India, Anna UniversIty, Chennai.

Raman, A. Singh, R. N. and Anna, M. N. (1997) Biology and karyology of the cecidogenous psylloid, *Trioza fletcheri minor* (Homoptera: psylloidea) and morphogenesis of their galls on the leaves of *Terminalia tomentosa* and *Terminalia arjuna* (combretaceae). *Insect Matsumurana* (Sappro, Japan) 53: 117-134

Ramani, S. (2000) Fortuitous introduction of an aphelinid parasitoid of the spiralling whitefly, Aleurodicus dispersus Russell (Homoptera: Aleyrodidae) into the Lakshadweep islands, with notes on host plants and other natural enemies. Journal of Biological Control, 14, 55–60.

Ramani, S., Poorani, J. and Bhumannavar, B.S. (2002) Spiralling whitefly, Aleurodicus dispersus, in India. Biocontrol News and Information, 23, 55–62.

Randhawa, M.S. (1980) A History of Agriculture in India, Indian Council of Agricultural Research, New Delhi, Vol.1, pp. 541.

Rani, P. U. (1992) Temperature induced effects on predation and growth of *Eocanthecona furcellata* (wolff.) (Pentatomidae : Heteroptera). *J. Biol. Control.* 6, 72-76.

Rani, P. U. and S. Wakamura (1993). Acceptance behaviour of predatory pentatomid *Eocanthecona furcellata* (wolff.) (Heteroptera : Pentatomidae) towards larvae of *Spodoptera litura* (Lepidoptera : Noctunidae). *Insect Sci.Applic.* 14, 141-147.

Ranjith, A.M., Rao, D.S. and Thomas, D.J. (1996) New host records of the mealy whitefly, Aleurodicus dispersus Russell in Kerala. Insect Environment, 2, 35-36.

Raychaudhury, S.P. (1964) Agriculture in Ancient India, Indian Council of Agricultural Research, New Delhi, pp. 167.

Read, D. P., P. D. Feeny and R. B. Root. (1970) Habitat selection by the aphid parasite *Diaeretiella rapae* (Hymenoptera : Braconidae) and hyperparasite *Charips brassicae* (Hymenoptera : Cynipidae). *Can. Ent.* 102, 1567-1578.

Redyy, D.N.R. and Kotikal, Y.K. (1988) Pest of mulberry and their management. Indian Silk, 10, 9–15.

Rembold,H., Forster,H., Czoppelt, Ch.Rao,J.P., Sieber,K.P. (1984) The azadiractins, a group of insect growth regulators from the neem tree. See Ref. 79, pp 153-62

Rembold.H., (1989) Isometric azadiractins and their mode of action. See Ref. 26, pp.47-67.

Rigway, R. L. and S. B. Vinson (1977) Biological control by augmentation of natural enemies. Plenum Press, New York, USA.

Rigway, R.L. and Vinson, S.B. (1977) Biological Control by Augmentation of Natural Enemies. Plenum Press, New York.

Roberts, D. A (1991) Using non-chemical methods to control diseases of plants. In : Handbook of Pest Management in Agriculture (2nd Edition) Vol. II (ed. David Pimental) CRC Press, USA pp. 13 –22.

Robinson, R. A. (1996) *Return to Resistance* : Breeding crops to Reduce Pesticides Dependence. Ottawa: AgAccess, Davis, California and International Development and Research Centre.

Rogers, D. J. and M. P. Hassell (1974) General models for insect parasite and predator searching behaviour interference. *J. Anim. Ecol.* 43, 239-253.

Rohwer, G. G. (1991) Regulatory Plant Pest Management. In : Handbook of Pest Management in Agriculture (2nd Edition) Vol. II (ed. David Pimental) CRC Press, USA pp. 285– 31022.

Rosen , D. and Huffaker, C.B. (1983) An over view of desired attributes of effective biological control agents, with particular emphasis on mites.In Biological Control of Pests and Mites. Edited by M. A. Hoy, G.L. Cunningham and L. Knutson pages 2 – 11. Div. Agric. Sci. Univ. Calif.. Publ.. 3304.

Rosen, D. (1978) The importance of cryptic species and specific identifications as related to biological control; in *Biosystematics in Agriculture*. Romberger, J. A. (ed.), pp. 23-35, Allanheld, Osmum & Co., Montclair, NJ.

Rosen, D. (1985) Biological Control. In : Kerkut, G.A. and L.I. Gilbert (eds) Biochemistry and Pharmacology, Pergamon Press, Oxford, New York, Frank pp.413 – 464. Vol 12.

Rosen, D. and C. B. Huffaker (1983) An overview of desired attributes of effective biological control agents with particular emphasis on mites; in *Biological Control of Pests by Mites*. Hoy, M. A., G. L. Cunningham and L. Knutsun (eds.), pp. 2-11, Div. Agric. Sci.. Univ. Calif. Publ. 3304.

Rosen, M. and DeBach, P. (1981) Citrus whitefly parasites established in California. California Agriculture, 35, 21–23.

Rosen, R. and Debach, P. (1979) Species of Aphytis of the world (Hymenoptera : Aphelinidae). Israel Universities Press and W. Junk, Jerusalem and The Hague

Ruesink, W. G. and M. Kogan (1982) The quantitative basis of pest management: sampling and measuring; in *Introduction to Insect Pest Management*. Metcalf, R.L and W. H. Luckmann(eds.), pp. 315-352, John Wiley & Sons, New York.

Ruesink, W. G. and M. Kogan. (1980) Introduction to sampling theory; in *Sampling methods for soyabean Entomology*. Kogan, M. and D. C. Herzog (eds.), pp. 587, Springer Verlag, New York.

Sahaf . K. A. and Munshi, N. A. *(2003) Trigoderma sp. (Coleoptera : Dermestidae), a new record of stored silk cocoon pest of mulberry silkworm Bombyx mori L.* Indian J. Seric. 42(2) : 169 – 170.

Sahaf, K. A and Chishti, M Z. and Khan, M. A. (1997) Morphological characteristics of *Tribolium freemani Hinton*(Coleptera : Tenebrionidae), a pest of stored silk cocoons. *J. Entomol. Res.* 21(2) : 165 – 168.

Samson, M. V. and O. K. Ramadevi (1985) A new record on the parasites of the uzi fly *Tricholyga bombycis* Beck. *Curr. Sci.*, 54, pp. 1144.

Santha Kumar, M.V., Chakraborty, N., Aswani Kumar, C., Bhattacharya, S.S. and Saha Kundu, A. K. (1995) New record of coccinellid predator on the pink mealybug, *Maconellicoccus hirsutus* (Green). *Sericologia* 35, 359-361.

Sauls, C. E., D. A. Norduland and W. J. Lewis (1979). Kairomones and their use for management of entomophagous insects. VIII Effect of diet on the kairomonal activity of frans from *Heliothis zea* (Boldie) larvae for *Mircoplitis croceipes* (Cresson). *J. Chem. Ecol.* 5, 363-369.

Saussure, H. de. (1892) *Hymenopteres*. In: Grandidier A. "Histoire physique naturelle et politique de Madagascar. 20." Paris. 590 pp.

Schluter.U. (1987) Effects of azadiractin on developing tissues of various insect larvae. See Ref. 80.pp 331-48.

Schmutterer, H. (1985) Which insect pests can be controlled by application of neem seed kernel extracts under field conditions. Z.Angew. Entonmol.100:468-75.

Schmutterer, H. (1990) Properties and potential of natural pesticides from the neem tree, Azadirachta indica. Annual Review of Entomology, 35, 271-297.

Schmutterer, H. (1988) Potential of azadiractin containing pesticides for integrated pest control in developing and industrialized countries. J.Insect Physiol.34:713-19.

Schmutterer, H., Ascher,K.R.S.ed. (1987) Natural Pesticides from the Neem Tree and Other Tropical Plants Proc.3rd Int.Neem Conf., Nairobi, 1986. Eschborn.GTZ.703 pp.

Schmutterer, H., Ascher.K.R.S.eds. (1984) Natural pesticides from the Neem Tree and Other Tropical Plants. Proc. 2nd Int.Neem Conf. Rauischolzhausen, 1983. Eschborn.GTZ.587 pp.

Schultz, J. C. (1983) Impact of variable plant defense chemistry on susceptibility of insects to natural enemies; in *Plant Resistance to Insects*. Hedin, P.A. (ed), pp. 37-53. American Chemical Society Symposium Series 208, American chemical society, Washington.

Schwalbe, C. P. (1993) Biologically based regulatory pest management. *Pest Management: Biologically Based Technologies*, (eds. R. D. Lumsden and J. L. Vaughn) pp. 404-409, American Chemical Society. Washington, D.C.

Sen, S. K., M.S. Jolly. and T. R Jammy (1971) Biology and life cycle of *Canthecona furcellata* Wolff. (Hemiptera : Pentatomidae) predator of tasar silkworm *Antheraea mylitta* Drury. *Indian J. Seric.* 10, 53-56.

Shahjahan, M. (1974) Erigeron flowers as food and attractive odour source for *Peristenus pseudopallipes* a braconid parasitoid of the tarnished plant bug. *Environ. Ent.* 3, 69-72.

Shepard, H. H. (1951) The Chemistry and Action of Insecticides. McGraw-Hill, New York.

Sheppard, C. A. and Smith, E. H. (1997) Entomologcial heritage. *American Entomologist* 142-146.

Shorey, H. H., L. K. Gaston, and C. A. Saerio (1967) Sex pheromones of Noctuid moths. XIV. Feasibility of behavioural control by disrupting pheromone communication in cabbage loopers. *J. Econ. Entomol.* 60, 1541-1545.

Shumakov, E. M., Gusev, G. V. and Fedorinchik, M. S. (1974). Biological Agents for Plant Protection. Kolos, Moscow. Translated from the Russian by Ruth L. Busbey.

Singh ,R. N. and B. Saratchandra (2002) An integrated approach in the pest management in sericulture. *Int. J. Indust. Entomol.* 5, 141 – 150.

Singh R. N. and K. Thangavelu (1999) Response and competition of the parasitoid *Trichomalopsis apanteloctena* Crawford at different host densities of uzi fly pupae. *Environ. and Ecology* 17, *45-48.*

Singh, R. N. and K. Thangavelu. (1995). First record of *Trichomalopsis apanteloctena* on the uzi fly to its parasitoid. *Indian J. Seric.* 34, 164-165.

Singh, R. N. (1998) Influence of male parasitoid on the fecundity, longevity and sex ratio of *Trichomalopsis apanteloctena* Crawford parasitising puparia of *Blepharipa zebina* (Walker). *J. Biol. Control* 12, 51-54.

Singh, R. N. and K. Thangavelu (1994) Pest management practices in tasar culture. *J. Insect Sci.* 7, 121 – 124.

Singh, R. N. and K. Thangavelu (1997) Effect of age of host and parasitoid on the parasitization of uzi fly *Blepharipa zebina* (Walker). *J. Insect Sci.* 10, *63-64.*

Singh, R. N. and S.S. Sinha (1995) The influence of *Eocanthecona furcellata* Wolff (Heteroptera : Pentatomidae) egg age on the progeny production of its parasitoid *Psix striaticeps*. *J. Biol. Control* 9, 99-101.

Singh, R. N. and K. Thangavelu (1991) Integrated pest management in tasar culuture. *Ann. Entomol.* 9, 59-65.

Singh, R. N. and K. Thangavelu (1994) Host discrimination ability in parasitoid wasp *Psix striaticeps* (Hymenoptera : Scelionidae). *Ann. Entomol.* 12, 19-23.

Singh, R. N. and B. Saratchandra (2002) An Integrated Approach in the Pest Management in Sericulture. *Int. J. Indust. Entomol.* 5, 141-151.

Singh, R. N. and B. Saratchandra (2005) The development of botanical products with special reference to seri- ecosystem. *Caspian J. Env. Sci.* 3, 1-8.

Singh, R. N. and K Thangavelu (1999) Interference among females of *Trichomalopsis apanteloctena* (Hymenoptera: Pteromalidae) and its effect on progeny production *J. Insect Sci* . 12, 77-78.

Singh, R. N. and K. Thangavelu (1996) Biological characteristic of *Trichomalopsis apanteloctena* Crawford, a parasitoid of *Blepharipa zebina* . *Indian J. Seric.* 35, 62-63.

Singh, R. N. and K. Thangavelu (1996) Parasitization behaviour of stinkbug parasitoid *Psix striaticeps* (Dodd.) (Hymenoptera : Scelionidae) at its varying host densities. *Ann. Entomol.* 14, 89-94.

Singh, R. N. and K. Thangavelu.(1995) Host discrimination ability in parasitic wasp, *Psix striaticeps* (Hymenoptera : Scelionidae). *Ann. Entomol.* 12, 27-31.

Singh, R. N. and S. S. Sinha (1995) Behaviour of parasitoid *Pediobius* sp. (Hymenoptera: Eulophidae) at varying host densities. *Proc. Nat. Acd. Sci.* India 65, 183-187.

Singh, R. N., A. K. Goel and K. Thangavelu (1992) Succession of insect pest in tasar ecosystem in Bihar. *Indian J. Ecol.* 19, 177-182.

Singh, R. N., S. S. Sinha and K. Thangavelu. (1995) Life-table studies of the scelionid egg parasitoid, *Psix striaticeps* (Hymenoptera : Scelionidae) on the silkworm predator *Canthecona furcellata* (Pentatomidae : Heteroptera). *J. Biol .Control* 9, 5-8.

Singh, R. N., T. Babu and S. S. Sinha (1995) Reproductive strategies of an ectophagous parasitoid *Trichomalopsis apanteloctena* (Hymenoptera: Pteromalidae) on the pupae of *Blepharipa zebina* (Diptera: Tachinidae). *Indian J. Ecol.* 22, 17-20.

Singh, R. N., B. R. R. P. Sinha and S. S. Sinha (1994) Host discrimination between parasitsed and unparasitised uzifly pupae by females of *Trichomalopsis apanteloctena* Crawford (Hymenoptera: Pteromalidae). *Trends Life Sci.* 9, 27-32.

Singh, R. N., J. V. Krishna Rao and M.V. Samson (2001) Role of parasitoids in Pest Management in Tasar culture; in *Biocontrol potential and its exploitation in sustainable agriculture.* Upadhaya, R. K., K.G. Mukerji and B. P. Chamola (eds.), pp. 379-388, Kluwer Academy, Plenum publisher, New York.

Singh, R. N., K. C. Mandal and S. S. Sinha (1993) Studies on the biology of *Blepharipa zebina* Walk.(Diptera : Tachinidae). *Indian J. Seric.* 32, 15-19.

Singh, R. N., K. C. Mandal and K. Thangavelu (1995) *Blepharipa zebina* Walker (Tachinidae : Diptera) a new host for hymenopteran parasitoids *Brachymeria lasus* and *Theronea maskeliya* Cameron. *Indian J. Seric.* 34, 72-73.

Singh, R. N., K. Thangavleu., P. K. Mishra and J. Jayaswal (1992) Reproductive strategy and sex ratio of the stink bug parasitoid *Psix striaticeps* (Dodd.) (Hymenoptera : Scelionidae) in tasar ecosystem. *G .it. Ent.* 6, 151-155.

Singh, R. N., Maheshwari, M. and Saratchandra, B. (2004) Prospects of biological control of mealybug in sericulture. *Modern J. Life Sci.* 3, 1-3.

Singh, R. N., Maheshwari, M. and Saratchandra, B. (2005) Biocoenology and control of whiteflies in sericulture. *Insect Sci.* 12, 401-412.

Singh, R. N., V. Kulreshtha and S.S. Sinha (1994) Sex ratio of stink bug parasitoid wasp in tasar culture. *Nat. Acad. Science* 64, 17-21.

Singh, R.N. (2002) Studies on the biological attributes of scelionid egg parasitoid *Psix striaticeps* (Dodd) for the control of stink bug *Canthecona furcellata* (Wolff) in sericulture. *Int. J. Indust. Entomol.* 4, 117-120.

Singh, R.N. and Saratchandra, B. (2002) Biological control of the pentatomid stink bug Eocanthecona furcellata (Wolff.) by using their parasitoid, Psix striaticeps Dodd. in sericulture. International Journal of Industrial Entomology, 5, 13–22.

Singh, R.N. and Saratchandra, B. (2004a) Properties and potential of natural pesticides against sericultural pests. New Horizon of Animal Science (eds. B.N. Pandey, N. Armugam, P. Natraj, S. Pranjith), pp. 200–206. Zoological Society of India, Bodh Gaya, India .

Singh, R.N. and Saratchandra, B. (2004b) Use of neem in plant protection in Vanya silk. Proceedings of national workshop on potential and strategies for sustainable development of Vanya silks in Himalayan states, pp. 125–130. Dehradun, India, November 8-9, 2004.

Singh, R.N., Krishnarao, J.V. and Samson M.V. (2000) Integrated pest management in non mulberry sericulture: An overview. International Journal of Wild Silkmoth and Silk, 5, 340-344.

Singh, R.N., Maheshwari, M. and Saratchandra, B. (2004) Sampling, surveillance and forecasting of insect population for integrated pest management in sericulture. International Journal of Industrial Entomolology, 8, 17–26.

Singh, R.N., Samson, M.V. and Datta, R.K. (2000) Pest Management in Sericulture. Indian Publishers Distributors, Delhi.

Singh, S. P. and S. K. Jalali.(1991) Trichogrammatid egg parasitoids. Biological control centre, National centre for integrated pest management, Bangalore, India.

Siswanto, M. Rita, O. Dzolkhifli and K. Elna, (2008) Population fluctuation of Helopeltis antoni Signoret on Cashew Anacarcium occidentalle L., in Java Indonesia. Pertanika J. Trop. Agric. Sci., 31: 191-196.

Smith, R. F. (1978). History and complexity of pest management. In Pest Control Strategies. Edited by E. H. Smith and D. Pimentel. Pages 41-52. Academic Press, New York.

Smith, R. F. and Reynolds, H. T. (1966) Principles, definition and scope of integrated pest control. *Proceeding of FAO (United Nation Food and Agricultural Organisation) symposium on Integrated Pest Control* 1, 11-17.

Smith, R. F., mittler, T. E. and smith, C. N. (1973). History of Entomology. Annual Reviews, Palo Alto, California.

Solomon, M.E (1949) The natural control of Animal population. *J. Insect Ecology.* 18, 1-35

Som prakash (1996) Chemical and biometric studies of selected terpenoids and phenolics. Ph. D. thesis, University of Delhi, Delhi, India.

Sorenson, W. C. (1995) Brethren of the Net, American Entomology 1840-1880. University of Alabama Press, Tuscaloosa, AL,USA.

Southwood, T.R.E. (1976) Ecological Methods with particular reference to the study of insect population, (2nd ed). John Wiley and Sons, New York, .

Southwood. T. R. E. (1978) Ecological methods. John Wiley & Sons, New York.

Srinivasa, M.V. (2000) Host plant of the spiraling whitefly Aleurodicus dispersus Russell (Hemiptera : Aleyrodidae). Pest Management in Horticulture Ecosystems, 6, 79–105.

Srinivasa, M.V., Viraktamath, C.A. and Reddy, C. (1999) A new parasitoid of the spiralling whitefly Aleurodicus dispersus Russell (Hemiptera: Aleyrodidae) in South India. Pest Management in Horticultural Ecosystems, 5, 59-61.

Srinivasan, G. and Mohanasundaram, M. (1997) A novel method to trap the spiralling whitefly, Aleurodicus dispersus Russell adults in the home gardens. Insect Environment, 3, 18.

Stern,V. M., Smith, R. F., Bosch, R. Van den and Hagen, K. S. (1959) The integrated control concept. *Hilgardia* 29, 81-101.

Stinner, R. E. (1976) Ovipositional response of *Ventturia canescens* (Grav.) (Hymenoptera : Ichneumonidae) to various host and parasite densities. *Res. Pop. Ecol.* 18, 57-73.

Stinner, R. E. and J. R. Jr. Bradley (1989) Habitat management to increase effectiveness of parasitoids and predators and parasites. *Proceedings of the workshop on biological control of Heliothis : Increasing the effectiveness of natural enemies.* Eastern Regional Research office, USDA, New Delhi, India.

Summers, C. G.(1992) Integrated pest management in the alfalfa agro ecosystem. Proceedings of the 11th Nevada Alfalfa Symposium, 5-6 February 1992, Sparks, Nevada, University of Nevada Co-operative Extension Service, Reno, Nevada, pp.119-126.

Suresh, S., S. Chandrasekaran and P.C. S. Babu (1989) Studies on pheromone and light trap for the attraction of diamond back moth, *Plutella xylostella. S. Indian Hort.* 37, 353 – 354.

Sutherst, R. W. (1991) Predicting the survival of immigrant insect pests new environments. *Crop protection* 10, 331-333.

Taneja, S. L. and A. P. Jayaswal (1983) Factors affecting pink ball worm moth cathes in gossyplure baited traps. Indian *J. Plant Protec.* 11, 78 –83.

Taylor, J. (1984) Assessing and interpreting the spatial distribution of insect population. *Ann. Rev. Entomol.* 29, 321 - 357.

Thangavelu, K. and R. N. Singh (1992) Record of new egg parasites of *Canthecona furcellata* (Pentatomidae : Heteroptera). *Entomon* 17, 65-66.

Thiaggarajan, V. and Govindaiah (1987) Menace of dermestid beetles in grainages. Indian Silk 26 : 26 –27.

Thomas, M. and Wagge, J . (1996) Integration of Biological control and Host Plant Resistance, Technical Centre for Agricultural and Rural Co-operation, Wageningen, The Netherlands; CAB International, Wallingford, UK.

Thomson, W. T. (1992) A world wide guide to beneficial animals. Thomson publications, Fresno, California, USA.

Thomson, W.T. (1992) A World Wide Guide to Beneficial Animals. Thomson Publications, Fresno.

Thorpe, W. A. and H. B. Caudle (1938) A study of the olfactory responses of insect parasites to the food plants of their host. *Parasitology* 30, 523-528.

Tikader, A. and Thangavelu, K (2003) Incidence of *O e c o p h y l l a smargdina* (Fabricious) (Hymenoptera : Formicidae) on mulberry (*Morus* species). Indian J. Seric. 42 (2): 186 – 187.

Tisdell, C. A. (1990) Economic impact of Biological Control Weeds and Insects; in *Critical Issues in Biological Control*. Mackauer, M., L. E., Ehler and J. Roland (eds.), pp. 301 – 316, Intercept, Andover, Hants, U.K.

Torii, T. (1967) Statistical methods in rice stem borer research; in *The Major Insect Pests of the Rice Plant*. pp.127-167. John Hopkins Press, Baltimore, Maryland, USA.

Townes, H. (1972) Ichneumonidae as biological control agents. Proceeding Tall Timber Conference and Ecological Animal Control by Habitat. Management. 3,235-248.

Tuskes, P.M., Tuttle, J.P. and Collins, M.M. (1996) *The wild silk moths of North America. Cornell University Press*, USA.

Ullal, S. R. and Narasimhanna, M. N. (1978) Handbook of practical sericulture, Central Silk Board Bangalore.

Upadhyay, R. K., K.G. Mukerji, and R. Rajat (1997) Integrated pest management system in agriculture, Vol.2. biocontrol in emerging biotechnology, Aditya books (P) limited, New Delhi, India.

Upadhyay, R..K., K. G. Mukerji, B. P. Chamola, and O. P. Dubey (1998) Integrated pest and disease management, APH Publishing Corp., New Delhi, India.

Van den Bosch, R. and F. H. Haramoto (1953) Competition among parasites of oriental fruit fly. *Proc. Hawaiian Entomol. Soc.* 35, 201-206.

Van den Bosch, R. (1978) The pesticide conspiracy. Doubleday, Garden city, NY, USA.

Van Driesche, R. G. and Bellows, T. S. Jr. (1996) *Biological Control.* Chapman & Hall, London, UK.

Van Lenteren, J. C. (1981) Host discrimination by parasitoids; in. *Semiochemicals : Their role in pest control.* Nordlund, D. L., R. L. Jones and W. J. Lewis, (eds.), pp. 153-179, John Wiley, New York.

Van Lenteren, J. C. (1983) Biological Pest Control : Passing fashion of here to stay, *Organorama* 20, 3-9.

Van Lenteren, J. C. (1986) Evaluation, mass production, quality control and release of entomophagous insects; in *Biological Plant and Health Protection*. Frang, J. M. (ed.), pp. 31-58, Fisher Veriag, Stutt.

Van Lenteren, J. C. (1986) Evaluation, mass production, quality control and release of entomophagous insects; *Biological Plant and Health Protection*. (ed. J. M. Frang), pp. 31-58, Fisher Veriag, Stutt.

Van Lenteren, J.C. (1983) Biological pest control : passing fashion of here to stay. *Organorama*. 20, 3-9.

van Lenteren, J.C. (1986) Evaluation, mass production, quality control and release of entomophagous insects. Biological Plant and Health Protection (ed. J. M. Frang), pp. 31-58. Fisher Veriag, Stuttgart.

Van Lenteren, J.C. and K. Bakker (1978) Behavioural aspect of the functional response of a parasite *Pseudeucoila bochei* to its host *Drosophila melanogaster. Neth. J. Zool.* 28, 213-233.

Van Lenteren, J.C., (1987) Environmental manipulation advantageous to natural enemies of pests. In: Delucchi, V. (ed.) Integrated Pest Management, Protection Integree : Quo Vadis ? An International Perspective. Parasites 86, Geneva, pp. 123 – 163.

Veer, V., Negi, B.K., and Rao, K.M. (1996) Dermestid beetles and some other insect pests associated with sotred silkworm cocoons in India, including a world list of dermestid species found attacking this species. J. Stored Prod. Res. 32 (1): 69- 80.

Venkatesan, T., Singh, S. P. and Jalali, S.K. (2001) Development of *Cryptolaemus montrouzieri* Mulsant (Coleoptera: Coccinellidae), a predator of mealybugs on freeze-dried artificial diet. *Jour. Biol. Cont.* 15, 139-142.

Vijayalakshmi, K., Radha, K.S. and Vandanashiva (1995) Neem ..A Users Manual, Centre for Indigenous knowledge systems, Chennai, pp.86.

Vinson, S. B. (1975) Biochemical co evolution between parasitoids and their hosts; in *Evolutionary Strategies of Parasitic Insects and Mites*. Price, P. W. (ed.), pp. 14-18, Plenum Press, New York and London.

Vinson, S. B. (1981) Habitat location; in *Semiochemicals : Their Role in Pest Control*. D. A. Norduland, R. L. Jones and W. J. Lewis (eds.), pp. 51-77, John Wiley & Sons , New York.

Voronin, K. E. (1981) Ecological aspects of the behaviour of *Telenomia* (Hymenoptera:Scelionidae) ; in *Insect behavior as a basis for developing control measures against pests of field crops and forests*. .Pristavko, V. P. (ed.), pp. 36-41. Amerind, New Delhi, India.

Waage, J. K. (1986) Family planning in parasitoids: Adaptive pattern of progeny and sex allocation; in *Insect parasitoids*. Waage, J. K. & D. J. Greathead (eds.), pp. 416, Academic press, London.

Waage, J. K. (1993) Making IPM work: developing country experiences and prospects. In : Srivastava, J. P. and Alderman, H. (eds) *Agriculture and Environmental Challenges*. Proceedings of the thirteenth Agricultural Sector Symposium. World Bank, Washington DC, USA, pp. 3 –11.

Waage, J. K. and J. A. Lane.(1984) The reproductive strategy of a parasitic wasp. II. sex allocation and local mate competition in *Trichogramma evanescens*. *J. Anim. Ecol.* 53, 417-426.

Waage, J. K. and Greathead, D. J. (eds) (1986) Insect Parasitoids. Academic Press, London, UK.

Waage, J.K. (1986) Family planning in parasitoids: Adaptive pattern of progeny and sex allocation. Insect Parasitoids (eds. J.K. Waage & D.J. Greathead), pp. 416. Academic Press, London.

Wall, C. (1989) Monitoring and spray timing; in *Insect Pheromones in plant protection*. Jutsum, A. R. and R. F. S. Gordan, pp. 39-66, John Wiley and Sons Ltd.

Wang, J. X. (1992) *Systematic Monitoring of Mulberry pests and its Methods*. Hangzhou: Hang Zhou University Press.

Wang, J.X. (1992) Systematic Monitoring of Mulberry Pests and its Methods. Hang Zhou University Press, Hangzhou.

Waterhouse, D.F. and Norris, K.R. (1989) *Aleurodicus dispersus* spiraling whitefly. Biological Control Pacific Prospects, Supplement 1, pp. 12-23. Australian Center for International Agriculture Research, Canberra.

Waters, W. E. (1955) Sequential sampling in forest insect survey. *For. Sci.* **1**, 68-79.

Way and K.L. Heong, (1994) The role of biodiversity in the dynamics and management of insect pests of tropical irrigated rice-a review. *Bull. Entomol.* Res., 84: 567-587.

Way, M. J. (1974) Studies on the life history and ecology of the ant *Oecophylla smaragdina* Lareille. *Bull. Entomol. Res.* 45 : 307 – 344.

Weatherston, I. (1989) Alternative dispensers for trapping and disruption; in *Insect Pheromones in plant protection*. Jutsum , A. R. and R. F. S. Gordan, pp. 249-278, John Wiley and Sons Ltd.

Wen, H.C., Hsu, T.C. and Chen, C.N. (1994) Supplementary description and host plants of the spiralling whitefly, Aleurodicus dispersus Russell. Chinese Journal of Entomology, 14, 147-161.

Wen, H.C., Tung, C.H. and Chen, C.N. (1995) Yield loss and control of spiralling whitefly (Aleurodicus dispersus Russell). *Journal of Agricultural Research of China*, 44, 147-156.

West, T. F. and Campbell, G. A. (1950). DDT and Newer Persistent Insecticides. 2nd edition. Chapman & Hall, London.

Wheatly, A. R. D., Wightman, J. A., Williams, J. H., and Wheatly, S. J. (1989) The influence of drought stress on the distribution of insects on four groundnut genotypes grown near Hyderabad, India. *Bulletin of Entomological Research* 79, 567 – 577.

Wigglesworth, V. B. (1976). Insects and the Life of Man. Chapman & Hall, London.

Wijesekera, G.A.W. and Kudagamage, C. (1990) Life history and control of 'spiralling' whitefly Aleurodicus dispersus (Homoptera: Aleyrodidae): fast spreading pest in Sri Lanka. Quarterly Newsletter, Asia and Pacific Plant Protection Commission, 33, 22-24.

Wilkes, A. (1963) Environmental causes of variation in the sex ratio of arrhenotokous insect, *Dahlbominus fuliginosus* (Nees) (Hymenoptera : Eulophidae). *Can. Ent.* 95, 183-202.

Williams DJ. 1996. A brief account of the hibiscus mealybug *Maconellicoccus hirsutus* (Hemiptera: Pseudococcidae), a pest of agriculture and horticulture, with descriptions of two related species from southern Asia. *Bulletin of Entomological Research.* 86: 617-628.

Williams, J. R. (2003) *Pink Hibiscus mealybug biologicalcontrol in Imperial valley*, California. California Deaprtment of Food and Agriculture, California.

Wilson, F. (1960) A review of the biological control of the insects and weeds in Australia and Australian New Guinea. Common Wealth Institute of Technical Communication, CAB, London.

Wilson, F. and Huffaker, C.B. (1976) The philosophy, scope, and importance of biological control. In Theory and Practice of Biological control. Edited by C.B. Huffaker and P.S. Messenger. Pages 3- 15. Academic press, New York, San Francisco and London.

Wilson, L. T., W. L. Sterling, D. R. Rummel and J. E. Devay (1989) Quantitative sampling principles in cotton IPM; in *IPM Systems and Cotton Production.* Frisbie, F., K. M. Elzin and L. T. Wilson), pp. 851-89 , John Wiley & Sons, New York.

Wood, D. L. (1982) The role of pheromones, kairomones and allomones in the host selection and colonizing behaviour of bark beetles. *A. Rev. Ent.* 27, 411- 446.

Wright, J. W. (1976). Insecticides in human health. In The Future for Insecticides: Needs and Prospects. Edited by R. L. Metcalf and J. J. McKelvey, Jr. Pages 17-34. John Wiley & Sons, New York.

Wylie, H. G. (1982) Reproductive capability and longevity of the parasitic wasps *Telenomus podisi* and *Trissolcus euschisti. Ann. Entomol. Soc. Amer.* 75, 181-183.

Wylie, H. G. (1976) Observation on the life history and sex ratio variability of *Eupterous dubius* (Hymenoiptera : Pteromalidae), A parasite of cyclorrhaphus diptera. *Can. Ent.* 108, 1267 –1274.

Yamada, Y. (1987) Factors determining the rate of parasitism by a parasitoid with low fecundity, *Chrysis shanghaiensis* (Hymenoptera : Chrysididae). *J. Anim. Ecol.* 56, 1029-1042.

Yang, Peng (1982) A preliminary study of the biology of the yellow citrus ant *Oecophylla smaragdina* Fabr. and its utilization against citrus insect pests: Acta Scientorium Naturalium Universitatis Sunyatseni, 3: 102 – 105.

Yeargon, K. V. (1982) Reproductive capability and longevity of the parasitic wasps, *Telenomus podisi* and *Trissolcus eus*chisti. *Ann. Entomol. Soc. Amer.* 75, 181-183.

Yoshimeki, M. (1978) Japanese forecasting for rice pests. Proceeding on Integrated Pest Control for irrigated Rice in South and Southeast Asia, Philippines.

Zadoks, J. C. (1983) An integrated disease and pest management scheme, EPIPRE for wheat; in *Better Crops for food,* pp.116-29, Ciba Foundation Symposium, Pitman, London.

Zebitz,C.P.W. (1987). Potential of neem seed kernel extracts in mosquito control. See Ref.80, pp.553-73.

Zhang, A.J. and Amalin, D. (2005) Sex pheromone of the female pink hibiscus mealybug, *Maconellicoccus hirsutus* (Green) (Homoptera: Pseudococcidae): biological activity evaluation. *Environmental Entomology* 34, 264-270.

Zhang, A.J., Amalin, D., Shirali, S., Serrano, M.S., Franqui, R.A., Oliver, J.E., Klun, J. A., Aldrich, J.R., Meyerdirk, D.E. and Lapointe, S. L. (2004) Sex pheromone of the pink hibiscus mealybug, *Maconellicoccus hirsutus*, contains an unusual cyclobutanoid monoterpene. *Proceedings of the National Academy of Sciences of the United States of America* 101, 9601-9606.

Zhu, Z.R., (1999) Population ecology and management strategy of the white backed planthoppers S. furcifera (Horvath) in subtropical rice. Ph.D. Thesis, Nanjing Agricultural University China.

GLOSSARY OF TERMS USED IN APPLIED ENTOMOLOGY AND SERICULTURE

Abiotic Components : Abiotic components are the non-living components of the environment such as space and weather. Pesticides could also be considered an abiotic factor.

Active ingredient : The biologically active chemical in pesticide preparation.

Acute toxicity : It refers to the toxic effect produced by a single dose of toxicant.

Alternate host: The other species of plant that is necessary for the completion of the life cycle of some insects and plant disease producing organisms.

Antibiotic : A substance produce by microorganisms which inhibit the growth of a poisonous or toxic chemical.

Antidote: A substance given to a patient to counteract the effect produced by the injestion of a poisonous or toxic chemical.

Antifeedant: A chemical possessing the property of inhibiting the feeding of certain insect pests.

Apodus : Without legs

Arthropoda : The phylum of animals to which insects and other group belongs

Asexual : Without sex, parthenogenetic

Attractant : A chemical or physical source which induces insects to move towards

Augmentation : Mass culture and local release of parasites, predators or pathogens to provide short term biological control of pest species.

Autocide : The control of a pest by the sterile male technique

Bait: Food stuff used for attracting pests, mixed with a poison which is usually injected by the pest.

Bioassay: Quantitative determination of strength of a biologically active substance from its effect on an organism.

Biocide : A general poison or toxicant used for killing.

Bionomics : The study of habits, life histories, and adaptation of living organisms.

Biotic Components : Biotic components of the environment are living organisms. Biotic components of the environment include primary producers, herbivores, natural enemies (e.g. parasitoids, predators, pathogens) and competitors (individuals of the same or different species competing for resources).

Biotic potential (reproductive potential) : The maximum possible rate of increase of an organism in absence of any limiting factor.

Biotype : A genetically distinct strain or sub group of a species distinguished by some behavioral or physiological difference but indistinguishable morphologically.

Bordeaux mixture : It consists of copper sulphate, lime and water in the ratio of 4-5-50.The solution of copper sulphate and finely ground quicklime are prepared separately and both mixed together in a third container with constant agitation.

Broad spectrum : Having an effect on wide range of insects.

Brood : Offspring of approximately same age arising from a single species of animal. .

Carbamate: A group name applied to insecticides which are based on the carbonic acid nucleus.

Chemical control: The use of chemicals to kill, deter or in any way suppress pest populations.

Chemosterilant : A chemical which has the property of sterilizing an insect without killing it.

Chlorosis : A Yellowing of foliage of plants due to lack of chlorophyll.

Cocoon: A silken covering produced by some insects which encloses and protects the pupa.

Coleoptera: (coleon= a sheath; ptera= wings): The order of insects comprising the beetle and weevils.

Community : Community is the species that occur together in space and time.

Compatibility: The ability to mix different pesticides without physical or chemical interactions which would lead to reduction in biological efficiency or increase in phytotoxicity

Concentrated solution (C.S.): Commercial pesticide preparation before dilution for use.

Contact insecticide: One which is able to kill insects by contact with the external body surface.

Control: To reduce damage or pest density to a level below the economic threshold.

Cross-resistance: The phenomenon of insect resistance to one type of insecticide providing resistance to other insecticides with similar mode of action.

Cultural control: Manipulation of cultural practices to provide control of a pest.

Damage threshold : Pest population above which crop loss occurs.

Density dependent (factor) : A factor regulating population of living organisms, whose influence varies with the population density of the organism concerned

Density independent (factor) : A factor regulating population of living organisms, whose influence is independent of the population density of the organism concerned.

Detoxication: The process by which a poison is rendered harmless.

Diapause: A state of suspended activity and metabolism for surviving unfavourable conditions and requires a particular stimulus for its termination.

Diptera (du=two; ptera=wings): The order of insects that comprises the true flies which are characterized by possession of only one pair of wings.

Dust (D): A dry pesticide formulation prepared by milling the pesticidal compound into a fine powder.

Ecological management: Purposeful manipulation of the cropping environment to reduce rates of pest increase or damage.

Ecological Niche : Ecological niche is all the components of the habitat with which an organism or population interacts.

Ecology : Ecology is the study of the interrelationships of organisms and their surrounding environment. In other words, ecology is the study of how organisms interact with their abiotic and biotic environment.

Ecology and Pest Management : IPM introduced ecological thinking into crop protection. Ecology gives context to pest management.

Economic control: Control of a pest such that expenditure on control measures is more than compensated by increased value of yield.

Economic damage: The injury done to a crop which will justify the cost of artificial control measures.

Economic injury level (EIL): Economic Injury Level (EIL) is the lowest population density of a pest that will cause economic damage to a crop. Sometimes referred to as the "damage threshold"

Economic pest: A pest causing a crop loss of about 5-10% according to definition.

Economic Threshold Level (ETL):Economic Threshold Level is the pest population level at which a control action, such as pesticide application, should be considered in order to prevent economic loss. ETL is synonymous with "action threshold" or "treatment threshold". At the ETL, the economic return of the control effort is greater than the cost of control.

Ecosystem: The interacting system of the living organisms in an area and their physical environment.

Ecotype: A specific strain of a species adapted to a particular set of environmental condition.

Emulsifier: Spray additive which permits formulation of a stable suspension of oil droplets in aqueous solution or of aqueous solution in oil.

Endoparasite : A parasite which lives internally within the body of its host.

Entomophagous: An animal or plant which feeds upon insects.

Environmental resistance: The physical and biological restraints that prevent a species from releasing its biotic potential.

Eradication: Complete elimination of a species from an area or country.

Exotic: Introduced from other country or continent.

Fecundity: Capacity to produce offspring (reproduce); power of species to multiply rapidly.

Fumigant: Pesticide exhibiting toxicity in the vapour phase.

Gall: A swelling or outgrowth produced by a plant as a result of attack by an insect, mite or micro-organisms.

Genetic control: Altering the genetic makeup of organisms to inhibit their reproduction and survival.

Genus : The next highest category to the species in the classification of an organism. A number of species which are similar in all but coloring or other unimportant characters are said to form a genus.

Granule: Coarse particle of inert material impregnated with a insecticide or on which the insecticide is coated.

Grubs (white) : A scarabaeiform larva, thick-bodied with a well developed head and thoracic legs, without abdominal pro-legs, usually sluggish in behavious; general term for larvae of coleopteran.

Habitat : Habitat is the environment in which the individual organism lives.

Hazard: The danger that injury will occur with use of a particular pesticide, depends on both toxicity and exposure.

Hemiptera (hemi=half; ptera=wings): The order of insects possessing piercing and sucking mouth parts.

Heteroptera: A sub-order of the insect order Hemiptera. All insects have sucking and piercing mouth parts.

Homoptera : A sub-order of insect or Hemiptera. All insects have piercing and sucking types of mouth parts.

Hormone: A chemical secreted into the blood stream which produces a specific effects on some bodily organ or process.

Host: The organism in or on which a parasite lives and the plant on which an insect feeds.

Hymenoptera (hymen= membrane; ptera=wings): The order of insects comprising sawflies, ants, bees and wasps.

Hyper parasite: A parasite whose host is another parasite.

Injury : The deleterious effects of pest activities on host physiology.

Insecticide: A chemical which is toxic to insects

Instar: The form of an insect between successive moults.

Integrated pest management (IPM): IPM is an ecosystem-based pest management strategy that focuses on long-term prevention of pests and their damage through a combination of techniques such as biological control, habitat manipulation, modification of cultural practices, and use of resistant cultivars. Pesticides are used only when needed as determined by established guidelines.

Isoptera (iso=similar; ptera=wings): The order of insects comprising termites or white ants.

Key pest:: The serious and persistent pests which attack a crop.

Larva: The immature stage of an insect between the egg and pupa having a complete metamorphosis.

Larvicide : Toxicant (poison) effective against insect larvae.

LC 50 : Lethal concentration of toxicant required to kill 50% of large group of individuals of one species.

LD 50: Lethal dose of toxicant required to kill 50% of a large group of individuals of one species.

Leaf minor: An insect which lives in and feeds upon the cells between the upper and lower epidermis.

Lepidoptera (lepis= a scale; ptera=wings) : The order of insects that comprises butterflies and moths.

Lethal: Deadly

Life-table : The separation of a pest population into its different age components such as eggs, larvae, pupae, adults.

Looper : A caterpillar with only one pair of abdominal pro legs in addition to the terminal claspers and which moves by looping its body.

Maggot: : A vermiform larva, legless, without a distinct head capsule (Diptera).

Mechanical control: Any means of pest control which involves mechanically trapping or killing insects or providing barriers to prevent insects gaining access to pants or other material.

Mite : A member of class Acari related to but distinct from the insects having single body region, four pairs of legs and absence of antennae.

Monophagous: An insect restricted to a single host plant species.

Natural control : The natural regulation of population of living organisms that takes place without human interference.

Natural enemy: Any living organism which is harmful to a species by way of its predatory, parasitic or disease inducing habit.

Nocturnal : Active in night

Nymph : The juvenile stage of an insect which undergoes incomplete metamorphosis.

Obligate parasite: A parasite which is obliged to follow a parasitic mode of life and which can not exist in any other way.

Occasional pest : : One which reaches significant levels only occasionally.

Orthoptera (orthos= straight; ptera=wings): The order of insects that includes crickets, grasshoppers and locusts

Oviposition: The act of egg laying

Ovipositor : Specialized egg laying organ possessed by most female insects.

Parasite: A parasite is an organism that is usually much small than its host and derives material essential for its existence while conferring no benefit in return. Generally, a single individual does not kill the host.

Parasitoid: Parasitoids have often been included in the parasite category but a parasitoid is a special kind of predator that often about the same size as its host, kills its host and require only one host for development into free living adult eg. Braconid wasps.

Pest : Pest is an organism that interferes with the availability, quality or value of a managed resource. In an agronomical sense a pest is any organism or microbe with the potential to lower the value of a crop by reducing crop yield, quality or reproductive ability (e.g. storage pests that reduce seed viability).

Pest management (IPM): An approach to pest control that blends biological and chemical control methods according to needs

Pesticide : A chemical which kills pest organisms by virtue of its toxicity.

Pheromone (Ecto hormone): A chemical substance produced by an insect or other animal which conveys a message to other members of the same species and usually results in a behavioural response.

Physical control: Pest control which involves either modification of some physical factor in the environment such as temperature and light (to attract insects).

Poison bait : An attractant food stuff for insects, molluscs, or rodents, mixed with an appropriate toxicants.

Polyphagous : An animal feeding upon a range of hosts.

Population : Population is a group of individuals of the same species.

Predator : A free living organisms throughout its life, it kills its prey, is usually larger than its prey and requires more than one prey to complete its development e.g. Mantids, spiders and many species of lady beetles.

Pupa : The quiescent stage in the life cycle between larva and adult in insect which undergoes complete metamorphosis.

Puparium : Specialized form of pupa in which the last larval skin is retained as an extra covering , characteristic of the higher Dipterans

Quarantine: All operations associated with the prevention of unwanted organisms into a territory or their exportation from it.

Repellant: A chemical or physiological source which induce insects to move away from it.

Resistance (of insects to insecticides): The developed ability of a strain within an insect species to withstand an insecticide to which it was formerly susceptible.

Resurgence: The recovery of pest population (sometimes to levels higher than before treatment) following application of an insecticide.

Secondary pest: Species whose numbers are usually controlled by biotic and abiotic factors which sometimes break down allowing the pest to increase in numbers.

Sex attractant: A volatile chemical substance produced generally by females to attract the opposite sex.

Soldier: A caste present in some social insects especially in termites and ants whose task is to defend the colony.

Species : A species is a group of reproductively isolated organisms. It is the basic unit of taxonomic classification of organisms, designated in scientific nomenclature by a unique Latin binomial with a genus and specific epithet.

Synergist: A chemical which when added to an insecticide increases its effectiveness, but which is not itself toxic.

Systemic insecticide : A insecticide that is taken up and translocated within plants or animals.

Tolerance: A mechanism of plant resistance to pests whereby the plant is able to grow and produce despite pest injury.

Toxicant : Poison or chemical exhibiting toxicity.

Toxicity : Ability to poison or to interfere adversely with vital process of the organism by physio-chemical means.

Trap crop: Crop of plants grown especially to attract insect pests.

Wettable powder : An insecticide formulation consisting of finely ground solid particles of active ingredient plus inert carrier, forms a suspension when added to water.

APPENDIX

Sl. No.	Pesticide	Formulation	Trade name*
1	2	3	4
1	Aldicarb	10g	Temik
2	Aldrin**	5% 30 EC	Aldrex, Octalane, Hexamar, Aldrite
3	BHC (HCH)*** (Benzene hexachloride)	5% 10% 50 WP	Hilbeech, Hexachlore
4	Carbaryl	5% 10% dust 4 G 50 WP	Sevin, Hexavin, Carbavin Ravgon, Killex
5	Carbon bisulphide (Carbon disulphide)	Fumigant	
6	Carbofuran	3 G	Furadan, Hexafuran
7	Carbosulfan	25 STD	
8	Chlordane**	5 % dust	Octachlor, Intox, Chlortox, Chlorexe, Micochlordance, Aspan, Coroden, Velsicol
9	Chlorpyriphos	5 G 10 G 20 EC 24 EC	Dursban, Lorsban, Coroban
10	Cypermethrin	10 EC 25 EC	Ripcord, Cymbush, Cyperkill
11	DDT (Dichlorodiphenyl- trichloroethane)***	5 % dust 10% dust 25 EC 50 WP	Didimax, Anodex
12	Decamethrin	2.8 EC	Decis
13	Diazinon	5 G 20 EC	Basudin, Suzon, Ditaf, Nicidol, Diazitol, Neocidol

1	2	3	4
14	Dicofol	85 EC	Kelthane
15	Dichlorvos	100 EC	DDVP, Nuvan, Dichlorphos, Nogos, Vapona, Suchlor, Marvex Super 100
16	Dieldrin***	20 EC	Dieldrex, Octalox, Dielmoth, Micodieldrin, Dieldrite
17	Dimethoate	30 EC	Rogor, Cygon, Hexagon, Tagor, Hilthoate, Tara-909
18	Disulfoton	3 G 5 G	Di-syston, Solvirex, Disystox, Thiodemeton
19	EDCT (Ethyline dichloride-carbon tetrachloride)	Fumigant	
20	Endosulfan	2% dust 5% EC 4 G 35 EC	Thiodan, Hexasulfan, Thionex, Thiotox, Hildan, Endocel, Cyclodon, Thimul, Tiovel
21	Fenitrothion	5% dust 50 EC	Folithion, sumithion, Accothion
22	Fenthion	1000 EC	Lebayeid, Baytex, Entex, Tiguvon
23	Fenvalerate	20 EC	Agrofen, Fenkill, Fenval Sumicidin
24	Formothion	25 EC	Anthio, Aflix
25	Heptachlor**	5% dust 20 EC 5 G	Heptagran, Eeptox, Hiltachlor, Helpaf, Heptamul, Micoheptachlor
26	Isofenfos	5 G	Oftanol
27	Lead Arsenate		
28	Indane***	2% dust 1 G 6 G 10 G 20 EC	
29	Malathion	2% dust 5% dust 50 WP 50 EC	Cythion, Malamar, Malatox, Hilthion, Micomalathion

1	2	3	4
30	Menazon	100 EC	Sayfos, Saphizon,
31	Mephosfolan	5 G	Cytrolane, Carbicron, Bidrin
32	Methyl-demeton	25 EC	Metasystox
33	Methyl-oxydemeton	20 EC	Metasystox R
34	Methyl Parathion***	25 EC 50 EC	Folidol-M Metacid, Paramar, Tagpar, Paratox, Metaphos
35	Monocrotophos	36 WSC 40 WSC 40 EC	Sufos, Monophos, Monocil
36	Nicotine Sulphate	40 EC	Black leaf 40
37	Oxithion	5% dust	
38	Paradichlorobenzene	Fumigant	
39	Permethrin	25 EC 50 EC	Ambush, Hilthrion, Permasect, Perthane
40	Phenthoate	50 EC	Elsan, Phendal, Tagsan, Cidial, Dimethoate, Papthion
41	Thiometon	25 EC	Ekatin, Hexatin
42	Phorate	10 G	Himet, Foratox-10
43	Phosalone	35 EC	Zolone, Rubitox
44	Phosphamidon	85 WSC 100 EC	Dimecron, Sumidon
45	Quinalphos	5 G 25 EC	Ekalux, Suguin, Agrophos, Kinalox
46	Toxaphene**	80 EC	Anatox, Camphechlor, Toxakil, Phentox
47	Trichlorphon	5 % dust 5 G 50 EC	Dipterex, Dylox, Neguvon, Tugon

* The proprietary names mentioned against each insecticide does not imply their commentation nor does omission of other proprietary names imply their condemnation.

** Pesticides have been banned by Government for use in India

*** The use of pesticides have been restricted by the Government of India

INDEX

OOO